Introduction to Internal Combustion Engines

Introduction to Internal Combustion Engines

Richard Stone

Brunel University
Uxbridge, Middlesex

Second Edition

Published by

Society of Automotive Engineers, Inc.
400 Commonwealth Drive
Warrendale, PA 15096–0001
USA

Phone: (412) 776–4841
Fax: (412) 776–5760

Library of Congress Cataloging-in-Publication Data

Stone, Richard, 1955–
    Introduction to internal combustion engines / Richard Stone. —
2nd ed.
       p.   cm.
    Includes bibliographical references (p.   ) and index.
    ISBN 1–56091–390–8: $39.00
    1. Internal combustion engines.   I. Title.
TJ755.S87   1993
621.43—dc20   93–14970
                                                 CIP

Fifth printing 1997
Published in the United States of America 1993 by
Society of Automotive Engineers, Inc.
400 Commonwealth Drive,
Warrendale, PA 15096–0001

SAE Order No. R–129

Printed in Hong Kong

Contents

Preface to the Second Edition

This book aims to provide for students and engineers the background that is pre-supposed in many articles, papers and advanced texts. Since the book is primarily aimed at students, it has sometimes been necessary to give only outline or simplified explanations. However, numerous references have been made to sources of further information.

Internal combustion engines form part of most thermodynamics courses at Polytechnics and Universities. This book should be useful to students who are following specialist options in internal combustion engines, and also to students at earlier stages in their courses — especially with regard to laboratory work.

Practising engineers should also find the book useful when they need an overview of the subject, or when they are working on particular aspects of internal combustion engines that are new to them.

The subject of internal combustion engines draws on many areas of engineering: thermodynamics and combustion, fluid mechanics and heat transfer, mechanics, stress analysis, materials science, electronics and computing. However, internal combustion engines are not just subject to thermodynamic or engineering considerations — the commercial (marketing, sales etc.) and economic aspects are also important, and these are discussed as they arise.

The second edition contains significant new material, both within the original chapters and in the new chapters. Preparing the second edition has also provided the opportunity to include further worked examples and many problems (with numerical answers).

The first 6 chapters still provide an introduction to internal combustion engines, and these can be read in sequence.

Chapter 1 provides an introduction, with definitions of engine types and operating principles. The essential thermodynamics is provided in chapter 2, while chapter 3 provides the background in combustion and fuel chemistry. The differing needs of spark ignition engines and compression ignition engines are discussed in chapters 4 and 5 respectively. Chapter 6 describes how the induction and exhaust processes are controlled.

Chapter 11 provides a discussion of some topics in the mechanical design of engines, and explains some of the criteria that influence material selection. Chapter 13 provides an overview of the experimental equipment that is an essential part of engine development. This chapter has substantial additions to describe gas analysis equipment and its use, and an introduction to computer-based combustion analysis.

The remaining chapters pre-suppose a knowledge of the material in the earlier chapters. Chapter 7 contains a treatment of two-stroke engines, a topic that has gained widened interest since the first edition; both spark ignition and compression ignition engines are discussed.

Chapter 8 is mainly concerned with turbulence: how it is measured, how it is defined, and how it affects combustion. Such knowledge is pre-supposed in technical papers and the contributions in specialist texts.

Chapter 9 discusses the application of turbochargers to spark ignition and compression ignition engines, and there is further material on turbo-charging in chapter 10 (Engine Modelling) as a turbocharged engine is used as an example.

The treatment of engine cooling in chapter 12 has been provided since it is a subject that is often overlooked, yet it is an area that offers scope for reductions in both fuel consumption and emissions.

The material in the book has been used by the author for teaching a final year course at Brunel, and contributing to the MSc in Automotive Product Engineering at Cranfield. These experiences have been useful in preparing the second edition, as have the written comments from readers. The material within the book has come from numerous sources. The published sources have been acknowledged, but of equal importance have been the conversations and discussion with colleagues and researchers in the area of internal combustion engines.

It is invidious to acknowledge individuals, but I must thank my colleague Dr Nicos Ladommatos for making a stimulating research environment in engines at Brunel. I must also thank Mrs June Phillips for her tireless and efficient typing of the manuscript.

I would again welcome any criticisms or comments on the book, either concerning the detail or the overall concept.

RICHARD STONE

Acknowledgements

The author and publisher wish to thank the following, who have kindly given permission for the use of copyright material:

The American Society of Mechanical Engineers, New York, for figures from technical publications.

Dr W. J. D. Annand (University of Manchester) for figures from W. J. D. Annand and G. E. Roe, *Gas Flow in the Internal Combustion Engine*, published by Foulis, Yeovil, 1974.

Atlantic Research Associates, Tunbridge Wells, and Martin H. Howarth for three figures from M. H. Howarth, *The Design of High Speed Diesel Engines*, published by Constable, London, 1966.

Blackie and Son Ltd, Glasgow, for three figures from H. R. Ricardo and J. G. G. Hempson, *The High Speed Internal Combustion Engine*, 1980.

Butterworth–Heinemann, Oxford, for figures from technical publications.

Butterworths, Guildford, for five figures from K. Newton, W. Steeds and T. K. Garrett, *The Motor Vehicle*, 10th edn, 1983.

The Engineer, London, for a figure from a technical publication.

Eureka Engineering Materials & Design, Horton Kirby, Kent, for a figure from a technical publication.

Ford of Europe, Inc., Brentwood, for eleven figures from Ford technical publications.

Froude Consine Ltd, Worcester, for two figures from *Publication No. 526/2*.

GKN Engine Parts Division, Maidenhead, for a figure from *Publication No. EPD 82100*.

Hutchinson Publishing Group Ltd, London, for a figure from A. Baker, *The Component Contribution* (co-sponsored by the AE Group), 1979.

Johnson Matthey Chemicals Ltd, Royston, for three figures from *Catalyst Systems for Exhaust Emission Control from Motor Vehicles*.

Longmans Group Ltd, Harlow, for four figures from G. F. C. Rogers and Y. R. Mayhew, *Engineering Thermodynamics*, 3rd edn, 1980; two figures from H. Cohen, G. F. C. Rogers and H. I. H. Saravanamuttoo, *Gas Turbine Theory*, 2nd edn, 1972.

Lucas CAV Ltd, London, for eight figures from *CAV Publications 586, 728, 730, 773* and *C2127E*, and a press release photograph.

Lucas Electrical Ltd, Birmingham, for two figures from publications *PLT 6339* (Electronic Fuel Injection) and *PLT 6176* (Ignition).

Mechanical Engineering Publications, Bury St Edmunds, for figures from technical publications.

MIT Press, Cambridge, Massachusetts, and Professor C. F. Taylor for a figure from C. F. Taylor, *The Internal Combustion Engine in Theory and Practice*, 1966/8.

Orbital Engine Corporation Ltd, Perth, Australia, for figures from technical publications.

Oxford University Press for two figures from Singer's *History of Technology*.

Patrick Stephens Ltd, Cambridge, for three figures from A. Allard, *Turbocharging and Supercharging*, 1982.

Pergamon Press Ltd, Oxford, for a figure from R. S. Benson and N. D. Whitehouse, *Internal Combustion Engines*, 1979; three figures from H. Daneshyar, *One-Dimensional Compressible Flow*, 1976.

Plenum Press, New York, and Professor John B. Heywood (MIT) for a figure from J. B. Heywood, *Combustion Modelling in Reciprocating Engines*, 1980.

Society of Automotive Engineers, Warrendale, Pennsylvania, for ten figures and copy from *SAE 790291, SAE 820167, SAE 821578* and *Prepr. No. 61A* (1968).

Sulzer Brothers Ltd, Winterthur, Switzerland, for three figures from the *Sulzer Technical Review*, Nos 1 and 3 (1982).

Verein Deutscher Ingenieure, Dusseldorf, for figures from technical publications.

Material is acknowledged individually throughout the text of the book.

Every effort has been made to trace all copyright holders but, if any have been inadvertently overlooked, the publisher will be pleased to make the necessary arrangements at the first opportunity.

Notation

abdc	after bottom dead centre
atdc	after top dead centre
A	piston area (m^2)
A_c	curtain area for poppet valve (m^2)
A_e	effective flow area (m^2)
A_f	flame front area (m^2)
A_o	orifice area (m^2)
A/F	air/fuel ratio
bbdc	before bottom dead centre
bdc	bottom dead centre
bmep	brake mean effective pressure (N/m^2)
bsac	break specific air consumption
bsfc	brake specific fuel consumption
btdc	before top dead centre
B	bore diameter (m)
BHP	brake horse power
c	sonic velocity (m/s)
c_p	specific heat capacity at constant pressure (kJ/kg K)
c_v	specific heat capacity at constant volume (kJ/kg K)
C_D	discharge coefficient
C_o	orifice discharge coefficient
C_p	molar heat capacity at constant pressure (kJ/kmol K)
C_v	molar heat capacity at constant volume (kJ/kmol K)
CI	compression ignition
CV	calorific value (kJ/kg)
dc	direct current
dohc	double overhead camshaft
D_v	valve diameter (m)
DI	direct injection (compression ignition engine)
E	absolute internal energy (kJ)
EGR	exhaust gas recirculation
f	fraction of exhaust gas residuals

ff	turbulent flame factor
fmep	frictional mean effective pressure
fwd	front wheel drive
g	gravitational acceleration (m/s^2)
G	Gibbs function (kJ)
h	specific enthalpy (kJ/kg); manometer height (m)
h_d	mean height of indicator diagram (m)
H	enthalpy (kJ)
imep	indicated mean effective pressure (N/m^2)
I	current (A)
IDI	indirect injection (compression ignition engine)
jv	just visible (exhaust smoke)
k	constant
K	equilibrium constant
KLSA	knock limited spark advance
l	length, connecting-rod length (m)
l_b	effective dynamometer lever arm length (m)
l_d	indicator diagram length (m)
L	stroke length (m); inductance (H)
L_D	duct length (m)
L_v	valve lift (m)
LBT	lean best torque
LDA	laser doppler anemometer
LDV	laser doppler velocimeter
m	mass (kg)
$\dot{m}_a$	air mass flow rate (kg/s)
$\dot{m}_f$	fuel mass flow rate (kg/s)
m_r	reciprocating mass (kg)
M	mutual inductance (H); molar mass (kg); moment of momentum flux (N/m)
MBT	minimum (ignition) advance for best torque (degrees)
n	number of moles/cylinders
N^*	rev./s for 2-stroke, rev./2s for 4-stroke engines
N'	total number of firing strokes/s ($\equiv n.N^*$)
Nu	Nusselt number (dimensionless heat transfer coefficient)
ohc	overhead camshaft
ohv	overhead valve
p	pressure (N/m^2)
p'	partial pressure (N/m^2)
$\bar{p}$	mean effective pressure (N/m^2)
Pr	Prandtl number (ratio of momentum and thermal diffusivities)
Q	heat flow (kJ)
r	crankshaft throw ($\equiv \frac{1}{2}$ engine stroke) (m)
r_p	pressure ratio

r_v	volumetric compression ratio
R	specific gas constant (kJ/kg K)
R_0	molar (or universal) gas constant (kJ/kmol K)
Re	Reynolds number
s	specific entropy (kJ/kg K)
sac	specific air consumption (kg/MJ)
sfc	specific fuel consumption (kg/MJ)
SI	spark ignition
t	time (s)
tdc	top dead centre
T	absolute temperature (K); torque (N m)
T_0	absolute temperature of the environment (K)
U_l	laminar flame front velocity (m/s)
U_t	turbulent flame front velocity (m/s)
v	velocity (m/s)
$\bar{v}_p$	mean piston velocity (m/s)
V	volume (m³)
$\dot{V}_a$	volumetric flow rate of air (m³/s)
V_s	engine swept volume (m³)
wmmp	weakest mixture for maximum power
W	work (kJ)
$\dot{W}$	power (kW)
W_b	brake work (kJ)
W_c	compressor work (kJ)
W_f	friction work (kJ)
W_i	indicated work (kJ)
W_{REV}	work output from a thermodynamically reversible process (kJ)
W_t	turbine work (kJ)
WOT	wide open throttle
x	a length (m); mass fraction
Z	Mach index
α	cut off (or load) ratio
γ	ratio of gas heat capacities, c_p/c_v or C_p/C_v
ΔH_0	enthalpy of reaction (combustion) $\equiv -CV$
Δ_p	pressure difference (N/m²)
$\Delta \theta_b$	combustion duration (crank angle, degrees)
ε	heat exchanger effectiveness = (actual heat transfer)/(max. possible heat transfer)
η	efficiency
η_b	brake thermal efficiency $\equiv \eta_o$
η_c	isentropic compressor efficiency
η_{Diesel}	ideal air standard Diesel cycle efficiency
η_{FA}	fuel–air cycle efficiency
η_i	indicated (arbitrary overall) efficiency

η_m	mechanical efficiency
η_o	arbitrary overall efficiency
η_{Otto}	ideal air standard Otto cycle efficiency
η_R	rational efficiency, W/W_{REV}
η_t	isentropic turbine efficiency
η_v	volumetric efficiency
θ	crank angle (degrees)
θ_0	crank angle at the start of combustion (degrees)
μ	dynamic viscosity (N s/m^2)
ρ	density (kg/m^3)
ρ_u	density of the unburnt gas (kg/m^3)
ϕ	equivalence ratio = (stoichiometric air/fuel ratio)/(actual air/ fuel ratio) (*note that* sometimes the reciprocal definition is used in other publications)
ω	specific humidity (kg water/kg dry air); angular velocity (rad/s)

Suffixes

a	air
b	brake
c	compressor
f	friction
i	indicated
m	mechanical
t	turbine
v	volumetric

1 Introduction

1.1 Fundamental operating principles

The reciprocating internal combustion engine must be by far the most common form of engine or prime mover. As with most engines, the usual aim is to achieve a high work output with a high efficiency; the means to these ends are developed throughout this book. The term 'internal combustion engine' should also include open circuit gas turbine plant where fuel is burnt in a combustion chamber. However, it is normal practice to omit the prefix 'reciprocating'; none the less this is the key principle that applies to both engines of different types and those utilising different operating principles. The divisions between engine types and between operating principles can be explained more clearly if stratified charge and Wankel-type engines are ignored initially; hence these are not discussed until section 1.4.

The two main types of internal combustion engine are: spark ignition (SI) engines, where the fuel is ignited by a spark; and compression ignition (CI) engines, where the rise in temperature and pressure during compression is sufficient to cause spontaneous ignition of the fuel. The spark ignition engine is also referred to as the petrol, gasoline or gas engine from its typical fuels, and the Otto engine, after the inventor. The compression ignition engine is also referred to as the Diesel or oil engine; the fuel is also named after the inventor.

During each crankshaft revolution there are two strokes of the piston, and both types of engine can be designed to operate in either four strokes or two strokes of the piston. The four-stroke operating cycle can be explained by reference to figure 1.1.

(1) The induction stroke. The inlet valve is open, and the piston travels down the cylinder, drawing in a charge of air. In the case of a spark ignition engine the fuel is usually pre-mixed with the air.
(2) The compression stroke. Both valves are closed, and the piston travels

1

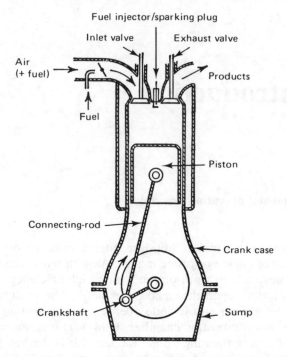

Figure 1.1   A four-stroke engine (reproduced with permission from Rogers and Mayhew (1980a))

up the cylinder. As the piston approaches top dead centre (tdc), ignition occurs. In the case of compression ignition engines, the fuel is injected towards the end of the compression stroke.

(3) The expansion, power or working stroke. Combustion propagates throughout the charge, raising the pressure and temperature, and forcing the piston down. At the end of the power stroke the exhaust valve opens, and the irreversible expansion of the exhaust gases is termed 'blow-down'.

(4) The exhaust stroke. The exhaust valve remains open, and as the piston travels up the cylinder the remaining gases are expelled. At the end of the exhaust stroke, when the exhaust valve closes some exhaust gas residuals will be left; these will dilute the next charge.

The four-stroke cycle is sometimes summarised as 'suck, squeeze, bang and blow'. Since the cycle is completed only once every two revolutions the valve gear (and fuel injection equipment) have to be driven by mechanisms operating at half engine speed. Some of the power from the expansion stroke is stored in a flywheel, to provide the energy for the other three strokes.

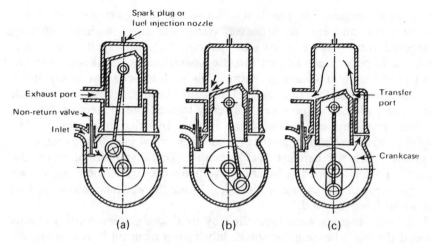

Figure 1.2   A two-stroke engine (reproduced with permission from Rogers and
            Mayhew (1980a))

The two-stroke cycle eliminates the separate induction and exhaust
strokes; and the operation can be explained with reference to figure 1.2.

(1) The compression stroke (figure 1.2a). The piston travels up the cylin-
    der, so compressing the trapped charge. If the fuel is not pre-mixed,
    the fuel is injected towards the end of the compression stroke; ignition
    should again occur before top dead centre. Simultaneously, the under-
    side of the piston is drawing in a charge through a spring-loaded
    non-return inlet valve.

(2) The power stroke. The burning mixture raises the temperature and
    pressure in the cylinder, and forces the piston down. The downward
    motion of the piston also compresses the charge in the crankcase. As
    the piston approaches the end of its stroke the exhaust port is un-
    covered (figure 1.2b) and blowdown occurs. When the piston is at
    bottom dead centre (figure 1.2c) the transfer port is also uncovered,
    and the compressed charge in the crankcase expands into the cylinder.
    Some of the remaining exhaust gases are displaced by the fresh charge;
    because of the flow mechanism this is called 'loop scavenging'. As the
    piston travels up the cylinder, first the transfer port is closed by the
    piston, and then the exhaust port is closed.

For a given size engine operating at a particular speed, the two-stroke
engine will be more powerful than a four-stroke engine since the two-
stroke engine has twice as many power strokes per unit time. Unfortunately the efficiency of a two-stroke engine is likely to be lower than that of

a four-stroke engine. The problem with two-stroke engines is ensuring that the induction and exhaust processes occur efficiently, without suffering charge dilution by the exhaust gas residuals. The spark ignition engine is particularly poor, since at part throttle operation the crankcase pressure can be less than atmospheric pressure. This leads to poor scavenging of the exhaust gases, and a rich air/fuel mixture becomes necessary for all conditions, with an ensuing low efficiency (see chapter 4, section 4.1).

These problems can be overcome in two-stroke compression ignition engines by supercharging, so that the air pressure at inlet to the crankcase is greater than the exhaust back-pressure. This ensures that when the transfer port is opened, efficient scavenging occurs; if some air passes straight through the engine, it does not lower the efficiency since no fuel has so far been injected.

Originally engines were lubricated by total loss systems with oil baths around the main bearings or splash lubrication from oil in the sump. As engine outputs increased a circulating high-pressure oil system became necessary; this also assisted the heat transfer. In two-stroke spark ignition engines a simple system can be used in which oil is pre-mixed with the fuel; this removes the need for an oil pump and filter system.

An example of an automotive four-stroke compression ignition engine is shown in figure 1.3, and a two-stroke spark ignition motor cycle engine is shown in figure 1.4.

The size range of internal combustion engines is very large, especially for compression ignition engines. Two-stroke compression ignition engines vary from engines for models with swept volumes of about 1 cm^3 and a fraction of a kilowatt output, to large marine engines with a cylinder bore of about 1 m, up to 12 cylinders in-line, and outputs of up to 50 MW.

An example of a large two-stroke engine is the Sulzer RTA engine (see figure 1.5) described by Wolf (1982). The efficiency increases with size because the effects of clearances and cooling losses diminish. As size increases the operating speed reduces and this leads to more efficient combustion; also the specific power demand from the auxiliaries reduces. The efficiency of such an engine can exceed 50 per cent, and it is also capable of burning low-quality residual fuels. A further advantage of a low-speed marine diesel engine is that it can be coupled directly to the propeller shaft. To run at low speeds an engine needs a long stroke, yet a stroke/bore ratio greater than 2 leads to poor loop scavenging.

The Sulzer RTA engine has a stroke/bore ratio of about 3 and uses uniflow scavenging. This requires the additional complication of an exhaust valve. Figure 1.6 shows arrangements for loop, cross and uniflow scavenging. The four-stroke compression ignition engines have a smaller size range, from about 400 cm^3 per cylinder to 60 litres per cylinder with an output of 600 kW per cylinder at about 600 rpm with an efficiency of over 45 per cent.

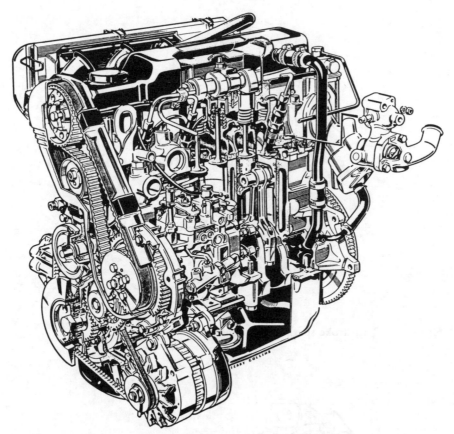

Figure 1.3   Ford 1.6 litre indirect injection Diesel (courtesy of Ford)

The size range of two-stroke spark ignition engines is small, with the total swept volumes rarely being greater than 1000 cm^3. The common applications are in motor cycles and outboard motors, where the high output, simplicity and low weight are more important than their poor fuel economy.

Since the technology of two-stroke engines is rather different from four-stroke engines, chapter 7 is devoted to two-stroke engines. The interest in two-stroke engines is because of the potential for higher specific power outputs and a more frequent firing interval. More detail can be found in the book published by Blair (1990).

Automotive four-stroke spark ignition engines usually have cylinder volumes in the range 50–500 cm^3, with the total swept volume rarely being greater than 5000 cm^3. Engine outputs are typically 45 kW/litre, a value that can be increased seven-fold by tuning and turbocharging.

The largest spark ignition engines are gas engines; these are usually

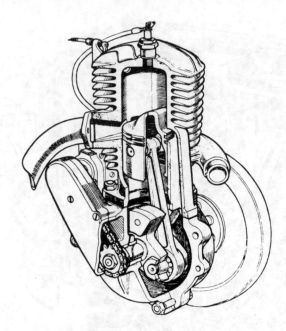

Figure 1.4   Two-stroke spark ignition engine

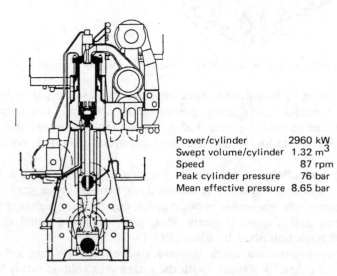

Power/cylinder	2960 kW
Swept volume/cylinder	1.32 m^3
Speed	87 rpm
Peak cylinder pressure	76 bar
Mean effective pressure	8.65 bar

Figure 1.5   Sulzer RTA two-stroke compression ignition engine (courtesy of the *Sulzer Technical Review*)

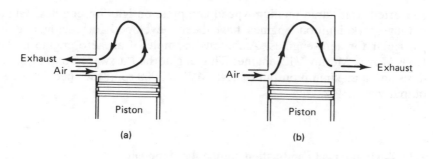

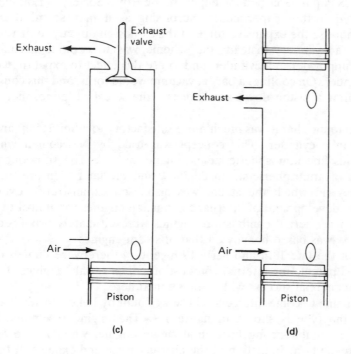

Figure 1.6   Two-stroke scavenging systems. (a) Loop scavenging; (b) cross scavenging; (c) uniflow scavenging with exhaust valve; (d) uniflow scavenging with opposed pistons

converted from large medium-speed compression ignition engines. High-output spark ignition engines have been developed for racing and in particular for aero-engines. A famous example of an aero-engine is the Rolls Royce Merlin V12 engine. This engine had a swept volume of 27 litres and a maximum output of 1.48 MW at 3000 rpm; the specific power output was 1.89 kW/kg.

1.2   Early internal combustion engine development

As early as 1680 Huygens proposed to use gunpowder for providing motive power. In 1688 Papin described the engine to the Royal Society of London, and conducted further experiments. Surprising as it may seem, these engines did not use the expansive force of the explosion directly to drive a piston down a cylinder. Instead, the scheme was to explode a small quantity of gunpowder in a cylinder, and to use this effect to expel the air from the cylinder. On cooling, a partial vacuum would form, and this could be used to draw a piston down a cylinder — the so-called 'atmospheric' principle.

Papin soon found that it was much more satisfactory to admit steam and condense it in a cylinder. This concept was used by Newcomen who constructed his first atmospheric steam engine in 1712. The subsequent development of atmospheric steam engines, and the later high-pressure steam engines (in which the steam was also used expansively), over-shadowed the development of internal combustion engines for almost two centuries. When internal combustion engines were ultimately produced, the technology was based heavily on that of steam engines.

Throughout the late 18th and early 19th century there were numerous proposals and patents for internal combustion engines; only engines that had some commercial success will be mentioned here.

The first engine to come into general use was built by Lenoir in 1860; an example of the type is shown in figure 1.7. The engine resembled a single-cylinder, double-acting horizontal steam engine, with two power strokes per revolution. Induction of the air/gas charge and exhaust of the burnt mixture were controlled by slide valves; the ignition was obtained by an electric spark. Combustion occurred on both sides of the piston, but considering just one combustion chamber the sequence was as follows.

(1) In the first part of the stroke, gas and air were drawn in. At about half stroke, the slide valves closed and the mixture was ignited; the explo-sion then drove the piston to the bottom of the stroke.

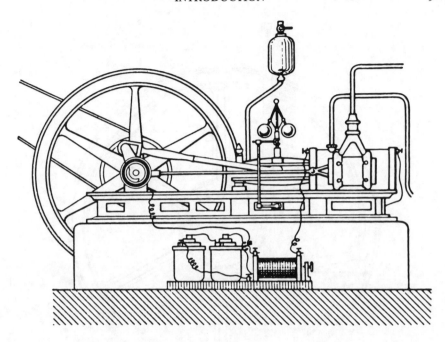

Figure 1.7   Lenoir gas-engine of 1860 (reproduced with permission from Singer, *History of Technology*, OUP, 1958)

(2) In the second stroke, the exhaust gases were expelled while combustion occurred on the other side of the piston.

The next significant step was the Otto and Langen atmospheric or free-piston engine of 1866; the fuel consumption was about half that of the Lenoir engine. The main features of the engine were a long vertical cylinder, a heavy piston and a racked piston rod (figure 1.8). The racked piston rod was engaged with a pinion connected to the output shaft by a ratchet. The ratchet was arranged to free-wheel on the upward stroke, but to engage on the downward stroke. Starting with the piston at the bottom of the stroke the operating sequence was as follows.

(1) During the first tenth or so of the stroke, a charge of gas and air was drawn into the cylinder. The charge was ignited by a flame transferred through a slide valve, and the piston was forced to the top of its stroke without delivering any work, the work being stored as potential energy in the heavy piston.
(2) As the cylinder contents cooled, the partial vacuum so formed and the weight of the piston transferred the work to the output shaft on the downward stroke. Exhaust occurred at the end of this stroke.

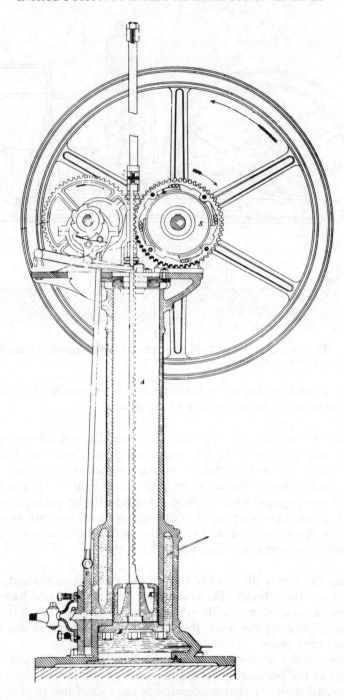

Figure 1.8   Otto and Langen free-piston engine

The piston had to weigh about 70 kg per kW of output, and by its nature the engine size was limited to outputs of a few kilowatts; none the less some 10 000 engines were produced within five years.

At the same time commercial exploitation of oil wells in the USA was occurring, as a result of the pioneer drilling by Drake in 1859. This led to the availability of liquid fuels that were much more convenient to use than gaseous fuels, since these often needed a dedicated gas-producing plant. Liquid fuels without doubt accelerated the development of internal combustion engines, and certainly increased the number of different types available, with oil products providing both the lubricant and the fuel. For the remainder of the 19th century any engine using a gaseous fuel was called a gas engine, and any engine using a liquid was called an oil engine; no reference was necessarily made to the mode of ignition or the different operational principles.

In 1876 the Otto silent engine using the four-stroke cycle was patented and produced. As well as being much quieter than the free-piston engine, the silent engine was about three times as efficient. Otto attributed the improved efficiency to a conjectured stratification of the charge. This erroneous idea was criticised by Sir Dugald Clerk, who appreciated that the improved efficiency was a result of the charge being compressed before ignition. Clerk subsequently provided the first analysis of the Otto cycle (see chapter 2, section 2.2.1).

The concept of compression before ignition can be traced back to Schmidt in 1861, but perhaps more remarkable is the work of Beau de Rochas. As well as advocating the four-stroke cycle, Beau de Rochas included the following points in 1862.

(1) There should be a high volume-to-surface ratio.
(2) The maximum expansion of the gases should be achieved.
(3) The highest possible mixture pressure should occur before ignition.

Beau de Rochas also pointed out that ignition could be achieved by sufficient compression of the charge.

Immediately following the Otto silent engine, two-stroke engines were developed. Patents by Robson in 1877 and 1879 describe the two-stroke cycle with under-piston scavenge, while patents of 1878 and 1881 by Clerk describe the two-stroke cycle with a separate pumping or scavenge cylinder.

The quest for self-propelled vehicles needed engines with better power-to-weight ratios. Daimler was the first person to realise that a light high-speed engine was needed, which would produce greater power by virtue of its higher speed of rotation, 500–1000 rpm. Daimler's patents date from 1884, but his twin cylinder 'V' engine of 1889 was the first to be produced in quantity. By the turn of the century the petrol engine was in a form that would be currently recognisable, but there was still much scope for development and refinement.

The modern compression ignition engine developed from the work of two people, Akroyd Stuart and Rudolf Diesel. Akroyd Stuart's engine, patented in 1890 and first produced in 1892, was a four-stroke compression ignition engine with a compression ratio of about 3 — too low to provide spontaneous ignition of the fuel. Instead, this engine had a large uncooled pre-chamber or vaporiser connected to the main cylinder by a short narrow passage. Initially the vaporiser was heated externally, and the fuel then ignited after it had been sprayed into the vaporiser at the end of the compression stroke. The turbulence generated by the throat to the vaporiser ensured rapid combustion. Once the engine had been started the external heat source could be removed. The fuel was typically a light petroleum distillate such as kerosene or fuel oil; the efficiency of about 15 per cent was comparable with that of the Otto silent engine. The key innovations with the Akroyd Stuart engine were the induction, being solely of air, and the injection of fuel into the combustion chamber.

Diesel's concept of compressing air to such an extent that the fuel would spontaneously ignite after injection was published in 1890, patented in 1892 and achieved in 1893; an early example is shown in figure 1.9. Some of Diesel's aims were unattainable, such as a compression pressure of 240 bar, the use of pulverised coal and an uncooled cylinder. None the less, the prototype ran with an efficiency of 26 per cent, about twice the efficiency of any contemporary power plant and a figure that steam power plant achieved only in the 1930s.

Diesel injected the fuel by means of a high-pressure (70 bar) air blast, since a liquid pump for 'solid' or airless injection was not devised until 1910 by McKechnie. Air-blast injection necessitated a costly high-pressure air pump and storage vessel; this restricted the use of diesel engines to large stationary and marine applications. Smaller high-speed compression ignition engines were not used for automotive applications until the 1920s. The development depended on experience gained from automotive spark ignition engines, the development of airless quantity-controlled fuel injection pumps by Bosch (chapter 5, section 5.5.2), and the development of suitable combustion systems by people such as Ricardo.

1.3 Characteristics of internal combustion engines

The purpose of this section is to discuss one particular engine the Ford 'Dover' direct injection in-line six-cylinder truck engine (figure 1.10). This turbocharged engine has a swept volume of 6 litres, weighs 488 kg and produces a power output of 114 kW at 2400 rpm.

Figure 1.9 Early (1898) Diesel engine (output: 45 kW at 180 rpm) (reproduced with permission from Singer, *History of Technology*, OUP, 1978)

Figure 1.10   Ford Dover compression ignition engine (courtesy of Ford)

In chapter 2 the criteria for judging engine efficiency are derived, along with other performance parameters, such as mechanical efficiency and volumetric efficiency, which provide insight into why a particular engine may or may not be efficient. Thermodynamic cycle analysis also indicates that, regardless of engine type, the cycle efficiency should improve with higher compression ratios. Furthermore, it is shown that for a given compression ratio, cycles that compress a fuel/air mixture have a lower efficiency than cycles that compress pure air. Cycle analysis also indicates the work that can be extracted by an exhaust gas turbine.

The type of fuel required for this engine and the mode of combustion are discussed in chapter 3. Since the fuel is injected into the engine towards the end of the compression stroke, the combustion is not pre-mixed but controlled by diffusion processes — the diffusion of the fuel into the air, the diffusion of the air into the fuel, and the diffusion of the combustion products away from the reaction zone. Turbulence is essential if these processes are to occur in the small time available. The main fuel requirement is that the fuel should readily self-ignite; this is the exact opposite to the requirements for the spark ignition engine. Fuel chemistry and combustion are discussed in chapter 3, along with additives, principally those that either inhibit or promote self-ignition. One of the factors limiting the output of this, and any other, diesel engine is the amount of fuel that can be injected before unburnt fuel leaves the engine as smoke (formed by agglomerated carbon particles). These and other engine emissions are discussed in sections 3.8 and 5.6.

A key factor in designing a successful compression ignition engine is the design of the combustion chamber, and the correct matching of the fuel injection to the in-cylinder air motion. These factors are discussed in chapter 5, the counterpart to chapter 4 which discusses spark ignition engines. The design and manufacture of fuel injection equipment is undertaken by specialist manufacturers; however, the final matching of the fuel injection equipment to the engine still has to be done experimentally. This turbocharged engine will have a lower compression ratio than its naturally aspirated counterpart, in order to limit the peak pressures during combustion. One effect of this is that starting is made more difficult; the compression ratio is often chosen to be the minimum that will give reliable starting. None the less, cold weather or poor fuel quality can lead to starting difficulties, and methods to improve starting are discussed in section 5.4.

The induction and exhaust processes are controlled by poppet valves in the cylinder head. The timing of events derives from the camshaft, but is modified by the clearances and the elastic properties of the valve gear. The valve timing for the turbocharged engine will keep the valves open for longer periods than in the naturally aspirated version, since appropriate turbocharger matching can cause the inlet pressure to be greater than the exhaust. These aspects, and the nature of the flow in the inlet and exhaust passages, are discussed in chapter 6. In the turbocharged engine the inlet and exhaust manifold volumes will have been minimised to help reduce the turbocharger response time. In addition, care will have been taken to ensure that the pressure pulses from each cylinder do not interfere with the exhaust system. The design of naturally aspirated induction and exhaust systems can be very involved if the engine performance is to be optimised. Pressure pulses can be reflected as rarefaction waves, and these can be used to improve the induction and exhaust processes. Since these are resonance effects, the engine speed at which maximum benefit occurs depends on the system design and the valve timing, see sections 6.4, 6.5 and 6.6.

Chapter 8 discusses the in-cylinder motion of the air in some detail. It provides definitions of turbulence and swirl, and explains how both can be measured. Chapter 5 should be read before chapter 8, as this is where the significance of swirl and turbulence is explained.

The turbocharger is another component that is designed and manufactured by specialists. Matching the turbocharger to the engine is difficult, since the flow characteristics of each machine are fundamentally different. The engine is a slow-running (but large) positive displacement machine, while the turbocharger is a high-speed (but small) non-positive displacement machine, which relies on dynamic flow effects. Turbochargers inevitably introduce a lag when speed or load is increased, since the flow rate can only increase as the rotor speed increases. These aspects, and the design of the compressor and turbine, are covered in chapter 9, along with applications to spark ignition engines. Turbochargers increase the efficiency of compression ignition engines since the power output of the engine is increased more than the mechanical losses.

Chapter 10 describes how computer modelling can be applied to internal combustion engines. Since the interaction of dynamic flow devices and positive displacement devices is complex, the chapter on computer modelling ends with an example on the computer modelling of a turbocharged diesel engine.

Some of the mechanical design considerations are dealt with in chapter 11. A six-cylinder in-line engine will have even firing intervals that produce a smooth torque output. In addition, there will be complete balance of all primary and secondary forces and moments that are generated by the reciprocating elements. The increased cycle temperatures in the turbocharged engine make the design and materials selection for the exhaust valve and the piston assembly particularly important. The increased pressures also raise the bearing loads and the role of the lubricant as a coolant will be more important. Computers are increasingly important in design work, as component weights are being reduced and engine outputs are being raised.

Chapter 12 is devoted to engine cooling systems. These are often taken for granted, but chapter 12 explains the principles, with particular emphasis on liquid cooled engines. There is also a description of how cooling systems can be designed to accelerate engine warm-up, and to provide higher efficiencies at steady-state operating points.

The use of computers is increasing in all aspects of engine work: modelling the engine to estimate performance, matching of the fuel injection equipment, selection and performance prediction of the turbocharger, estimation of vehicle performance (speed, fuel consumption etc.) for different vehicles, transmission and usage combinations.

None the less, when an engine such as the Dover Diesel is being developed there is still a need to test engines. The designers will want to know about its performance at all loads and speeds. Some measurements, such as the fuel efficiency are comparatively simple to make, and these techniques are introduced at the start of chapter 13. Chapter 13 also describes how exhaust emissions are measured, and the procedures needed for in-cylinder combustion analysis. The chapter ends with a description of computer-controlled test facilities.

1.4 Additional types of internal combustion engine

Two types of engine that fall outside the simple classification of reciprocating spark ignition or compression ignition engine are the Wankel engine and stratified charge engine.

1.4.1 The Wankel engine

The Wankel engine is a rotary combustion engine, developed from the work of Felix Wankel. The mode of operation is best explained with reference to figure 1.11. The triangular rotor has a centrally placed internal

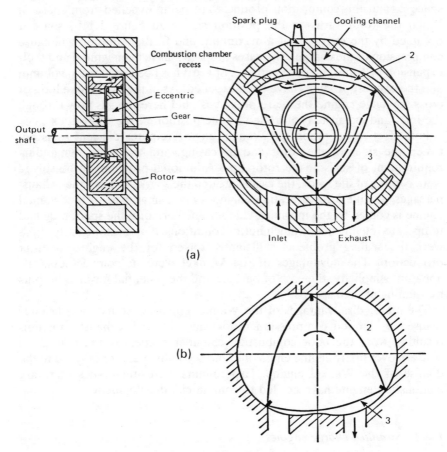

Figure 1.11   The Wankel engine (reproduced with permission from Rogers and Mayhew (1980a))

gear that meshes with a sun gear that is part of the engine casing. An eccentric that is an integral part of the output shaft constrains the rotor to follow a planetary motion about the output shaft. The gear ratios are such that the output shaft rotates at three times the speed of the rotor, and the tips of the rotor trace out the two-lobe epitrochoidal shape of the casing. The compression ratio is dictated geometrically by the eccentricity of the rotor and the shape of its curved surfaces. The convex surfaces shown in the diagram maximise and minimise the sealed volumes, to give the highest compression ratio and optimum gas exchange. A recess in the combustion chamber provides a better-shaped combustion chamber.

The sequence of events that produces the four-stroke cycle is as follows. In figure 1.11a with the rotor turning in a clockwise direction a charge is drawn into space 1, the preceding charge is at maximum compression in space 2, and the combustion products are being expelled from space 3. When the rotor turns to the position shown in figure 1.11b, space 1 occupied by the charge is at a maximum, and further rotation will cause compression of the charge. The gases in space 2 have been ignited and their expansion provides the power stroke. Space 3 has been reduced in volume, and the exhaust products have been expelled. As in the two-stroke loop or cross scavenge engine there are no valves, and here the gas flow through the inlet and exhaust is controlled by the position of the rotor apex.

For effective operation the Wankel engine requires efficient seals between the sides of the rotor and its casing, and the more demanding requirement of seals at the rotor tips. Additional problems to be solved were cooling of the rotor, the casing around the spark plug and the exhaust passages. Unlike a reciprocating engine, only a small part of the Wankel engine is cooled by the incoming charge. Furthermore, the spark plug had to operate reliably under much hotter conditions. Not until the early 1970s were the sealing problems sufficiently solved for the engine to enter production. The advantages of the Wankel were its compactness, the apparent simplicity, the ease of balance and the potential for high outputs by running at high speeds.

The major disadvantages of the Wankel engine were its low efficiency (caused by limited compression ratios) and the high exhaust emissions resulting from the poor combustion chamber shape. By the mid 1970s concern over firstly engine emissions, and secondly fuel economy led to the demise of the Wankel engine. Experiments with other types of rotary combustion engine have not led to commercial development.

1.4.2 Stratified charge engines

The principle behind stratified charge engines is to have a readily ignitable mixture in the vicinity of the spark plug, and a weaker (normally nonignitable) mixture in the remainder of the combustion chamber. The purpose of this arrangement is to control the power output of the engine by varying only the fuel supply without throttling the air, thereby eliminating the throttling pressure-drop losses. The stratification of the charge is commonly arranged by division of the combustion chamber to produce a pre-chamber that contains the spark plug. Typically fuel would also be injected into the pre-chamber, so that charge stratification is controlled by the timing and rate of fuel injection. Thus fuel supply is controlled in the same manner as compression ignition engines, yet the ignition timing of the spark controls the start of combustion.

An alternative means of preparing a stratified charge was to provide an extra valve to the pre-chamber, which controlled a separate air/fuel mixture. This was the method used in the Honda CVCC engine, figure 1.12, the first stratified charge engine in regular production.

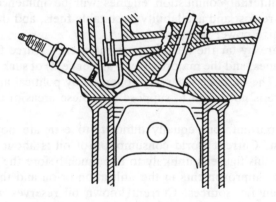

Figure 1.12   Honda CVCC engine, the first in regular production (from
                     Campbell (1978))

The advantages claimed for stratified charge engines include:

(1) Lower exhaust emissions than conventional spark ignition engines, since there is greater control over combustion.
(2) Improved efficiency, since throttling losses are eliminated.
(3) Greater fuel tolerance — suitable fuels should range from petrol to diesel.

Unfortunately stratified charge engines have not met the initial expectations. Compression ignition and spark ignition engines are the result of much development work, and instead of stratified charge engines having the best characteristics of each engine type, they can all too easily end up with the worst characteristics of each engine type, namely:

(1) high exhaust emissions
(2) unimproved efficiency
(3) low power output
(4) high expense (primarily fuel injection equipment and/or extra manu-
     facturing costs).

A proper discussion of stratified charge engines requires a knowledge of
chapters 2–5.

1.5 Prospects for internal combustion engines

The future of internal combustion engines will be influenced by two
factors: the future cost and availability of suitable fuels, and the development of alternative power plants.

Liquid fuels are by far the most convenient energy source for internal
combustion engines, and the majority (over 99 per cent) of such fuels come
from crude oil. The oil price is largely governed by political and taxation
policy, and there is no reason to suppose that these areas of control will
change.

It is very important, but equally difficult, to estimate how long oil
supplies will last. Current world consumption of oil is about 65 million
barrels per day; this figure is unlikely to rise much before the end of the
century owing to improvements in the utilisation of oil and the development of other energy sources. Current known oil reserves would then
imply a supply of crude oil for another 25–30 years. However, exploration
for oil continues and new reserves are being found; it must also be
remembered that oil companies cannot justify expensive exploration work
to demonstrate reserves for, say, the next 100 years. An alternative
approach is to look at the ratio of oil reserves to the rate of production, see
figure 1.13 from Ford (1982). This shows that as the cost of oil has
increased, and the cost of prospecting and producing in more difficult fields
has risen, then the reserves/production ratio for the Middle East has
stabilised at about 50 years. This suggests a continuing equilibrium between supply and demand.

Internal combustion engines can also be fuelled from renewable energy
sources. Spark ignition engines run satisfactorily on alcohol-based fuels,
and compression ignition engines can operate on vegetable oils. Countries
such as Brazil, with no oil reserves but plentiful sources of vegetation, are
already operating an alcohol-fuelled policy.

The other major source of hydrocarbons is coal, and even conservative
estimates show a 200-year supply from known reserves. One approach is to

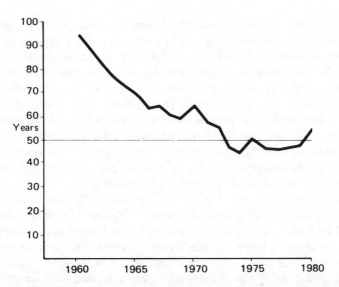

Figure 1.13   Middle East reserves/production ratio (from Ford (1982))

introduce a suspension of coal particles into the heavy fuel oil used by large compression ignition engines. A more generally applicable alternative is the preparation of fuels by 'liquefaction' or 'gasification' of coal. A comprehensive review of alternative processes is given by Davies (1983), along with the yields of different fuels and their characteristics.

The preceding remarks indicate that the future fuel supply is assured for internal combustion engines, but that other types of power plant may supersede them; the following is a discussion of some of the possibilities.

Steam engines have been used in the past, and would have the advantages of external combustion of any fuel, with readily controlled emissions. However, if it is possible to overcome the low efficiency and other disadvantages it is probable that this would have already been achieved and they would already have been adopted; their future use is thus unlikely.

Stirling engines have been developed, with some units for automotive applications built by United Stirling of Sweden. The fuel economy at full and part load is comparable to compression ignition engines, but the cost of building the complex engine is about 50 per cent greater. As the Stirling engine has external combustion it too has the capability of using a wide range of fuels with readily controlled emissions.

Gas turbines present another alternative: conventionally they use internal combustion, but external combustion is also possible. For efficient operation a gas turbine would require a high efficiency compressor and turbine, high pressure ratio, high combustion temperatures and an effective heat exchanger. These problems have been solved for large aero and industrial gas turbines, but scaling down to even truck engine size changes the design philosophy. The smaller size would dictate the less efficient radial flow compressor, with perhaps a pressure ratio of 5:1 and a regenerative heat exchanger to preserve the efficiency. All these problems could be solved, along with reductions in manufacturing cost, by the development of ceramic materials. However, part load efficiency is likely to remain poor, although this is of less significance in truck applications. The application to private passenger vehicles is even more remote because of the importance of part load efficiency, and the reductions in efficiency that would follow from the smaller size. Marine application has been limited because of the low efficiency. In addition, this can deteriorate rapidly if the turbine blades corrode as a result of the combustion of salt-laden air.

Electric vehicles present an interesting possibility which is currently restricted by the lack of a suitable battery. Lead/acid batteries are widely used, but 1 tonne of batteries only stores the same amount of energy that is available in about 5 litres of fuel. Typical performance figures are still a maximum speed of about 75 km per hour and a range of 75 km. None the less, this would meet the majority of personal transport needs, and is already economically viable for local delivery vehicles.

This leads to the possibility of a hybrid vehicle that has both an internal combustion engine and an electric motor. This is obviously an expensive solution but one that is versatile and efficient by using the motor and/or engine. Particularly ingenious systems have been developed by Volkswagen, in which the electric motor/generator is integrated with the flywheel; see Walzer (1990).

2   Thermodynamic Principles

2.1   Introduction and definitions of efficiency

This chapter provides criteria by which to judge the performance of internal combustion engines. Most important are the thermodynamic cycles based on ideal gases undergoing ideal processes. However, internal combustion engines follow a mechanical cycle, not a thermodynamic cycle. The start and end points are mechanically the same in the cycle for an internal combustion engine, whether it is a two-stroke or four-stroke mechanical cycle.

The internal combustion engine is a non-cyclic, open-circuit, quasi steady-flow, work-producing device. None the less it is very convenient to compare internal combustion engines with the ideal air standard cycles, as they are a simple basis for comparison. This can be justified by arguing that the main constituent of the working fluid, nitrogen, remains virtually unchanged in the processes. The internal combustion engine is usually treated as a steady-flow device since most engines are multi-cylindered, with the flow pulsations smoothed at inlet by air filters and at exhaust by silencers.

Air standard cycles have limitations as air and, in particular, air/fuel mixtures do not behave as ideal gases. These effects are discussed at the end of this chapter where computer modelling is introduced. Despite this, the simple air standard cycles are very useful, as they indicate trends. Most important is the trend that as compression ratio increases cycle efficiency should also increase.

At this stage it is necessary to define engine efficiency. This is perhaps the most important parameter to an engineer, although it is often very carelessly defined.

The fuel and air (that is, the reactants) enter the power plant at the temperature T_0 and pressure p_0 of the environment (that is, under ambient conditions). The discharge is usually at p_0, but in general the exhaust products from an internal combustion engine are at a temperature in excess

23

of T_0, the environment temperature. This represents a potential for producing further work if the exhaust products are used as the heat source for an additional cyclic heat power plant. For maximum work production all processes must be reversible and the products must leave the plant at T_0, as well as p_0.

Availability studies show that the work output from an ideal, reversible, non-cyclic, steady-flow, work-producing device is given by

$$W_{REV} = B_{in} - B_{out}$$

where the steady-flow availability function $B \equiv H - T_0S$.

In the case of an internal combustion engine the ideal case would be when both reactants and products enter and leave the power plant at the temperature and pressure of the environment, albeit with different composition. Thus

$$W_{REV} = (G_{R0} - G_{P0}) \equiv -\Delta G_0 \qquad (2.1)$$

where the Gibbs function G is defined by

$$G \equiv H - TS$$

and

$$G_{R0} = G \text{ of the reactants at } p_0, T_0$$
$$G_{P0} = G \text{ of the products at } p_0, T_0$$

Equation (2.1) defines the maximum amount of work (W_{REV}) that can be obtained from a given chemical reaction. As such it can be used as a basis for comparing the actual output of an internal combustion plant. This leads to a definition of rational efficiency, η_R.

$$\eta_R = \frac{W}{W_{REV}} \qquad (2.2)$$

where W is the actual work output.

It is misleading to refer to the rational efficiency as a thermal efficiency, as this term should be reserved for the cycle efficiency of a cyclic device. The upper limit to the rational efficiency can be seen to be 100 per cent, unlike the thermal efficiency of a cyclic device.

This definition is not widely used, since G_0 cannot be determined from simple experiments. Instead calorific value (CV) of a fuel is used

$$CV \equiv (H_{R0} - H_{P0}) \equiv - \Delta H_0 \qquad (2.3)$$

where $\qquad\qquad\qquad\qquad$ $H_{R0} \equiv H$ of the reactants at p_0, T_0
$\qquad\qquad\qquad\qquad\qquad$ $H_{P0} \equiv H$ of the products at p_0, T_0

and $\qquad\qquad\qquad\qquad\qquad$ $\Delta H_0 \equiv H_{P0} - H_{R0}$.

The difference in the enthalpies of the products and reactants can be readily found from the heat transfer in a steady-flow combustion calorimeter. This leads to a convenient, but arbitrary, definition of efficiency as

$$\eta_0 \equiv \frac{W}{\mathrm{CV}} \equiv \frac{W}{-\Delta H_0} \qquad\qquad (2.4)$$

where $\quad \eta_0 =$ arbitrary overall efficiency
and $\qquad W =$ work output.

While the arbitrary overall efficiency is often of the same order as the thermal efficiency of cyclic plant, it is misleading to refer to it as a thermal efficiency.

Table 2.1, derived from Haywood (1980), shows the difference between $-\Delta G_0$ and $-\Delta H_0$.

Table 2.1

Fuel	Reaction	$-\Delta G_0$	$-\Delta H_0$ (calorific value)
		MJ/kg of fuel	
C	$C + O_2 \rightarrow CO_2$	32.84	32.77
CO	$CO + \frac{1}{2} O_2 \rightarrow CO_2$	9.19	10.11
H_2	$H_2 + \frac{1}{2} O_2 \rightarrow H_2O$ liq.	117.6	142.0
	$H_2 + \frac{1}{2} O_2 \rightarrow H_2O$ vap.	113.4	120.0

The differences in table 2.1 are not particularly significant, especially as the arbitrary overall efficiency is typically about 30 per cent.

In practice, engineers are more concerned with the fuel consumption of an engine for a given output rather than with its efficiency. This leads to the use of specific fuel consumption (sfc), the rate of fuel consumption per unit power output:

$$\mathrm{sfc} = \frac{\dot{m}_f}{\dot{W}} \ \mathrm{kg/J} \qquad\qquad (2.5)$$

where $\dot{m}_f$ = mass flow rate of fuel
$\dot{W}$ = power output.

It can be seen that this is inversely proportional to arbitrary overall efficiency, and is related by the calorific value of the fuel:

$$\text{sfc} = \frac{1}{-\Delta H_0 \, \eta_0} \qquad (2.6)$$

It is important also to specify the fuel used. The specific fuel consumption should be quoted in SI units (kg/MJ), although it is often quoted in metric units such as (kg/kW h) or in British Units such as (lbs/BHP h).

Sometimes a volumetric basis is used, but this should be avoided as there is a much greater variation in fuel density than calorific value.

2.2   Ideal air standard cycles

Whether an internal combustion engine operates on a two-stroke or four-stroke cycle and whether it uses spark ignition or compression ignition, it follows a mechanical cycle not a thermodynamic cycle. However, the thermal efficiency of such an engine is assessed by comparison with the thermal efficiency of air standard cycles, because of the similarity between the engine indicator diagram and the state diagram of the corresponding hypothetical cycle. The engine indicator diagram is the record of pressure against cylinder volume, recorded from an actual engine. Pressure/volume diagrams are very useful, as the enclosed area equates to the work in the cycle.

2.2.1   The ideal air standard Otto cycle

The Otto cycle is usually used as a basis of comparison for spark ignition and high-speed compression ignition engines. The cycle consists of four non-flow processes, as shown in figure 2.1. The compression and expansion processes are assumed to be adiabatic (no heat transfer) and reversible, and thus isentropic. The processes are as follows:

1–2   isentropic compression of air through a volume ratio V_1/V_2, the compression ratio r_v
2–3   addition of heat Q_{23} at constant volume
3–4   isentropic expansion of air to the original volume

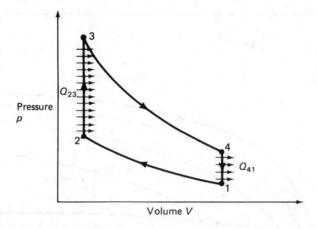

Figure 2.1   Ideal air standard Otto cycle

4–1   rejection of heat Q_{41} at constant volume to complete the cycle.

The efficiency of the Otto cycle, η_{Otto} is

$$\eta_{\text{Otto}} = \frac{W}{Q_{23}} = \frac{Q_{23} - Q_{41}}{Q_{23}} = 1 - \frac{Q_{41}}{Q_{23}}$$

By considering air as a perfect gas we have constant specific heat capacities, and for mass m of air the heat transfers are

$$Q_{23} = m\, c_v(T_3 - T_2)$$

$$Q_{41} = m\, c_v(T_4 - T_1)$$

thus

$$\eta_{\text{Otto}} = 1 - \frac{T_4 - T_1}{T_3 - T_2} \qquad (2.7)$$

For the two isentropic processes 1–2 and 3–4, $TV^{\gamma-1}$ is a constant. Thus

$$\frac{T_2}{T_1} = \frac{T_3}{T_4} = r_v^{\gamma-1}$$

where γ is the ratio of gas specific heat capacities, c_p/c_v. Thus

$$T_3 = T_4\, r_v^{\gamma-1} \quad \text{and} \quad T_2 = T_1\, r_v^{\gamma-1}$$

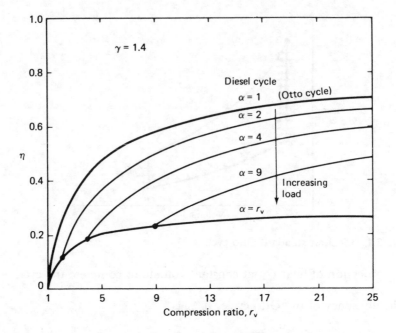

Figure 2.2   Diesel cycle efficiency for different load ratios, α

and substituting into equation (2.7) gives

$$\eta_{\text{Otto}} = 1 - \frac{T_4 - T_1}{r_v^{\gamma-1}(T_4 - T_1)} = 1 - \frac{1}{r_v^{\gamma-1}} \tag{2.8}$$

The value of η_{Otto} depends on the compression ratio, r_v, and not the temperatures in the cycle. To make a comparison with a real engine, only the compression ratio needs to be specified. The variation in η_{Otto} with compression ratio is shown in figure 2.2 along with that of η_{Diesel}.

2.2.2 The ideal air standard Diesel cycle

The Diesel cycle has heat addition at constant pressure, instead of heat addition at constant volume as in the Otto cycle. With the combination of high compression ratio, to cause self-ignition of the fuel, and constant-volume combustion the peak pressures can be very high. In large compression ignition engines, such as marine engines, fuel injection is arranged so that combustion occurs at approximately constant pressure in order to limit the peak pressures.

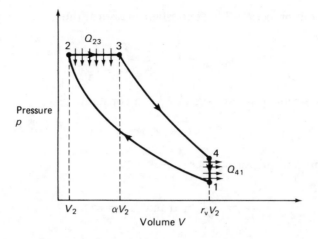

Figure 2.3   Ideal air standard Diesel cycle

The four non-flow processes constituting the cycle are shown in the state diagram (figure 2.3). Again, the best way to calculate the cycle efficiency is to calculate the temperatures around the cycle. To do this it is necessary to specify the cut-off ratio or load ratio:

$$\alpha \equiv V_3/V_2$$

The processes are all reversible, and are as follows:

1–2   isentropic compression of air through a volume ratio V_1/V_2, the compression ratio r_v
2–3   addition of heat Q_{23} at constant pressure while the volume expands through a ratio V_3/V_2, the load or cut-off ratio α
3–4   isentropic expansion of air to the original volume
4–1   rejection of heat Q_{41} at constant volume to complete the cycle. The efficiency of the Diesel cycle, η_{Diesel}, is

$$\eta_{\text{Diesel}} = \frac{W}{Q_{23}} = \frac{Q_{23} - Q_{41}}{Q_{23}} = 1 - \frac{Q_{41}}{Q_{23}}$$

By treating air as a perfect gas we have constant specific heat capacities, and for mass m of air the heat tranfers are

$$Q_{23} = m\,c_p(T_3 - T_2) \qquad \text{Const. pressure}$$

$$Q_{41} = m\,c_v(T_4 - T_1) \qquad \text{const. volume}$$

Note that the process 2–3 is at constant pressure, thus

$$\eta_{\text{Diesel}} = 1 - \frac{1}{\gamma} \frac{T_4 - T_1}{T_3 - T_2} \tag{2.9}$$

For the isentropic process 1–2, $TV^{\gamma-1}$ is a constant:

$$T_2 = T_1 r_v^{\gamma-1}$$

For the constant pressure process 2–3

$$\frac{T_3}{T_2} = \frac{V_3}{V_2} = \alpha \text{ thus } T_3 = \alpha r_v^{\gamma-1} T_1$$

For the isentropic process 3–4, $TV^{\gamma-1}$ is a constant:

$$\frac{T_4}{T_3} = \left(\frac{V_3}{V_4} \right)^{\gamma-1} = \left(\frac{\alpha}{r_v} \right)^{\gamma-1}$$

thus

$$T_4 = \left(\frac{\alpha}{r_v} \right)^{\gamma-1} T_3 = \alpha r_v^{\gamma-1} \left(\frac{\alpha}{r_v} \right)^{\gamma-1} T_1 = \alpha^{\gamma} T_1$$

Substituting for all the temperatures in equation (2.9) in terms of T_1 gives

$$\eta_{\text{Diesel}} = 1 - \frac{1}{\gamma} \cdot \frac{\alpha^{\gamma} - 1}{\alpha r_v^{\gamma-1} - r_v^{\gamma-1}} = 1 - \frac{1}{r_v^{\gamma-1}} \left[\frac{\alpha^{\gamma} - 1}{\gamma(\alpha - 1)} \right] \tag{2.10}$$

At this stage it is worth making a comparison between the air standard Otto cycle efficiency (equation 2.8) and the air standard Diesel cycle efficiency (equation 2.10).

The Diesel cycle efficiency is less convenient; it is not solely dependent on compression ratio, r_v, but is also dependent on the load ratio α. The two expressions are the same, except for the term in square brackets

$$\left[\frac{\alpha^{\gamma} - 1}{\gamma(\alpha - 1)} \right]$$

The load ratio lies in the range $1 < \alpha < r_v$, and is thus always greater than unity. Consequently the expression in square brackets is always greater than unity, and the Diesel cycle efficiency is less than the Otto cycle efficiency *for the same compression ratio*. This is shown in figure 2.2 where efficiencies have been calculated for a variety of compression ratios and load ratios. There are two limiting cases. The first is, as $\alpha \to 1$, then $\eta_{\text{Diesel}} \to \eta_{\text{Otto}}$. This can be shown by rewriting the term in square brackets, and using a binomial expansion:

$$\left[\frac{\alpha^{\gamma} - 1}{\gamma(\alpha - 1)} \right] = \left[\frac{\{1 + (\alpha - 1)\}^{\gamma} - 1}{\gamma(\alpha - 1)} \right]$$

$$= \left[\frac{1}{\gamma(\alpha - 1)} \left\{ 1 + [\gamma(\alpha - 1)] \right. \right.$$

$$\left. \left. + \frac{\gamma(\gamma - 1)}{2!}(\alpha - 1)^2 + \ldots - 1 \right\} \right] \qquad (2.11)$$

As $\alpha \to 1$, then $(\alpha - 1) \to 0$ and the $(\alpha - 1)^2$ terms and higher can be neglected, thus the term in square brackets tends to unity.

The second limiting case is when $\alpha \to r_{\text{v}}$ and point $3 \to 4$ in the cycle, and the expansion is wholly at constant pressure; this corresponds to maximum work output in the cycle. Figure 2.2 also shows that as load increases, with a fixed compression ratio the efficiency reduces. The compression ratio of a compression ignition engine is usually greater than for a spark ignition engine, so the Diesel engine is usually more efficient.

2.2.3   The ideal air standard Dual cycle

In practice, combustion occurs neither at constant volume nor at constant pressure. This leads to the Dual, Limited Pressure, or Mixed cycle which has heat addition in two stages, firstly at constant volume, and secondly at constant pressure. The state diagram is shown in figure 2.4; again all processes are assumed to be reversible. As might be expected, the efficiency lies between that of the Otto cycle and the Diesel cycle. An analysis can be found in several sources such as Taylor (1966) or Benson and Whitehouse (1979), but is merely quoted here as the extra complication does not give results significantly closer to reality.

$$\eta = 1 - \frac{1}{r_{\text{v}}^{\gamma-1}} \left[\frac{r_{\text{p}} \alpha^{\gamma} - 1}{(r_{\text{p}} - 1) + \gamma r_{\text{p}} (\alpha - 1)} \right] \qquad (2.12)$$

where r_{p} is the pressure ratio during constant-volume heat addition; $r_{\text{p}} = p_3/p_2$. As before, $\alpha = V_4/V_3$.

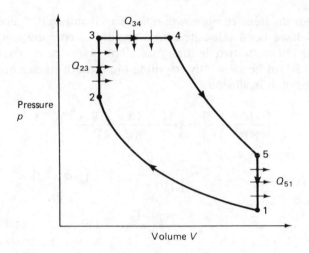

Figure 2.4   Ideal air standard Dual cycle

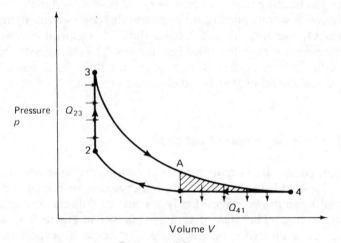

Figure 2.5   Ideal air standard Atkinson cycle

2.2.4   The ideal air standard Atkinson cycle

This is commonly used to describe any cycle in which the expansion stroke is greater than the compression stroke; the analysis is presented in example 2.3. Figure 2.5 shows the limiting case for the Atkinson cycle in which expansion is down to pressure p_1. All processes are reversible, and processes 1–2 and 3–4 are also adiabatic.

The shaded area (1A4) represents the increased work (or reduced heat rejection) when the Atkinson cycle is compared to the Otto cycle. The mechanical difficulties of arranging unequal compression and expansion strokes have prevented the development of engines working on the Atkinson cycle. However, expansion A4 can be arranged in a separate process, for example an exhaust turbine. This subject is treated more fully in chapter 9. The Atkinson cycle analysis is illustrated by example 2.3.

2.3 Comparison between thermodynamic and mechanical cycles

Internal combustion engines operate on a mechanical cycle, not a thermodynamic cycle. Although it is an arbitrary procedure it is very convenient to compare the performance of real non-cyclic engines with thermal efficiencies of hypothetical cycles. This approach arises from the similarity of the engine indicator diagram, figure 2.6, and the state diagram of a hypothetical cycle, figure 2.1 or figure 2.3.

Figure 2.6 is a stylised indicator diagram for a high-speed four-stroke engine, with exaggerated pressure difference between induction and exhaust strokes. As before, r_v is the compression ratio, and V_c is the clearance volume with the piston at top dead centre (tdc). The diagram could be from either a spark ignition or a compression ignition engine, as in both cases combustion occurs neither at constant pressure nor at constant volume. For simplicity it can be idealised as constant-volume combustion, figure 2.7, and then compared with the Otto cycle, figure 2.1.

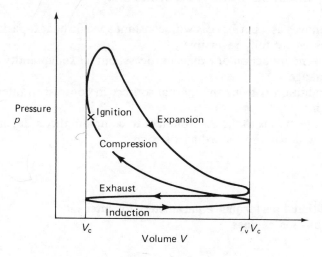

Figure 2.6   Stylised indicator diagram for four-stroke engine

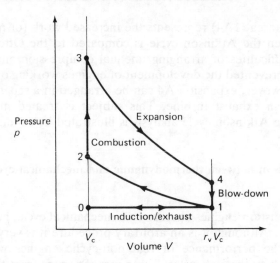

Figure 2.7   Idealised indicator diagram for four-stroke engine

In the idealised indicator diagram, induction 0–1 is assumed to occur with no pressure drop. The compression and expansion (1–2, 3–4) are not adiabatic, so neither are they isentropic. Combustion is assumed to occur instantaneously at constant volume, 2–3. Finally, when the exhaust valve opens blow-down occurs instantaneously, with the exhaust expanding into the manifold 4–1, and the exhaust stroke occurring with no pressure drop, 1–0. The idealised indicator diagram is used as a basis for the simplest computer models.

In comparison, the Otto cycle assumes that

(1) air behaves as a perfect gas with constant specific heat capacity, and all processes are fully reversible
(2) there is no induction or exhaust process, but a fixed quantity of air and no leakage
(3) heat addition is from an external source, in contrast to internal combustion
(4) heat rejection is to the environment to complete the cycle, as opposed to blow-down and the exhaust stroke.

2.4   Additional performance parameters for internal combustion engines

The rational efficiency and arbitrary overall efficiency have already been defined in section 2.1. The specific fuel consumption which has a much

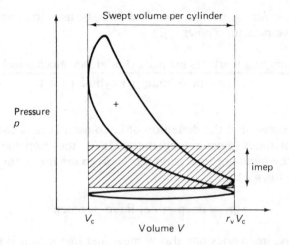

Figure 2.8   Indicated mean effective pressure

greater practical significance has also been defined. The additional parameters relate to the work output per unit swept volume in terms of a mean
effective pressure, and the effectiveness of the induction and exhaust
strokes.

There are two types of mean effective pressure, based on either the work
done by the gas on the piston or the work available as output from the
engine.

(a) Indicated mean effective pressure (imep)
The area enclosed on the p–V trace or indicator diagram from an engine is
the indicated work (W_I) done by the gas on the piston. The imep is a
measure of the indicated work output per unit swept volume, in a form
independent of the size and number of cylinders in the engine and engine
speed.

The imep is defined as

$$\text{imep (N/m}^2) = \frac{\text{indicated work output (N m) per cylinder per mechanical cycle}}{\text{swept volume per cylinder (m}^3)}$$

(2.13)

Figure 2.8 shows an indicator diagram with a shaded area, equal to the net
area of the indicator diagram. In a four-stroke cycle the negative work
occurring during the induction and exhaust strokes is termed the *pumping
loss*, and has to be subtracted from the positive indicated work of the other
two strokes. When an engine is throttled the pumping loss increases,
thereby reducing the engine efficiency. In figure 2.8 the shaded area has the
same volume scale as the indicator diagram, so the height of the shaded
area must correspond to the imep.

The *pumping loss* or *pumping work* can also be used to define a pumping mean effective pressure (*pmep*), $\bar{p}_p$

$$\bar{p}_p = \frac{\text{pumping work (N m) per cylinder per mechanical cycle}}{\text{swept volume per cylinder (m}^3\text{)}}$$

(2.14)

It should be noted that the definition of imep used here is not universally adopted. Sometimes (most notably in the USA), the imep does not always incorporate the pumping work, and this leads to the use of the terms *gross imep* and *net imep*:

$$\text{gross imep} = \text{net imep} + \text{pmep}$$

(2.15)

Unfortunately, imep does not always mean net imep, so it is necessary to check the context to ensure that it is not a gross imep.

The imep bears no relation to the peak pressure in an engine, but is a characteristic of engine type. The imep in naturally aspirated four-stroke spark ignition engines will be smaller than the imep of a similar turbocharged engine. This is mainly because the turbocharged engine has greater air density at the start of compression, so more fuel can be burnt.

(b) *Brake mean effective pressure (bmep)*
The work output of an engine, as measured by a brake or dynamometer, is more important than the indicated work output. This leads to a definition of bmep, $\bar{p}_b$, very similar to equation (2.13):

$$\bar{p}_b \text{ (N/m}^2\text{)} = \frac{\text{brake work output (N m) per cylinder per mechanical cycle}}{\text{swept volume per cylinder (m}^3\text{)}}$$

(2.16)

or in terms of the engine brake power

$$\text{brake power} = \bar{p}_b \, L \, A \, N'$$
$$= \bar{p}_b \, (L \, A \, n)N^* = \bar{p}_b \, V_s \, N^*$$

(2.17)

where L = piston stroke (m)
        A = piston area (m^2)
        n = number of cylinders
        V_s = engine swept volume (m^3)

and N' = number of mechanical cycles of operation per second
       $N^* = N'/n$

$$= \begin{cases} \text{rev./s for two-stroke engines} \\ \dfrac{\text{rev./s}}{2} \text{ for four-stroke engines.} \end{cases}$$

The bmep is a measure of work output from an engine, and not of pressures in the engine. The name arises because its unit is that of pressure.

(c) *Mechanical efficiency, η_m and Frictional mean effective pressure, fmep*
The difference between indicated work and brake work is accounted for by friction, and work done in driving essential items such as the lubricating oil pump. Frictional mean effective pressure (fmep), $\bar{p}_f$, is the difference between the imep and the bmep:

$$\text{fmep} = \text{imep} - \text{bmep}$$
$$\bar{p}_f = \bar{p}_i - \bar{p}_b \qquad (2.18)$$

Mechanical efficiency is defined as

$$\eta_m = \frac{\text{brake power}}{\text{indicated power}} = \frac{\text{bmep}}{\text{imep}} \qquad (2.19)$$

(d) *Indicated efficiency, η_i*
When comparing the performance of engines it is sometimes useful to isolate the mechanical losses. This leads to the use of indicated (arbitrary overall) efficiency as a means of examining the thermodynamic processes in an engine:

$$\eta_i = \frac{\dot{W}_i}{\dot{m}_f.CV} = \frac{\dot{W}}{\dot{m}_f.CV.\eta_m} = \frac{\eta_0}{\eta_m} \qquad (2.20)$$

(e) *Volumetric efficiency, η_v*
Volumetric efficiency is a measure of the effectiveness of the induction and exhaust processes. Even though some engines inhale a mixture of fuel and air it is convenient, but arbitrary, to define volumetric efficiency as

$$\eta_v = \frac{\text{mass of air inhaled per cylinder per cycle}}{\text{mass of air to occupy swept volume per cylinder at ambient } p \text{ and } T}$$

Assuming air obeys the Gas Laws, this can be rewritten as

$$\eta_v = \frac{\text{volume of ambient density air inhaled per cylinder per cycle}}{\text{cylinder swept volume}}$$

$$= \frac{\dot{V}_a}{V_s N^*} \qquad (2.21)$$

where $\dot{V}_a$ = volumetric flow rate of air with ambient density
$\quad\quad V_s$ = engine swept volume

$$N^* = \begin{cases} \text{rev./s for two-stroke engines} \\ \dfrac{\text{rev./s}}{2} \text{ for four-stroke engines} \end{cases}$$

The output of an engine is related to its volumetric efficiency, and this can be shown by considering: the mass of fuel that is burnt (m_f), its calorific value (CV), and the brake efficiency of the engine.

$$m_f = \frac{\eta_v \, V_s \, \rho_a}{\text{AFR}} \tag{2.22}$$

where AFR is the gravimetric air/fuel ratio and ρ_a is the ambient air density.

The amount of brake work produced per cycle (W_b) is given by

$$W_b = \bar{p}_b V_s \tag{2.23}$$

$$= m_f \, \eta_b \, CV$$

Combining equations (2.22) and (2.23) and then rearranging gives

$$\bar{p}_b = \eta_v \, \eta_b \, \rho_a \, CV/\text{AFR} \tag{2.24}$$

and similarly

$$\bar{p}_i = \eta_v \, \eta_i \, \rho_a \, CV/\text{AFR} \tag{2.25}$$

In the case of supercharged engines, compressor or intercooler delivery conditions should be used instead of ambient conditions. Volumetric efficiency has a direct effect on power output, as the mass of air in a cylinder determines the amount of fuel that can be burnt. In a well-designed, naturally aspirated engine the volumetric efficiency can be over 90 per cent, and over 100 per cent for a tuned induction system.

Volumetric efficiency depends on the density of the gases at the end of the induction process; this depends on the temperature and pressure of the charge. There will be pressure drops in the inlet passages and at the inlet valve owing to viscous effects. The charge temperature will be raised by heat transfer from the induction manifold, mixing with residual gases, and heat transfer from the piston, valves and cylinder. In a petrol engine, fuel evaporation can cool the charge by as much as 25 K, and alcohol fuels have much greater cooling effects; this improves the volumetric efficiency.

In an idealised process, with charge and residuals having the same specific heat capacity and molar mass, the temperature of the residual gases does not affect volumetric efficiency. This is because in the idealised process induction and exhaust occur at the same constant pressure, and when the two gases mix the contraction on cooling of the residual gases is exactly balanced by the expansion of the charge.

In practice, induction and exhaust processes do not occur at the same pressure and the effect this has on volumetric efficiency for different compression ratios is discussed by Taylor (1985a). The design factors influencing volumetric efficiency are discussed in chapter 6, section 6.3.

(f) *Specific air consumption (sac)*

As with specific fuel consumption, specific air consumption can be evaluated as a brake specific air consumption (bsac) or as an indicated specific air consumption (isac). The specific air consumption indicates how good a combustion system is at utilising the air trapped within the cylinder, and it is most likely to be encountered with Diesel engines. This is because the output of Diesel engines (especially when naturally aspirated) is often limited by the fuelling level that causes smoke. In other words, when the specific air consumption is evaluated at full load, it tells the designer how well the air and fuel have been mixed.

When the gravimetric air/fuel ratio is known (AFR), the relationship with brake specific fuel consumption (bsfc) and indicated specific fuel consumption (isfc) is

$$isac = AFR \times isfc \qquad (2.26)$$

$$isac = AFR \times bsfc \qquad (2.27)$$

Equations (2.26) and (2.27) show that the specific air consumption is also dependent on the engine efficiency. Indicated specific air consumption thus provides an insight into the combustion and thermodynamic performance; brake specific air consumption also includes the mechanical performance.

2.5 Fuel–air cycle

The simple ideal air standard cycles overestimate the engine efficiency by a factor of about 2. A significant simplification in the air standard cycles is the assumption of constant specific heat capacities. Heat capacities of gases are strongly temperature-dependent, as shown by figure 2.9.

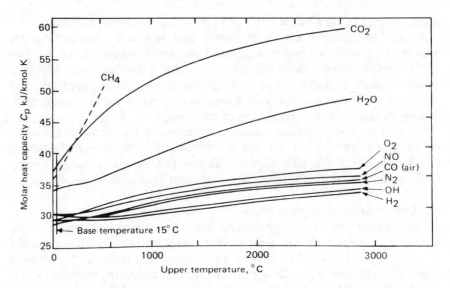

Figure 2.9   Molar heat capacity at constant pressure of gases above 15°C
quoted as averages between 15°C and abscissa temperature (adapted
from Taylor (1985a))

The molar constant-volume heat capacity will also vary, as will γ the
ratio of heat capacities:

$$C_p - C_v = R_0, \ \gamma = C_p/C_v$$

If this is allowed for, air standard Otto cycle efficiency falls from 57 per
cent to 49.4 per cent for a compression ratio of 8.

When allowance is made for the presence of fuel and combustion
products, there is an even greater reduction in cycle efficiency. This leads
to the concept of a fuel-air cycle which is the same as the ideal air standard
Otto cycle, except that allowance is made for the real thermodynamic
behaviour of the gases. The cycle assumes instantaneous complete combustion, no heat transfer, and reversible compression and expansion.
Taylor (1985a) discusses these matters in detail and provides results in
graphical form. Figures 2.10 and 2.11 show the variation in fuel-air cycle
efficiency as a function of equivalence ratio for a range of compression
ratios. Equivalence ratio ϕ is defined as the chemically correct (stoichiometric) air/fuel ratio divided by the actual air/fuel ratio. The datum
conditions at the start of the compression stroke are pressure (p_1) 1.013
bar, temperature (T_1) 115°C, mass fraction of combustion residuals
(f) 0.05, and specific humidity (ω) 0.02 – the mass fraction of water
vapour.

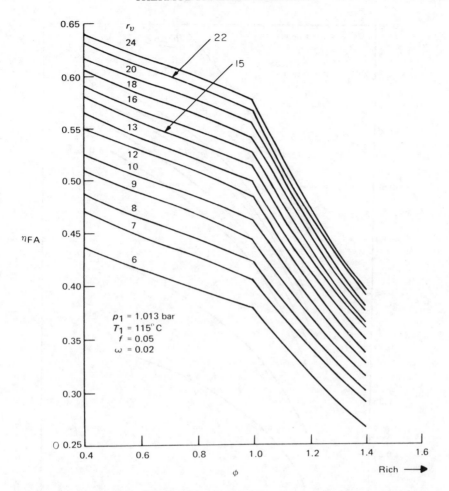

Figure 2.10   Variation of efficiency with equivalence ratio for a constant-volume
fuel–air cycle with 1-octene fuel for different compression ratios
(adapted from Taylor (1985a))

The fuel 1-octene has the formula $C_8 H_{16}$, and structure

Figure 2.10 shows the pronounced reduction in efficiency of the fuel-air
cycle for rich mixtures. The improvement in cycle efficiency with increasing
compression ratio is shown in figure 2.11, where the ideal air standard Otto
cycle efficiency has been included for comparison.

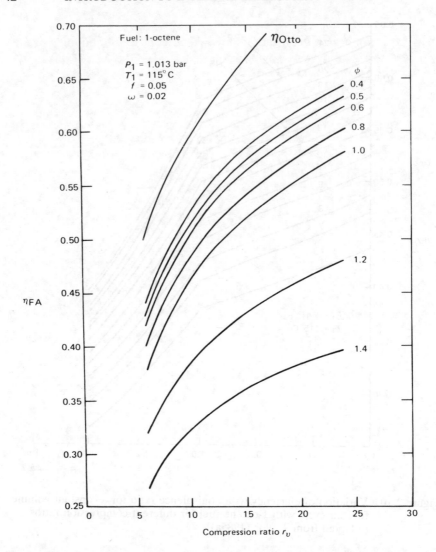

Figure 2.11   Variation of efficiency with compression ratio for a
constant-volume fuel–air cycle with 1-octene fuel for different
equivalence ratios (adapted from Taylor (1985a))

In order to make allowances for the losses due to phenomena such as
heat transfer and finite combustion time, it is necessary to develop
computer models.

Prior to the development of computer models, estimates were made for
the various losses that occur in real operating cycles. Again considering the
Otto cycle, these are as follows:

(a) 'Finite piston speed losses' occur since combustion takes a finite time and cannot occur at constant volume. This leads to the rounding of the indicator diagram and Taylor (1985a) estimates these losses as being about 6 per cent.

(b) 'Heat losses', in particular between the end of the compression stroke and the beginning of the expansion stroke. Estimates of up to 12 per cent have been made by both Taylor (1985a) and Ricardo and Hempson (1968). However, with no heat transfer the cycle temperatures would be raised and the fuel-air cycle efficiencies would be reduced slightly because of increasing gas specific heats and dissociation.

(c) Exhaust losses due to the exhaust valve opening before the end of the expansion stroke. This promotes gas exchange but reduces the expansion work. Taylor (1985a) estimates these losses as 2 per cent.

Since the fuel is injected towards the end of the compression stroke in compression ignition engines (unlike the spark ignition engine where it is pre-mixed with the air) the compression process will be closer to ideal in the compression ignition engine than in the spark ignition engine. This is another reason for the better fuel economy of the compression ignition engine.

2.6 Computer models

In internal combustion engines the induction, compression, expansion and exhaust strokes are all influenced by the combustion process. In any engine model it is necessary to include all processes, of which combustion is the most complex. Combustion models are discussed in chapter 3, section 3.9.

Benson and Whitehouse (1979) provide a useful introduction to engine modelling by giving Fortran programs for spark ignition and compression ignition engine cycle calculations. The working of the programs is explained and typical results are presented. These models have now been superseded but none the less they provide a good introduction.

The use of engine models is increasing as engine testing becomes more expensive and computing becomes cheaper. Additionally, once an engine model has been set up, results can be produced more quickly. However, it is still necessary to check model results against engine results to calibrate the model. Engine manufacturers and research organisations either develop their own models or buy-in the expertise from specialists.

The aspects of engine models discussed in this section are the compression and expansion processes. The main difference between spark ignition and compression ignition cycles is in the combustion process. The other significant difference is that spark ignition engines usually induct and compress a fuel/air mixture.

This section considers the compression and expansion strokes in the approach adopted by Benson and Whitehouse (1979).

The 1st Law of Thermodynamics expressed in differential form is

$$dQ = dE + dW \qquad (2.28)$$

The heat transfer, dQ, will be taken as zero in this simple model. Heat transfer in internal combustion engines is still very poorly understood, and there is a shortage of experimental data. A widely used correlation for heat transfer inside an engine cylinder is due to Annand (1963); this and other correlations are discussed in chapter 10, section 10.2.4.

The change in absolute energy of the cylinder contents, dE, is a complex function of temperature, which arises because of the variation of the gas specific heat capacities with temperature. Equation (2.24) cannot be solved analytically, so a numerical solution is used, which breaks each process into a series of steps. Consider the ith to the $(i + 1)^{th}$ steps, for which the values of E can be found as functions of the temperatures

$$dE = E(T_{i+1}) - E(T_i)$$

The work term, dW, equals pdV for an infinitesimal change. If the change is sufficiently small the work term can be approximated by

$$dW = \frac{(p_i + p_{i+1})}{2} (V_{i+1} - V_i)$$

These results can be substituted into equation (2.17) to give

$$0 = E(T_{i+1}) - E(T_i) + \frac{(p_{i+1} + p_i)}{2} (V_{i+1} - V_i) \qquad (2.29)$$

To find p_{i+1}, the state law is applied

$$pV = nR_0T$$

If the gas composition is unchanged, n the number of moles is constant, thus

$$p_{i+1} = \frac{(V_i)}{(V_{i+1})} \frac{(T_{i+1})}{(T_i)} P_i$$

For each step change in volume the temperature (T_{i+1}) can be found, but because of the complex nature of equation (2.25) an iterative solution is needed. The Newton-Raphson method is used because of its rapid convergence.

For the $(n + 1)^{th}$ iteration

$$(T_{i+1})_{n+1} = (T_{i+1})_n - \frac{f(E)_n}{f'(E)_n} \qquad (2.30)$$

where

$$f'(E) = \frac{d}{dT} (f(E)).$$

For each volume step the first estimate of T_{i+1} is made by assuming an isentropic process with constant specific heat capacities calculated at temperature T_i.

This model is equivalent to that used in section 2.5 on the fuel-air cycle; the efficiencies quoted from this model are slightly higher than those of Taylor (1985a). The differences can be attributed to Taylor considering starting conditions of 115°C with 5 per cent exhaust residuals and 2 per cent water vapour, and fuel of octene ($C_8 H_{16}$) as opposed to octane ($C_8 H_{18}$). As an example, figure 2.12 shows the variation of efficiency with equivalence ratio for a compression ratio of 8 using octane as fuel; this can be compared with figure 2.10.

The simple computer model is comparable to the results from the fuel-air cycle as no account is taken of either heat transfer or finite combustion times. More complex computer models are discussed in chapter 10, and these make allowance for finite combustion time, heat transfer and reaction rate kinetics. Complex models have close agreement to results that might be obtained from an experimental determination of indicated arbitrary overall efficiency. This is not surprising, as the complex models will have empirical constants derived from experiments to calibrate the model.

2.7 Conclusions

This chapter has devised criteria by which to judge the performance of actual engines, and it has also identified some of the means of increasing both efficiency and power output, namely:

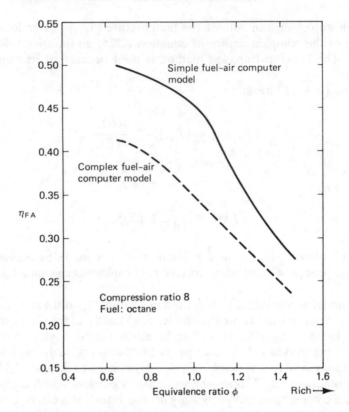

Figure 2.12   Variation of efficiency with equivalence ratio for simple fuel–air computer model and complex fuel–air model with allowance for heat transfer and combustion time (reprinted with permission from Benson and Whitehouse (1979), © Pergamon Press Ltd)

raising the compression ratio
minimising the combustion time (cut-off ratio, $\alpha \to 1$)

Reciprocating internal combustion engines follow a mechanical cycle – not a thermodynamic cycle since the start and end points are thermodynamically different. For this reason their performance should be assessed using a rational (or exergetic) efficiency. Unlike the thermal efficiency of a cyclic plant, the upper limit to the rational efficiency is 100 per cent. None the less, the arbitrary overall efficiency based on calorific value (equation 2.4) is widely used because of its convenience. Furthermore, arbitrary overall efficiencies are often compared with thermal efficiencies, since both are typically in the range 30–40 per cent. The problem of how to quote efficiency is often side-stepped by quoting specific fuel consumption, in which case the fuel ought to be fully specified.

The use of the air standard cycles arises from the similarity between engine indication (p–V) diagrams and the state diagrams of the corresponding air standard cycle. These hypothetical cycles show that, as the volumetric compression ratio is increased, the efficiency also increases (equations 2.8 and 2.10). The air standard Diesel cycle also shows that, when combustion occurs over a greater fraction of the cycle, then the efficiency reduces. No allowance is made in these simple analyses for the real properties of the working fluid, or for the changes in composition of the working fluid. These shortcomings are overcome in the fuel–air cycle and in computer models.

Despite combustion occurring more slowly in compression ignition engines, they are more efficient than spark ignition engines because of several factors:

(1) their higher compression ratios
(2) their power output is controlled by the quantity of fuel injected, not by throttling, with its associated losses
(3) during compression the behaviour of air is closer to ideal than the behaviour of a fuel/air mixture.

Additional performance parameters have also been defined: imep, bmep, fmep, pmep, mechanical efficiency, indicated efficiency and volumetric efficiency. The indicated performance parameters are particularly useful since they measure the thermodynamic performance, as distinct from brake performance which includes the associated mechanical losses.

2.8 Examples

Example 2.1

(i) The Rolls Royce CV12 turbocharged four-stroke direct injection diesel engine has a displacement of 26.1 litres. The engine has a maximum output of 900 kW at 2300 rpm and is derated to 397.5 kW at 1800 rpm for industrial use. What is the bmep for each of these types?
(ii) The high-performance version of the CV12 has an sfc of 0.063 kg/MJ at maximum power, and a minimum sfc of 0.057 kg/MJ. Calculate the arbitrary overall efficiencies for both conditions and the fuel flow rate at maximum power. The calorific value of the fuel is 42 MJ/kg.

Solution:

(i) Using equation (2.16)

$$\text{brake power} = \bar{p}_b\, V_s\, N^*$$

remembering that for a four-stroke engine $N^* = (\text{rev./s})/2$.
Rearranging gives

$$\text{bmep, } \bar{p}_b = \frac{\text{brake power}}{V_s\, \dfrac{\text{rev./s}}{2}}$$

For the high-performance engine

$$\bar{p}_b = \frac{900 \times 10^3}{26.1 \times 10^{-3} \times (2300/120)} = 18.0 \times 10^5 \text{ N/m}^2$$

$$= \underline{18 \text{ bar}}$$

For the industrial engine

$$\bar{p}_b = \frac{397.5 \times 10^3}{26.1 \times 10^{-3} \times (1800/120)} = 10.15 \times 10^5 \text{ N/m}^2$$

$$= \underline{10.15 \text{ bar}}$$

(ii) Equation (2.6) relates specific fuel consumption to arbitrary overall efficiency:

$$\text{sfc} = \frac{1}{-\Delta H_0 \eta_0} \quad \text{or} \quad \eta_0 = \frac{1}{\text{CV.sfc}}$$

At maximum power

$$\eta_0 = \frac{1}{0.063 \times 42} = 37.8 \text{ per cent}$$

At maximum economy

$$\eta_0 = \frac{1}{0.057 \times 42} = 41.8 \text{ per cent}$$

Finally calculate the maximum flow rate of fuel, $\dot{m}_f$:

$\dot{m}_f$ = brake power.sfc

$$= (900 \times 10^3) \times (0.063 \times 10^{-6}) = \underline{0.0567 \text{ kg/s}}$$

Example 2.2

A high-performance four-stroke SI engine has a swept volume of 875 cm³ and a compression ratio of 10:1. The indicated efficiency is 55 per cent of the corresponding ideal air standard Otto cycle. At 8000 rpm, the mechanical efficiency is 85 per cent, and the volumetric efficiency is 90 per cent. The air/fuel ratio (gravimetric, that is, by mass) is 13:1 and the calorific value of the fuel is 44 MJ/kg. The air is inducted at 20°C and 1 bar.
Calculate: (i) the arbitrary overall efficiency and the sfc
             (ii) the air mass flow rate, power output and bmep

(i) The first step is to find the arbitrary overall efficiency, η_0: $\eta_0 = \eta_m.\eta_i$, and the question states that $\eta_i = 0.55\, \eta_{\text{Otto}}$

$$\eta_{\text{Otto}} = 1 - \frac{1}{r_v^{\gamma-1}} \tag{2.8}$$

$$= 1 - \frac{1}{10^{(1.4-1)}} = 0.602$$

Thus $\eta_0 = 0.85 \times 0.55 \times 0.602 = \underline{28.1 \text{ per cent}}$

From equation (2.6)

$$\text{sfc} = \frac{1}{CV.\eta_0} = \frac{1}{44 \times 0.281} = \underline{0.0809 \text{ kg/MJ}}$$

(ii) The air mass flow rate is found from the volume flow rate of air using the equation of state.

$$p\dot{V}_a = m_a R_a T, \quad R_a = 287 \text{ J/kg K}$$

From equation (2.21) $\dot{V}_a = V_s\, \eta_v N^*$, where $N^* = (\text{rev.}/\text{s})/2$ for a four-stroke engine. Combining and rearranging gives

$$\dot{m}_a = \frac{pV_s \eta_v N^*}{R_a T} = \frac{10^5 \times 875 \times 10^{-6} \times 0.9 \times (8000/120)}{287 \times 293}$$

$$= 0.0624 \text{ kg/s}$$

The power output can be found for the fuel flow rate, since the efficiency (or sfc) is known. The fuel flow rate is then found for the air flow rate, from the air/fuel ratio (A/F).

$$\dot{m}_f = \dot{m}_a/A/F$$

$$\text{Brake power output} = \frac{\dot{m}_f}{\text{sfc}} = \frac{0.0624}{13.0809 \times 10^{-6}} = \underline{59.3 \text{ kW}}$$

Finally, from equation (2.13)

$$\bar{p}_b = \frac{\dot{W}_b}{V_s N^*} = \frac{59.3 \times 10^3}{875 \times 10^{-6} \times (8000/120)} = \underline{10.2 \text{ bar}}$$

Example 2.3

Reciprocating internal combustion engines have been fitted with ingenious mechanisms that allow the expansion ratio (r_e) to be greater than the compression ratio (r_c). When such a system is modelled by an ideal gas cycle there is:

(i)   heat addition at constant volume (as in the Otto cycle)
(ii) some heat rejection at constant volume (as in the Otto cycle)
(iii) and some heat rejection at constant pressure, to complete the cycle.

The processes are shown in figure 2.13 on the *T–s* plane. Depict the processes on the *p–V* plane.

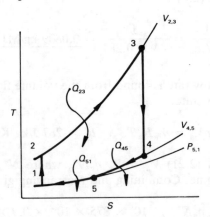

Figure 2.13   The temperature/entropy state diagram for the Atkinson cycle

The constant volume temperature rise (2–3) is θT_1. Derive an expression for the cycle efficiency in terms of r_c, r_e and θ; state any assumptions.

The limiting case is the Atkinson cycle, in which the expansion ratio is sufficient to provide full expansion to the initial pressure, and all heat rejection is then at constant pressure. Determine the limiting value of the expansion ratio (r_e) in terms of θ and r_c, and the Atkinson Cycle Efficiency.

Solution:

Assume

$1 \rightarrow 2$	isentropic compression
$2 \rightarrow 3$	constant-volume heat input
$3 \rightarrow 4$	isentropic expansion
$4 \rightarrow 5$	constant-volume heat rejection
$5 \rightarrow 1$	isobaric heat rejection

Perfect gas

with r_c = compression ratio
               r_e = expansion ratio
               v_c = clearance volume.

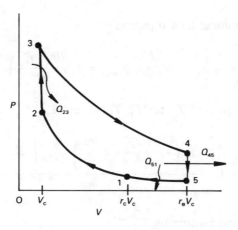

Figure 2.14   The pressure/volume state diagram for the Atkinson cyle

$$\eta = 1 - \frac{Q_0}{Q_i} = 1 - \frac{Q_{45} + Q_{51}}{Q_{23}}$$

$$\left. \begin{array}{l} Q_{23} = c_v m(T_3 - T_2) \\ Q_{45} = c_v m(T_4 - T_5) \\ Q_{51} = c_p m(T_5 - T_1) \end{array} \right\}$$

giving

$$\eta = \frac{c_v(T_4 - T_5) + c_p(T_5 - T_1)}{c_v(T_3 - T_2)}$$

It is possible to find all the temperatures in the cycle by working around the cycle, starting from state 1:

for isentropic compression

$$1 \rightarrow 2, \qquad T_2 = T_1 \, r_c^{\gamma-1}$$

for constant-volume heat addition

$$2 \rightarrow 3, \qquad T_3 = T_2 + \theta T_1 = T_1(r_c^{\gamma-1} + \theta)$$

Now working from state 1 in the opposite direction:

for constant-pressure heat rejection

$$5 \rightarrow 1, \qquad T_5 = (V_5/V_1) \times T_1 = T_1 \times (r_e/r_c)$$

for constant-volume heat rejection

$$4 \rightarrow 5, \qquad T_4 = T_5/r_e^{\gamma-1} = T_1\left(\frac{\theta + r_c^{\gamma-1}}{r_e^{\gamma-1}}\right)$$

Substitute in terms of T_1, for T_2, T_3, T_4 and T_5:

$$\eta = 1 - \frac{c_v\left(\dfrac{\theta + r_c^{\gamma-1}}{r_e^{\gamma-1}} - \dfrac{r_e}{r_c}\right) + c_p\left(\dfrac{r_e}{r_c} - 1\right)}{\theta c_v}$$

Multiply top and bottom by $\dfrac{r_c r_e^{\gamma-1}}{c_v}$:

$$\eta = 1 - \frac{\theta r_c + r_c^\gamma - r_e^\gamma + \gamma\,(r_e - r_c) \times r_e^{\gamma-1}}{\theta r_c \, r_e^{\gamma-1}}$$

or

$$\eta = 1 - \frac{(\gamma - 1)r_e^{\gamma} + r_c(\theta - \gamma r_e^{\gamma-1}) + r_c^{\gamma}}{\theta \, r_c \, r_e^{\gamma-1}}$$

<u>Check:</u> as $r_c \to r_e$, $r_c = r_e = r_v$, $\eta \to \eta_{\text{Otto}} = 1 - \dfrac{1}{r_v^{\gamma-1}}$

The limiting case (the Atkinson Cycle) is when full expansion occurs, that is, state 4 merges with state 5. Thus $T_4 = T_5$, and

$$T_1 \left(\frac{\theta + r_c^{\gamma-1}}{r_e^{\gamma-1}} \right) = T_1 \left(\frac{r_e}{r_c} \right)$$

Rearranging to give an expression for r_e:

$$r_e^{\gamma} = \theta r_c + r_c^{\gamma}, \qquad r_e = \sqrt[\gamma]{(\theta r_c + r_c^{\gamma})}$$

$$\eta_{\text{Atkinson}} = 1 - \frac{Q_{51}}{Q_{23}} = 1 - \frac{c_p(r_e/r_c - 1)T_1}{c_v \theta T_1}$$

$$\eta_{\text{Atkinson}} = 1 - \frac{\gamma(r_e - r_c)}{\theta r_c} = 1 - \frac{\gamma(r_e - r_c)}{r_e^{\gamma} - r_c^{\gamma}}$$

The limiting compression ratio for a SI engine is around 10; suppose $T_1 = 300$ K.

Consider the combustion of a stoichiometric mixture, with a gravimetric air/fuel ratio of 15, and a calorific value (CV) of 44 MJ/kg of fuel:

$$\theta = \frac{T_{23}}{T_1} \frac{\text{CV}/(\text{AFR} + 1)}{C_v/T_1} = \frac{44 \times 10^3/16}{0.75 \times 300}$$

$$\theta \sim 12$$

This enables us to find the expansion ratio corresponding to a compression ratio of 10:

$$r_e = \sqrt[\gamma]{(\theta r_c + r_c^{\gamma})}$$
$$= \sqrt[1.4]{(2 \times 10 + 10^{1.4})}$$
$$= 35$$

An expansion ratio 3.5 times the compression ratio will be very difficult to obtain:

$$\eta_{\text{Atkinson}} = 1 - \frac{\gamma(r_e - r_c)}{\theta r_c} = 1 - \frac{1.4(35 - 10)}{12 \times 10} = 70.8 \text{ per cent}$$

$$\eta_{\text{Otto}} = 1 - \frac{1}{r_v^{\gamma-1}} = 60.2 \text{ per cent}$$

Thus, at most, η_{Atkinson} gives a 16 per cent improvement in cycle efficiency and work output. However, in practice, this is unlikely to be obtained, because of the increased frictional losses. Also, the engine would be very bulky, about 3 times the size of a conventional engine that has the same output.

2.9 Problems

2.1 For the ideal air standard Diesel cycle with a volumetric compression ratio of 17:1 calculate the efficiencies for cut-off rates of 1, 2, 4, 9. Take $\gamma = 1.4$. The answers can be checked with figure 2.2.

2.2 Outline the shortcomings of the simple ideal cycles, and explain how the fuel-air cycle and computer models overcome these problems.

2.3 A 2 litre four-stroke indirect injection diesel engine is designed to run at 4500 rpm with a power output of 45 kW; the volumetric efficiency is found to be 80 per cent. The sfc is 0.071 kg/MJ and the fuel has a calorific value of 42 MJ/kg. The ambient conditions for the test were 20°C and 1 bar. Calculate the bmep, the arbitrary overall efficiency, and the air/fuel ratio.

2.4 A twin-cylinder two-stroke engine has a swept volume of 150 cm³. The maximum power output is 19 kW at 11 000 rpm. At this condition the sfc is 0.11 kg/MJ, and the gravimetric air/fuel ratio is 12:1. If ambient test conditions were 10°C and 1.03 bar, and the fuel has a calorific value of 44 MJ/kg, calculate: the bmep, the arbitrary overall efficiency and the volumetric efficiency.

2.5 A four-stroke 3 litre V6 spark ignition petrol engine has a maximum power output of 100 kW at 5500 rpm, and a maximum torque of 236 N m at 3000 rpm. The minimum sfc is 0.090 kg/MJ at 3000 rpm, and the air flow rate is 0.068 m³/s. The compression ratio is 8.9:1 and the mechanical efficiency is 90 per cent. The engine was tested under ambient conditions of 20°C and 1 bar; take the calorific value of the fuel to be 44 MJ/kg.

(a) Calculate the power output at 3000 rpm and the torque output at 5500 rpm.
(b) Calculate for both speeds the bmep and the imep.
(c) How does the arbitrary overall efficiency at 3000 rpm compare with the corresponding air standard Otto cycle efficiency?
(d) What is the volumetric efficiency and air/fuel ratio at 3000 rpm?

2.6 Show that the air standard Dual cycle efficiency is given by

$$\eta = 1 - \frac{1}{r_v^{\gamma-1}} \left[\frac{r_p\, \alpha^\gamma - 1}{(r_p - 1) + \gamma r_p(\alpha - 1)} \right] \qquad (2.12)$$

where r_v = volumetric compression ratio
  r_p = pressure ratio during constant-volume heat addition
  α = volumetric expansion ratio during constant-pressure, heat addition.

3 Combustion and Fuels

3.1 Introduction

The fundamental difference between spark ignition (SI) and compression ignition (CI) engines lies in the type of combustion that occurs, and not in whether the process is idealised as an Otto cycle or a Diesel cycle. The combustion process occurs at neither constant volume (Otto cycle), nor constant pressure (Diesel cycle). The difference between the two combustion processes is that spark ignition engines usually have pre-mixed flames while compression ignition engines have diffusion flames. With pre-mixed combustion the fuel/air mixture must always be close to stoichiometric (chemically correct) for reliable ignition and combustion. To control the power output a spark ignition engine is throttled, thus reducing the mass of fuel and air in the combustion chamber; this reduces the cycle efficiency. In contrast, for compression ignition engines with fuel injection the mixture is close to stoichiometric only at the flame front. The output of compression ignition engines can thus be varied by controlling the amount of fuel injected; this accounts for their superior part load fuel economy.

With pre-mixed reactants the flame moves relative to the reactants, so separating the reactants and products. An example of pre-mixed combustion is with oxy-acetylene equipment; for welding, the flame is fuel-rich to prevent oxidation of the metal, while, for metal cutting, the flame is oxygen-rich in order to burn as well as to melt the metal.

With diffusion flames, the flame occurs at the interface between fuel and oxidant. The products of combustion diffuse into the oxidant, and the oxidant diffuses through the products. Similar processes occur on the fuel side of the flame, and the burning rate is controlled by diffusion. A common example of a diffusion flame is the candle. The fuel is melted and evaporated by radiation from the flame, and then oxidised by air; the process is evidently one governed by diffusion as the reactants are not pre-mixed.

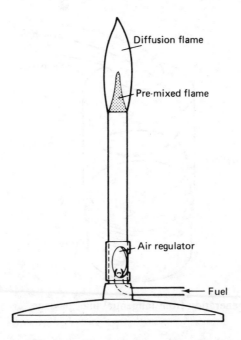

Figure 3.1   Bunsen burner with pre-mixed and diffusion flames

The Bunsen burner, shown in figure 3.1, has both a pre-mixed flame and a diffusion flame. The air entrained at the base of the burner is not sufficient for complete combustion with a single pre-mixed flame. Consequently, a second flame front is established at the interface where the air is diffusing into the unburnt fuel.

The physics and chemistry of combustion are covered in some detail by both Gaydon and Wolfhard (1979) and Lewis and von Elbe (1961), but neither book devotes much attention to combustion in internal combustion engines. Hydrocarbon/air mixtures have a maximum laminar burning velocity of about 0.5 m/s, a notable exception being acetylene/air with a value of 1.58 m/s.

An order of magnitude calculation for the burning time in a cylinder of 100 mm diameter with central ignition gives about 100 ms. However, for an engine running at 3000 rpm the combustion time can only be about 10 ms. This shows the importance of turbulence in speeding combustion by at least an order of magnitude.

Turbulence is generated as a result of the induction and compression processes, and the geometry of the combustion chamber. In addition there can be an ordered air motion such as swirl which is particularly important

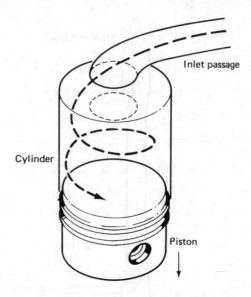

Figure 3.2   Swirl generation from tangential inlet passage

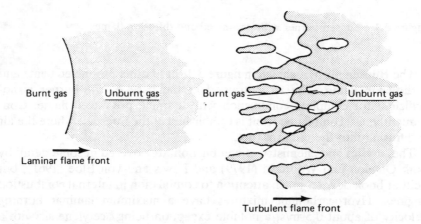

Figure 3.3   Comparison between laminar and turbulent flame fronts for
              pre-mixed combustion

in diesel engines. This is obtained from the tangential component of
velocity during induction, figure 3.2.

For pre-mixed combustion the effect of turbulence is to break up, or
wrinkle the flame front. There can be pockets of burnt gas in the unburnt
gas and vice versa. This increases the flame front area and speeds up
combustion. Figure 3.3 shows a comparison of laminar and turbulent flame
fronts.

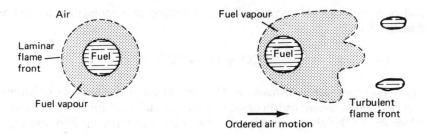

Figure 3.4　Comparison between laminar flame front in stagnant air with turbulent flame front and ordered air motion for diffusion-controlled combustion

For diffusion-controlled combustion the turbulence again enhances the burning velocity. Fuel is injected as a fine spray into air which is hot enough to vaporise and ignite the fuel. The ordered air motion is also important because it sweeps away the vaporised fuel and combustion products from the fuel droplets; this is shown in figure 3.4.

Finally, it should be noted that there are compression ignition engines with pre-mixed combustion, for example 'model diesel' engines that use a carburetted ether-based fuel. Conversely, there are spark ignition engines such as some stratified charge engines that have diffusion processes; neither of these exceptions will be considered separately.

Before discussing combustion in spark ignition engines and compression ignition engines in any greater detail, it is necessary to study combustion chemistry and dissociation.

3.2 Combustion chemistry and fuel chemistry

Only an outline of combustion chemistry will be given here as the subject is treated in general thermodynamics books such as Rogers and Mayhew (1980a) or more specialised works like Goodger (1979). However, neither of these books emphasises the difference between rational efficiency (η_R) and arbitrary overall efficiency (η_0).

For reacting mixtures the use of molar quantities is invaluable since reactions occur between integral numbers of molecules, and the mole is a unit quantity of molecules. The mole is the amount of substance in a system that contains as many elementary entities as there are atoms in 0.012 kg of carbon 12. The normal SI unit is the kilomole (kmol), and the molar number (Avogadro constant) is 6.023×10^{26} entities/kmol.

Consider the reaction between two molecules of carbon monoxide (CO)

and one molecule of oxygen (O_2) to produce two molecules of carbon dioxide (CO_2):

$$2CO + O_2 \rightarrow 2CO_2$$

There is conservation of mass, and conservation of the number of atoms.

It is often convenient to consider a unit quantity of fuel, for instance a kilomole, so the above reaction can be written in terms of kilomoles as

$$CO + \tfrac{1}{2}O_2 \rightarrow CO_2$$

A further advantage of using kilomoles is that, for the same conditions of temperature and pressure, equal volumes of gas contain the same number of moles. Thus the volumetric composition of a gas mixture is also the molar composition of the gas mixture. This is obviously not the case if liquid or solids are also present.

As well as the forward reaction of carbon monoxide oxidising to carbon dioxide, there will be a reverse reaction of carbon dioxide dissociating to carbon monoxide and oxygen:

$$CO + \tfrac{1}{2}O_2 \rightleftharpoons CO_2$$

When equilibrium is attained the mixture will contain all possible species from the reaction. In addition the oxygen can also dissociate

$$O_2 \rightleftharpoons 2O$$

With internal combustion engines, dissociation is important. Furthermore, there is not always sufficient time for equilibrium to be attained. Initially, complete combustion will be assumed.

Fuels are usually burnt with air, which has the following composition:

Molar   21.0 per cent O_2 79 per cent N_2^*

Gravimetric   23.2 per cent O_2 76.8 per cent N_2^*

Atmospheric nitrogen, N_2^*, represents all the constituents of air except oxygen and water vapour. Its thermodynamic properties are usually taken to be those of pure nitrogen. The molar masses (that is, the masses containing a kilomole of molecules) for these substances are:

$$O_2 \; 31.999 \text{ kg/kmol}; \; N_2 \; 28.013 \text{ kg/kmol}$$

$$N_2^* \; 28.15 \text{ kg/kmol}; \; \text{air } 28.96 \text{ kg/kmol}$$

When carbon monoxide is burnt with air the reaction (in kmols) is

$$CO + \tfrac{1}{2}\left(O_2 + \frac{79}{21}\,N_2^*\right) \rightarrow CO_2 + \tfrac{1}{2}\,\frac{79}{21}N_2^*$$

The nitrogen must be kept in the equation even though it does not take part in the reaction; it affects the volumetric composition of the products and the combustion temperature. The molar or volumetric air/fuel ratio is

$$1: \tfrac{1}{2}\left(1 + \frac{79}{21}\right)$$

or   1: 2.38

The gravimetric air/fuel ratio is found by multiplying the number of moles by the respective molar masses – (12 + 16) kg/kmol for carbon monoxide, and 29 kg/kmol for air:

$$1.(12 + 16): 2.38\,(29)$$

or   1: 2.47

So far the reacting mixtures have been assumed to be chemically correct for complete combustion (that is, stoichiometric). In general, reacting mixtures are non-stoichiometric; these mixtures can be defined in terms of the excess air, theoretical air or equivalence ratio. Consider the same reaction as before with 25 per cent excess air, or 125 per cent theoretical air:

$$CO + \frac{1.25}{2}\left(O_2 + \frac{79}{21}\,N_2^*\right) \rightarrow CO_2 + \frac{0.25}{2}\,O_2 + \frac{1.25}{2}\cdot\frac{79}{21}N_2^*$$

The equivalence ratio

$$\phi = \frac{\text{stoichiometric air/fuel ratio}}{\text{actual air/fuel ratio}} = \frac{1}{1.25} = 0.8$$

The air/fuel ratio can be either gravimetric or molar; the usual form is gravimetric and this is often implicit.

Fuels are often mixtures of hydrocarbons, with bonds between carbon atoms, and between hydrogen and carbon atoms. During combustion these bonds are broken, and new bonds are formed with oxygen atoms, accompanied by a release of chemical energy. The principal products are carbon dioxide and water.

Table 3.1   Alkane family of compounds

Formula	Name	Comments	
CH_4	methane	'Natural gas' ⎫	LPG
C_2H_6	ethane	⎪	Liquefied
C_3H_8	propane ⎫	⎬	Petroleum
C_4H_{10}	butane ⎭	'Bottled gas' ⎭	Gases
C_5H_{12}	pentane		
C_6H_{14}	hexane	Liquids at room	
C_7H_{16}	heptane	temperature	
C_8H_{18}	octane		
.			
.			
.			
$C_{16}H_{34}$	cetane		
.			
.			
etc.			

As combustion does not pass through a succession of equilibrium states it is irreversible, and the equilibrium position will be such that entropy is a maximum. The different compounds in fuels are classified according to the number of carbon atoms in the molecules. The size and geometry of the molecule have a profound effect on the chemical properties. Each carbon atom requires four bonds; these can be single bonds or combinations of single, double and triple bonds. Hydrogen atoms require a single bond.

An important family of compounds in petroleum (that is, petrol or diesel fuel) are the alkanes, formerly called the paraffins. Table 3.1 lists some of the alkanes; the different prefixes indicate the number of carbon atoms.

The alkanes have a general formula C_nH_{2n+2}, where n is the number of carbon atoms. Inspection shows that all the carbon bonds are single bonds, so the alkanes are termed 'saturated'. For example, propane has the structural formula

$$
\begin{array}{c}
\quad H \quad\ H \quad\ H \\
\quad | \quad\ | \quad\ | \\
H-C-C-C-H \\
\quad | \quad\ | \quad\ | \\
\quad H \quad\ H \quad\ H
\end{array}
$$

When four or more carbon atoms are in a chain molecule it is possible to form isomers. Isomers have the same chemical formula but different structures, which often leads to very different chemical properties. Iso-octane is of particular significance for spark ignition engines; although it should be called 2, 2, 4-trimethylpentane, the isomer implied in petroleum technology is

```
        |
      — C —
   |   |   |       |
 — C — C — C — C — C —        (Hydrogen atoms not shown)
   |   |   |       |
      — C —   — C —
        |       |
```

Compounds that have straight chains with a single double bond are termed alkenes (formerly olefines); the general formula is C_nH_{2n}. An example is propylene, C_3H_6:

$$
\begin{array}{c}
H \\
|H \\
H-C-\;C=C \\
||H \\
HH
\end{array}
$$

Such compounds are termed 'unsaturated' as the double bond can be split and extra hydrogen atoms added, a process termed 'hydrogenation'.

Most of the alkene content in fuels comes from catalytic cracking. In this process the less volatile alkanes are passed under pressure through catalysts such as silica or alumina at about 500°C. The large molecule are decomposed, or cracked, to form smaller more volatile molecules. A hypothetical example might be

$$C_{20}H_{42} \rightarrow C_4H_8 + C_5H_{10} + C_4H_{10} + C_6H_{14} + C$$

alkane   alkenes   alkanes   carbon

A disadvantage of alkenes is that they can oxidise when the fuel is stored in contact with air. The oxidation products reduce the quality of the fuel and leave gum deposits in the engine. Oxidation can be inhibited by the addition of alkyl phenol, typically 100 ppm (parts per million by weight).

Hydrocarbons can also form ring structures, which can be saturated or unsaturated. Cyclo-alkanes are saturated and have a general formula C_nH_{2n}; in petroleum technology they are called naphthenes. An example is cyclopropane.

```
        H       H
         \     /
          \   /
           C
          / \
         /   \
  H — C  —  C — H
     /         \
    /           \
   H             H
```

Aromatic compounds are unsaturated and based on the benzene molecule, C_6H_6. This has an unsaturated ring best represented as

The inner circle signifies two groups of three electrons, which form molecular bonds each side of the carbon atom plane. The structure accounts for the distinct properties of the aromatic compounds. Benzene and its derivatives occur in many crude oils but in particular they come from the distillation of coal.

The final class of fuels that have significance for internal combustion engines are the alcohols, in particular methanol (CH_3OH) and ethanol (C_2H_5OH):

The resurgent interest in alcohols is due to their manufacture from renewable energy sources, such as the destructive distillation of wood and fermentation of sugar respectively.

The calculations for the combustion of fuel mixtures are not complicated, but are best shown by worked solutions; see examples 3.1 and 3.2. At this stage it is sufficient to note that the mean composition of molecules will be close to C_xH_{2x} and that most of the bonds will be single carbon–carbon or carbon–hydrogen. This implies that the stoichiometric (gravimetric) air/fuel ratio is always close to 14.8:1 (see problem 3.1). Furthermore, because of the similarity in bond structure, the calorific value and the density will vary only slightly for a range of fuels; this is shown in figure 3.5 using data from Blackmore and Thomas (1977) and Taylor (1985a).

3.3   Combustion thermodynamics

Only combustion under the simple conditions of constant volume or constant pressure will be considered here. The gases, both reactants and

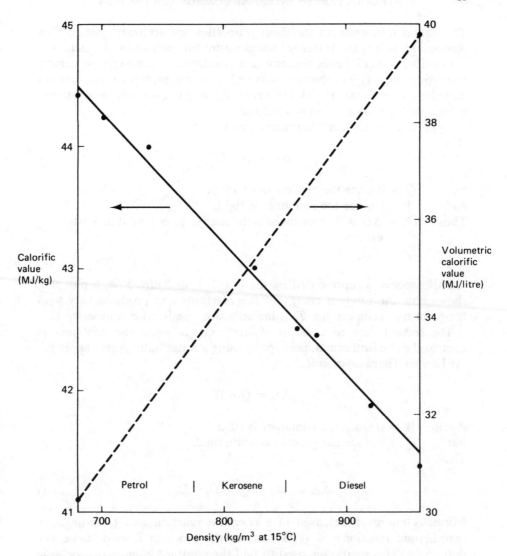

Figure 3.5 Variation in calorific value for different fuel mixtures (with acknowledgement to Blackmore and Thomas (1977))

products, will be considered as ideal (a gas that obeys the equation of state $pV = RT$, but has specific heat capacities that vary with temperature but not pressure). The molar enthalpy and molar internal energy of gases can be tabulated, Rogers and Mayhew (1980b), but if only molar enthalpy is tabulated, Haywood (1972), molar internal energy has to be deduced:

$$H \equiv U + pV \quad \text{or} \quad H \equiv U + R_0 T$$

The datum temperatures for these properties are arbitrary, but 25°C is convenient as it is the reference temperature for enthalpies of reaction.

Consider a rigid vessel containing a stoichiometric mixture of carbon monoxide and oxygen; the reactants (R) react completely to form carbon dioxide, the products (P). The first case, figure 3.6a, is when the container is insulated, so the process is adiabatic.

Apply the 1st Law of Thermodynamics:

$$\Delta E = Q - W$$

but $Q = 0$ since the process is adiabatic
and $W = 0$ since the container is rigid.
Thus $\Delta E = \Delta U = 0$ since there is no change in potential or kinetic energy
and $U_R = U_P$.

This process is represented by the line 1–2 on figure 3.6c, a plot that shows how the internal energy of the reactants and products vary with temperature. Temperature T_2 is the adiabatic combustion temperature.

The second case to consider, figure 3.6b, is when the container is contrived to be isothermal, perhaps by using a water bath. Again apply the 1st Law of Thermodynamics:

$$\Delta E = Q - W$$

Again $W = 0$ since the container is rigid
but $Q \neq 0$ since the process is isothermal.
Thus

$$\Delta E = \Delta U = Q = U_P - U_R \qquad (3.1)$$

Normally energy is released in a chemical reaction and Q is negative (exothermic reaction); if Q is positive the reaction is said to be endothermic. This method is used to find the constant volume or isochoric calorific value of a fuel in a combustion bomb.

Isochoric calorific value $= (CV)_{T,V} = - (\Delta U)_{T,V} = (U_R - U_P)_{T,V}$. The process is shown by the line 1–3 in figure 3.6c.

Combustion at constant pressure can be analysed by considering a rigid cylinder closed by a free-moving perfectly sealed piston, figure 3.7a. As before, the combustion can be contrived to be adiabatic or isothermal.

Again apply the 1st Law of Thermodynamics:

$$Q = \Delta U + W$$

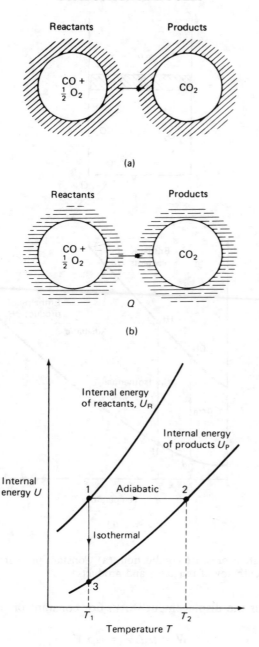

Figure 3.6   Constant-volume combustion. (a) Rigid insulated bomb
calorimeter; (b) rigid isothermal bomb calorimeter; (c) internal
energy of products and reactants

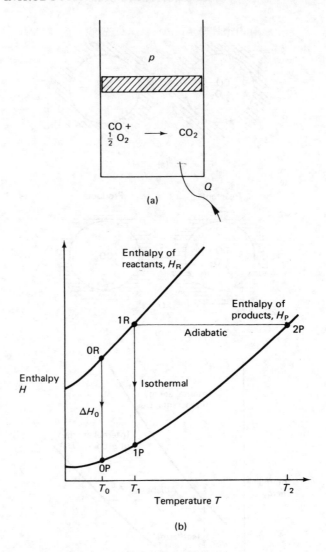

Figure 3.7   Constant-pressure combustion. (a) Constant-pressure calorimeter; (b) enthalpy of reactants and products

In this case W is the displacement work. For constant pressure:

$$W = p \, \Delta V = \Delta p \, V$$

Thus

$$Q = \Delta U + \Delta p \, V = \Delta H = H_P - H_R \tag{3.2}$$

Adiabatic and isothermal combustion processes are shown on figure 3.7b. Calorific value normally refers to the isobaric isothermal calorific value, $CV = -\Delta H_{Tp} = (H_R - H_P)_{T,p}$. The datum temperature T_0 for tabulating calorific value is usually 298.15 K (25°C); this is denoted by $-\Delta H_0$. To evaluate calorific values at other temperatures, say T_1, recall that the calorific value at this temperature, $-\Delta H_{T_1p}$, will be independent of the path taken.

Referring to figure 3.7b, 1R–0R–0P–1P will be equivalent to 1R–1P. Thus

$$\Delta H_{T_1P} = (H_{0P} - H_{1P}) + \Delta H_{T_0P} + (H_{1R} - H_{0R}) \qquad (3.3)$$

This principle is demonstrated by example 3.3. These calorific values can be found using steady-flow combustion calorimeters.

Since most combustion occurs at constant pressure (boilers, furnaces and gas turbines) isobaric calorific value ($-\Delta H_0$) is tabulated instead of isochoric value ($-\Delta U_0$); consequently there is a need to be able to convert from isobaric to isochoric calorific value.

Consider the reactants at the same temperature, pressure and volume (T, p, V_R). The difference between the calorific values is given by

$$(CV)_{p,T} - (CV)_{V,T} = (H_R - H_P)_{T,p} - (U_R - U_P)_{T,V_R}$$

Recalling that $H \equiv U + pV$

$$(CV)_{p,T} - (CV)_{V,T} = (U_R - U_P)_{T,p} + p(V_R - V_P)_{T,p} - (U_R - U_P)_{T,V_R}$$

and assuming that internal energy is a function only of temperature and not of pressure, that is

$$(U_P)_{p,T} = (U_P)_{V_R,T}$$

thus

$$(CV)_{p,T} - (CV)_{V,T} = p[V_R - V_P]_{p,T}$$

Neglect the volumes of any liquids or solids, and assume ideal gas behaviour, $pV = n R_0 T$; so

$$(CV)_{p,T} - (CV)_{V,T} = (n_{GR} - n_{GP})R_0 T \qquad (3.4)$$

where n_{GR} number of moles of gaseous reactants/unit of fuel
 n_{GP} number of moles of gaseous products/unit of fuel.

The difference between the calorific values is usually very small.

In most combustion problems any water produced by the reaction will be in the vapour state. If the water were condensed, the calorific value would be increased and then be referred to as the higher calorific value (HCV). The relationship between higher and lower calorific value (LCV) is

$$HCV - LCV = yh_{fg} \qquad (3.5)$$

where y is the mass of water per unit quantity of fuel and h_{fg} is the enthalpy of evaporation of water at the temperature under consideration.

For hydrocarbon fuels the difference is significant, but lower calorific value is invariably used or implied. See example 3.3 for the use of calorific values. Similarly the state of the fuel must be specified, particularly if it could be liquid or gas. However, the enthalpy of vaporisation for fuels is usually small compared to their calorific value. For example, for octane at 298.15 K (Rogers and Mayhew (1980b)):

$$H_{fg} = 41\ 500 \text{ kJ/kmol}$$
$$\Delta H_0 = 5\ 116\ 180 \text{ kJ/kmol}$$

Empirical equations for evaluating the internal energy are presented in chapter 10, section 10.2.2.

3.4 Dissociation

Dissociation has already been introduced in this chapter by discussing the dissociation of carbon dioxide

$$CO + \tfrac{1}{2} O_2 \rightleftharpoons CO_2$$

According to Le Châtelier's Principle, an equilibrium will always be displaced in such a way as to minimise any changes imposed from outside the system. This equilibrium can be affected in three ways: a change in concentration of a constituent, a change in system pressure, or a change in temperature. Considering one change at a time

(i) Suppose excess oxygen were added to the system.
     The reaction would move in the forwards direction, as this would reduce the concentration of oxygen.
(ii) Suppose the system pressure were increased.

Again the reaction would move in the forwards direction, as this reduces the total number of moles (n), and the pressure reduces since

$$p V = n R_0 T$$

(iii) Suppose the temperature of the system were raised.
The reaction will move in a direction that absorbs heat; for this particular reaction, that will be the reverse direction.

Care must be taken in defining the forward and reverse directions as these are not always self-evident. For example, take the water gas reaction

$$CO_2 + H_2 \rightleftharpoons CO + H_2O$$

To study this matter more rigorously it is necessary to introduce the concept of equilibrium constants, K; these are also called dissociation constants.

Consider a general reaction of

$$a \text{ kmols } A + b \text{ kmols } B \rightleftharpoons c \text{ kmols } C + d \text{ kmols } D$$

It can be shown by use of a hypothetical device, the Van t'Hoff equilibrium box, that

$$K = \frac{(p'_c)^c (p'_d)^d}{(p'_a)^a (p'_b)^b} \tag{3.6}$$

The derivation can be found in thermodynamics texts such as Rogers and Mayhew (1980a). K is a function only of temperature and will have units of pressure to the power $(c + d - a - b)$. p', the partial pressure of a component is defined as

$$p'_a = \frac{a}{a + b + c + d} \, p, \text{ where } p \text{ is the system pressure}$$

To solve problems involving dissociation introduce a variable x

$$CO + \tfrac{1}{2}O_2 \rightarrow (1 - x)CO_2 + xCO + \frac{x}{2} O_2$$

$$K = \frac{p'_{CO_2}}{(p'_{CO})(p'_{O_2})^{\frac{1}{2}}} \text{ where } p'_{CO_2} = \frac{1 - x}{(1 - x) + x + \dfrac{x}{2}} \, p, \text{ etc.} \tag{3.7}$$

so

$$K = \frac{1-x}{x(x/2)^{\frac{1}{2}}} \left(\frac{1 + \dfrac{x}{2}}{p} \right)^{\frac{1}{2}}$$

As the equilibrium constant varies strongly with temperature, it is most convenient to tabulate $\log_{10}(K)$; the numerical value will also depend on the pressure units adopted (Rogers and Mayhew (1980b) or Haywood (1972)). Again, the use of equilibrium constants is best shown by a worked problem, example 3.5. With internal combustion engines there will be several dissociation mechanisms occurring, and the simultaneous solution is best performed by computer.

However, the most significant dissociation reactions are

$$CO + \tfrac{1}{2}O_2 \rightleftharpoons CO_2$$

and the water gas reaction

$$CO_2 + H_2 \rightleftharpoons CO + H_2O$$

The dissociation of carbon dioxide means that carbon monoxide will be present, even for the combustion of weak mixtures. Of course, carbon monoxide is most significant when rich mixtures are being burned, and there is insufficient oxygen for full oxidation of the fuel. When carbon monoxide is present from the combustion of hydrocarbons, then there will also be water vapour present. The water gas reaction implies that the water vapour can be dissociated to produce hydrogen, and this will be seen later in figure 3.15. The water gas equilibrium can be assumed to be the sole determinant of the burned gas composition, and this is illustrated by example 3.6.

3.5 Combustion in spark ignition engines

Combustion either occurs normally — with ignition from a spark and the flame front propagating steadily throughout the mixture — or abnormally. Abnormal combustion can take several forms, principally pre-ignition and self-ignition. Pre-ignition is when the fuel is ignited by a hot spot, such as the exhaust valve or incandescent carbon combustion deposits. Self-ignition is when the pressure and temperature of the fuel/air mixture are such that the remaining unburnt gas ignites spontaneously. Pre-ignition can lead to self-ignition and vice versa; these processes will be discussed in more detail after normal combustion has been considered.

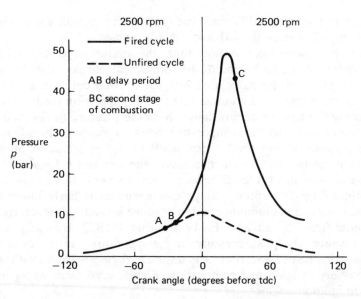

Figure 3.8   Hypothetical pressure diagram for a spark ignition engine

3.5.1   Normal combustion

When the piston approaches the end of the compression stroke, a spark is discharged between the sparking plug electrodes. The spark leaves a small nucleus of flame that propagates into the unburnt gas. Until the nucleus is of the same order of size as the turbulence scale, the flame propagation cannot be enhanced by the turbulence. This causes a delay period of approximately constant duration. Figure 3.8 compares the pressure diagrams for the cases where a mixture is ignited, and where it is not ignited. The point at which the pressure traces diverge is ill-defined, but is used to denote the end of the delay period. The delay period is typically of about 0.5 ms duration, which corresponds to about $7\frac{1}{2}°$ of crank angle at 2500 rpm. The delay period depends on the temperature, pressure and composition of the fuel/air mixture, but it is a minimum for slightly richer than stoichiometric mixtures, in other words, when the laminar flame speed is highest.

The end of the second stage of combustion is also ill-defined on the pressure diagram, but occurs shortly after the peak pressure. The second stage of combustion is affected in the same way as delay period, and also by the turbulence. This is very fortunate since turbulence increases as the engine speed increases, and the time for the second stage of combustion reduces almost in proportion. In other words, the second stage of combustion occupies an approximately constant number of crank angle degrees. In practice, the maximum cylinder pressure usually occurs 5–20° after top dead centre (Benson and Whitehouse (1979)). It is normal for combustion to be complete before the exhaust valve is opened.

Since combustion takes a finite time, the mixture is ignited before top dead centre (btdc), at the end of the compression stroke. This means that there is a pressure rise associated with combustion before the end of the compression stroke, and an increase in the compression (negative) work. Advancing the ignition timing causes both the pressure to rise before top dead centre and also the compression work to increase. In addition the higher pressure at top dead centre leads to higher pressures during the expansion stroke, and to an increase in the expansion (positive) work. Obviously there is a trade-off between these two effects, and this leads to an optimum ignition timing. Since the maximum is fairly insensitive to ignition timing, the minimum ignition advance is used; this is referred to as 'minimum (ignition) advance for best torque' (MBT). By using the minimum advance, the peak pressures and temperatures in the cylinder are reduced; this helps to restrict heat transfer, engine noise, emissions, and susceptibility to abnormal combustion. Similar arguments apply to compression ignition engines.

During the early stages of combustion, while the flame nucleus is still small, it can be displaced from the sparking plug region by large-scale flows in the cylinder. This can occur in a random way, and can have a significant effect on the subsequent propagation of combustion. This is readily shown by the non-repeatability of consecutive indicator diagrams from an engine, and is called variously 'cyclic dispersion', 'cycle-by-cycle variation' or 'cyclic variation'. (This is illustrated later by figure 4.15 where there is further discussion of cyclic dispersion.)

3.5.2 Abnormal combustion

Pre-ignition is caused by the mixture igniting as a result of contact with a hot surface, such as an exhaust valve. Pre-ignition is often characterised by 'running-on'; that is, the engine continues to fire after the ignition has been switched off.

If the engine is operating with the correct mixture strength, ignition timing and adequate cooling, yet there is pre-ignition, the usual explanation is a build-up of combustion deposits, or 'coke'. The early ignition causes an increase in the compression work and this causes a reduction in power. In a multi-cylinder engine, with pre-ignition in just one cylinder, the consequences can be particularly serious as the other cylinders continue to operate normally. Pre-ignition leads to higher peak pressures, and this in turn can cause self-ignition.

Self-ignition occurs when the pressure and temperature of the unburnt gas are such as to cause spontaneous ignition, figure 3.9. The flame front

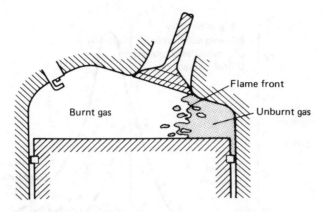

Figure 3.9   Combustion in a spark ignition engine

propagates away from the sparking plug, and the unburnt (or 'end') gas is heated by radiation from the flame front and compressed as a result of the combustion process. If spontaneous ignition of the unburnt gas occurs, there is a rapid pressure rise which is characterised by a 'knocking' The 'knock' is probably caused by resonances of the combustion chamber contents. As a result of knocking, the thermal boundary layer at the combustion chamber walls can be destroyed. This causes increased heat transfer which might then lead to certain surfaces causing pre-ignition.

In chapter 2 it was shown how increasing the compression ratio should improve engine performance. Unfortunately, raising the compression ratio also increases the susceptibility to knocking. For this reason, much research has centred on the fundamental processes occurring with knock. These mechanisms are discussed in section 3.7.

3.6   Combustion in compression ignition engines

Near the end of the compression stroke, liquid fuel is injected as one or more jets. The injector receives fuel at very high pressures in order to produce rapid injection, with high velocity jet(s) of small cross-sectional area; in all but the largest engines there is a single injector. The fuel jets entrain air and break up into droplets; this provides rapid mixing which is essential if the combustion is to occur sufficiently fast. Sometimes the fuel jet is designed to impinge on to the combustion chamber wall; this can help to vaporise the fuel and break up the jet. There will be large variations in fuel/air mixtures on both a large and small scale within the combustion

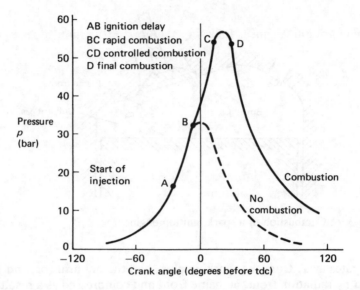

Figure 3.10   Hypothetical pressure diagram for a compression ignition engine

chamber. Figure 3.10 shows the pressure diagram for a compression ignition engine; when compared to figure 3.8 (spark ignition engine) it can be seen immediately that the pressures are higher, especially for the unfired cycle. Referring to figure 3.10, there are several stages of combustion, not distinctly separated:

(i) Ignition delay, AB. After injection there is initially no apparent deviation from the unfired cycle. During this period the fuel is breaking up into droplets being vaporised, and mixing with air. Chemical reactions will be starting, albeit slowly.

(ii) Rapid or uncontrolled combustion, BC. A very rapid rise in pressure caused by ignition of the fuel/air mixture prepared during the ignition delay period.

(iii) Controlled combustion, CD. Combustion occurs at a rate determined by the preparation of fresh air/fuel mixture.

(iv) Final combustion D. As with controlled combustion the rate of combustion is governed by diffusion until all the fuel or air is utilised.

As with spark ignition engines the initial period is independent of speed, while the subsequent combustion occupies an approximately constant number of crank angle degrees. In order to avoid too large a rapid combustion period, the initial fuel injection should be carefully controlled. The 'rapid' combustion period can produce the characteristic 'diesel knock'. Again this is caused by a sudden pressure rise, but is due to self-ignition occurring too slowly. Its cure is the exact opposite to that used in spark ignition engines; fuels in compression ignition engines should self-ignite readily. For a given fuel and engine, diesel knock can be reduced by avoiding injection of too much fuel too quickly. Some systems inject a small quantity of fuel before the main injection, a system known as pilot injection. Alternatively, the engine can be modified to operate with a higher compression ratio. This increases the temperature and pressure during the compression stroke, and this will reduce the ignition delay period. The modelling of the ignition delay and the subsequent combustion are treated in section 10.2.

To obtain maximum output the peak pressure should occur about 10–20° after top dead centre. Sometimes the injection is later, in order to retard and to reduce the peak pressure.

Combustion photography in compression ignition engines can be very useful, as it shows the effectiveness of the injection process. The fuel ignites spontaneously at many sites, and produces an intense flame. There is significant radiation from the flame front, and this is important for vaporising the fuel. Since the combustion occurs from many sites, compression ignition engines are not susceptible to cyclic variation or cyclic dispersion.

Compression ignition engines can operate over a wide range of air/fuel mixtures with equivalence ratios in the range 0.14–0.90. The power output of the engine is controlled by the amount of fuel injected, as in the combustion region the mixture is always approximately stoichiometric. This ensures good part load fuel economy as there are no throttling losses. The fuel/air mixture is always weaker than stoichiometric, as it is not possible to utilise all the air. At a given speed the power output of an engine is usually limited by the amount of fuel that causes the exhaust to become smoky.

3.7 Fuels and additives

The performance and in particular fuel economy of internal combustion engines should not be considered in isolation, but also in the context of the oil refinery. Oil refining and distribution currently has an overall efficiency of about 88 per cent (Francis and Woollacott (1981)). This efficiency would be changed if the demand for the current balance of products was changed. Oil refining can be compared to the work of a butcher; each process has a raw material (a barrel of oil or a carcass) that has to be cut in such a way that all the products can be sold at a competitive price. Just as there are different animals, there are also different types of crude oil, depending on the source. However, the oil refinery can change the type of product in additional ways, such as cracking, although there is an energy cost associated with this.

The energy content of a typical petrol is 44 MJ/kg or 31.8 MJ/litre and of a typical diesel fuel 42 MJ/kg or 38.15 MJ/litre; associated with these is an energy content at the refinery of typically 2.7 MJ/kg and 1.6 MJ/kg (Francis and Woollacott (1981)). This gives an effective primary energy of 46.7 MJ/kg or 35.5 MJ/litre for petrol and 43.65 MJ/kg or 39.95 MJ/litre for diesel fuel. There may be circumstances in which it is more appropriate to use primary energy density than to use calorific value. These figures also highlight the difference in energy content of unit volume of fuel. This is often overlooked when comparing the fuel economy of vehicles on a volumetric basis.

3.7.1   Characteristics of petrol

The properties of petrol are discussed thoroughly by Blackmore and Thomas (1977). The two most important characteristics of petrol are its volatility and octane number (its resistance to self-ignition).

Volatility is expressed in terms of the volume percentage that is distilled at or below fixed temperatures. If a petrol is too volatile, when it is used at high ambient temperatures the petrol is liable to vaporise in the fuel lines and form vapour locks. This problem is most pronounced in vehicles that are being restarted, since under these conditions the engine compartment is hottest. If the fuel is not sufficiently volatile the engine will be difficult to start, especially at low ambient temperatures. The volatility also influences the cold start fuel economy. Spark ignition engines are started on very rich mixtures, and continue to operate on rich mixtures until they reach their normal operating temperature; this is to ensure adequate vaporisation of fuel. Increasing the volatility of the petrol at low temperatures will evidently improve the fuel economy during and after starting. Blackmore and Thomas (1977), point out that in the USA as much as 50 per cent of all petrol is consumed on trips of 10 miles or less. Short journeys have a profound effect on vehicle fuel economy, yet fuel consumption figures are invariably quoted for steady-state conditions.

Fuel volatility is specified in British Standard 4040: 1978, and these data are compared with typical fuel specifications from Blackmore and Thomas (1977) in table 3.2. This is plotted with further data in figure 3.11.

Table 3.2 shows how the specification of petrol varies to suit climatic conditions. Petrol stored for a long time in vented tanks is said to go stale; this refers to the loss of the more volatile components that are necessary for easy engine starting.

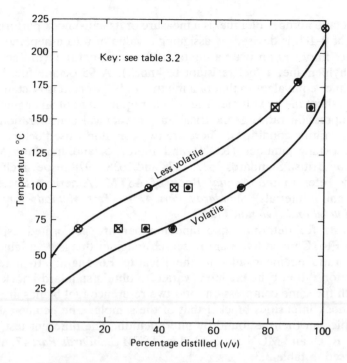

Figure 3.11 Distillation curves for petrol (with acknowledgement to Blackmore and Thomas (1977))

Table 3.2 Volatility of different petrol blends

	BS4040		Less volatile	Volatile	North-west Europe		Central Africa
	Min.	Max.			Summer	Winter	
Distillate evaporated at 70°C (per cent V/V)	10	45	10	42	25	35	10
Distillate evaporated at 100°C (per cent V/V)	36	70	38	70	45	50	38
Distillate evaporated at 160°C (per cent V/V)			80	98	80	95	80
Distillate evaporated at 180°C (per cent V/V)	90						
Final boiling point °C	—	220					
Residue (per cent V/V)		2					
Symbol used in figure 3.11	⊗	⊙	—	—	☒	⊡	—

The octane number of a fuel is a measure of its anti-knock performance. A scale of 0–100 is devised by assigning a value of 0 to n-heptane (a fuel prone to knock), and a value of 100 to iso-octane (in fact 2, 2, 4-trimethylpentane, a fuel resistant to knock). A 95 octane fuel has the performance equivalent to that of a mixture of 95 per cent iso-octane and 5 per cent n-heptane by volume. The octane requirement of an engine varies with compression ratio, geometrical and mechanical considerations, and also its operating conditions. There are two commonly used octane scales, research octane number (RON) and motor octane number (MON), covered by British Standards 2637: 1978 and 2638: 1978 respectively. Both standards refer to the *Annual Book of ASTM* (American Society for Testing and Materials) *Standards Part 47 — Test Methods for Rating Motor, Diesel and Aviation Fuels*.

The tests for determining octane number are performed using the ASTM-CFR (Cooperative Fuel Research) engine; this is a variable compression ratio engine similar to the Ricardo E6 engine. In a test the compression ratio of the engine is varied to obtain standard knock intensity. With the same compression ratio two reference fuel blends are found whose knock intensities bracket that of the sample. The octane rating of the sample can then be found by interpolation. The different test conditions for RON and MON are quoted in *ASTM Standards Part 47*, and are summarised in table 3.3.

Table 3.3 shows that the motor octane number has more severe test conditions since the mixture temperature is greater and the ignition occurs earlier. There is not necessarily any correlation between MON and RON as the way fuel components of different volatility contribute to the octane rating will vary. Furthermore, when a carburetted engine has a transient increase in load, excess fuel is supplied. Under these conditions it is the octane rating of the more volatile components that determines whether or not knock occurs. The minimum octane requirements for different grades of petrol are given by BS4040, see table 3.4. A worldwide summary of octane ratings is published by the Associated Octel Co. Ltd, London.

The attraction of high octane fuels is that they enable high compression ratios to be used. Higher compression ratios give increased power output and improved economy. This is shown in figure 3.12 using data from Blackmore and Thomas (1977). The octane number requirements for a given compression ratio vary widely, but typically a compression ratio of 7.5 requires 85 octane fuel, while a compression ratio of 10.0 would require 100 octane fuel. There are even wide variations in octane number requirements between supposedly identical engines.

Of the various fuel additives, those that increase octane numbers have greatest significance. In 1922 Midgely and Boyd discovered that lead-based compounds improved the octane rating of fuels. By adding 0.5 grams of lead per litre, the octane rating of the fuel is increased by about 5 units.

Table 3.3   Summary of RON and MON test conditions

Test conditions	Research octane number	Motor octane number
Engine speed, rpm	600 ± 6	900 ± 9
Crankcase oil, SAE grade	30	30
Oil pressure at operating temperature, psi	25–30	25–30
Crankcase oil temperature	$135 \pm 15°F$ ($57 \pm 8.5°C$)	$135 \pm 15°F$ ($57 \pm 8.5°C$)
Coolant temperature		
Range	$212 \pm 3°F$ ($100 \pm 1.5°C$)	$212 \pm 3°F$ ($100 \pm 1.5°C$)
Constant within	$\pm 1°F$ ($0.5°C$)	$\pm 1°F$ ($0.5°C$)
Intake air humidity, grains of water per lb. of dry air	25–50	25–50
Intake air temperature	See *ASTM Standard Part 47*	$100 \pm 5°F$ ($38 \pm 2.8°C$)
Mixture temperature		$300 \pm 2°F$ ($149 \pm 1.1°C$)
Spark advance, deg. btdc	13	14–26 depending on compression ratio
Spark plug gap, in.	0.020 ± 0.005	0.020 ± 0.005
Breaker point, gap, in.	0.020	0.020
Valve clearances, in.		
Intake	0.008	0.008
Exhaust	0.008	0.008
Fuel/air ratio	Adjusted for maximum knock	

Table 3.4   Octane number requirements for different fuel grades

Grade designation	RON	MON
5 star	100.0	86.0
4 star	97.0	86.0
3 star	94.0	82.0
2 star	90.0	80.0

The lead additives take the form of lead alkyls, either tetramethyl lead $(CH_3)_4$ Pb, or tetraethyl lead $(C_2H_5)_4$ Pb. Since the active ingredient is lead, the concentration of the additives is expressed in terms of the lead content. Thus

$$0.5 \text{ g Pb/1} = 0.645 \text{ g } (CH_3)_4 \text{ Pb/1}$$

and

$$0.5 \text{ g Pb/1} = 0.780 \text{ g } (C_2H_5)_4 \text{ Pb/1}$$

Most countries now have restrictions on the use of lead in automotive fuels for environmental reasons. However, the use of lead additives in

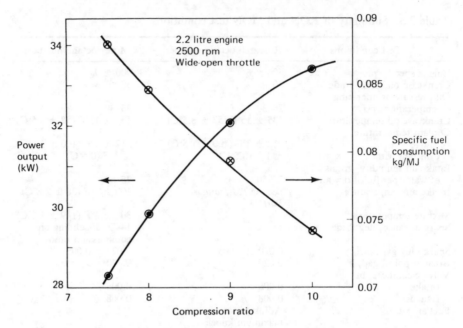

Figure 3.12   Effect of changing compression ratio on engine power output and
              fuel economy (with acknowledgement to Blackmore and Thomas
              (1977))

aviation gasoline is still very significant as in low lead fuel there is 0.5 g
Pb/litre, giving a RON of about 106. As well as the possible dangers of lead
pollution, catalysts for the conversion of other engine pollutants are made
inactive by lead. However, manufacturers of lead additives claim that
suitable filters could be installed in exhaust systems to remove the lead
particulates.

To understand how lead alkyls can inhibit knocking, the chemical
mechanism involved in knocking must be considered in more detail. Two
possible causes of knocking are cool flames or low-temperature auto-
ignition, and high-temperature auto-ignition. Cool flames can occur in
many hydrocarbon fuels and are studied by experiments with fuel/oxygen
mixtures in heated vessels. If some mixtures are left for a sufficient time, a
flame is observed at temperatures that are below those for normal self-
ignition; the flames are characterised by the presence of peroxide and
aldehyde species. Engine experiments have been conducted in which the
concentrations of peroxide and aldehyde species have been measured,
giving results that imply the presence of a cool flame. However, knock was
obtained only with higher compression ratios, which implied that it is a
subsequent high-temperature auto-ignition that causes the rapid pressure
rise and knock. Cool flames have not been observed with methane and

benzene, so when knock occurs with these fuels it is a single-stage high-temperature auto-ignition effect.

Downs and Wheeler (1951–52) and Downs *et al.* (1961) discuss the possible chemical mechanisms of knock and how tetraethyl lead might inhibit knock. In the combustion of, say, heptane, it is unlikely that the complete reaction

$$C_7H_{16} + 11O_2 \rightarrow 7CO_2 + 8H_2O$$

can occur in one step. A gradual degradation through collisions with oxygen molecules is much more likely, finally ending up with CO_2 and H_2O. This is a chain reaction where oxygenated hydrocarbons such as aldehydes and peroxides will be among the possible intermediate compounds. A possible scheme for propane starts with the propyl radical $(C_3H_7)^-$, involves peroxide and aldehyde intermediate compounds, and finally produces a propyl radical so that the chain reaction can then repeat:

$$CH_3CH^-CH_3 + O_2 \rightarrow CH_3CH(OO)CH_3$$

propyl radical   hydroperoxide

$$\rightarrow CH_3CHO + CH_3O^- \xrightarrow{+C_3H_8} CH_3OH + CH_3CH^-CH_3 + CH_3CHO$$

aldehyde   alcohol

or

$$\rightarrow C_2H_5CHO + OH^-$$

hydroxyl

radical

There are many similar chain reactions that can occur and some of the possibilities are discussed by Lewis and von Elbe (1961) in greater detail.

Tetraethyl lead improves the octane rating of the fuel by modifying the chain reactions. During the compression stroke, the lead alkyl decomposes and reacts with oxygenated intermediary compounds to form lead oxide, thereby combining with radicals that might otherwise cause knock.

The suggested mechanism is as follows:

$$PbO + OH \rightarrow PbO(OH)$$
$$PbO(OH) + OH \rightarrow PbO_2 + H_2O$$
$$PbO_2 + R \rightarrow PbO + RO$$

where R is a radical such as the propyl radical, C_3H_7.

One disadvantage with the lead alkyls is that lead compounds are deposited in the combustion chamber. These can be converted to more volatile lead halides by alkyl halide additives such as dichloroethane, $(C_2Cl_2H_4)$ or dibromoethane $(C_2Br_2H_4)$. However, some lead halides remain in the combustion chamber and these deposits can impair the insulation of spark plugs, and thus lead to misfiring. By adding aryl phosphates to petrol, lead halides are converted to phosphates, which have greater electrical resistivity. A second benefit is that lead phosphates are less prone to cause pre-ignition by surface ignition. These and other additives are discussed more fully by Blackmore and Thomas (1977).

Alcohols have certain advantages as fuels, particularly in countries without oil resources, or where there are sources of the renewable raw materials for producing methanol (CH_3OH) or ethanol (C_2H_5OH). Car manufacturers have extensive programmes for developing alcohol-fuelled vehicles (Ford (1982)). Alcohols can also be blended with oil-derived fuels and this improves the octane ratings. Both alcohols have high octane ratings (ethanol has a RON of 106) and high enthalphy of vaporisation; this improves the volumetric efficiency but can cause starting problems. For cold ambient conditions it may be necessary to start engines with petrol. The other main disadvantages are the lower energy densities (about half that of petrol for methanol and two-thirds for ethanol), and the miscibility with water.

3.7.2 In-vehicle performance of fuels, and the potential of alcohols

In Western Europe the lead content of fuel is now mostly limited to 0.15 grams of lead per litre. The premium leaded fuel (4 star fuel in the UK with its minimum RON of 97) is from the same feedstock as the premium unleaded fuel. In the UK, British Standard 7070 defines the specification of unleaded fuel, see table 3.5.

In addition, the fuel companies have introduced unleaded fuels with an octane rating performance equivalent to that of 4 star fuel, but the anti-knock performance of this fuel is not defined by BS7070. The improvement in the anti-knock rating of the unleaded fuel (and leaded fuel with its reduced lead content) has been achieved by adding fuel components that have high octane ratings. Notable in this respect are oxygenates and benzene. Since oxygenate fuels have: a lower calorific value, a different density, and a different stoichiometric air fuel ratio, they can lead to engines operating away from design conditions. Furthermore, oxygenate fuels can absorb moisture, and this can lead to corrosion and fuel separation problems. BS7070 limits the total amount of oxygenates by limiting the oxygen content of the fuel to 2.5 per cent of mass. There are also limits on the individual oxygenate components and on benzene (for reasons of carcinogenity). Table 3.6 summarises these limits, and indicates their octane rating.

Table 3.5 Octane number requirements of different fuel grades

Description	RON	MON
Regular unleaded	90	80
Premium unleaded	95	85

Table 3.6 The limits imposed by BS7070 on oxygenates and benzene (BV denotes Blending Value)

Component	Limit per cent (V/V)	RON	MON	BV (RON)
Benzene C_6H_6	5.0		115	
Methanol CH_3OH	3.0	106	92	112
Ethanol C_2H_5OH	5.0	107	89	110
Isopropyl alcohol (propan-2-ol) $CH_3CHOHCH_3$	5.0	—	—	118
Tertiary butyl alcohol C_3H_7CHOH	7.0	—	—	107
Ethers containing five or more carbon atoms	10.0	—	—	—
Other oxygenates specified by BS7070	7.0	—	—	—

The improvement that oxygenate fuels give to the resulting fuel blend depends on the oxygenate fuel and the fuel that it is being blended with. By definition, when mixtures of iso-octane and n-heptane are prepared, there is a linear relationship between composition and the octane rating of the blend. However, many fuel components exhibit non-linear mixing properties, and this is why a blending value has been indicated in table 3.6. For a given component, the blending value will depend on what it is being mixed with, but it is convenient to assume a linear mixing relation and a blending value for oxygenates, to estimate the octane rating of a fuel blend.

Most fuels have different values of RON and MON (the difference is known as the sensitivity), and this illustrates that the anti-knock performance of the fuel depends on the test conditions. Furthermore, there are several limitations associated with the standard tests in the CFR engine. Firstly, both MON and RON are evaluated at low engine speeds, and secondly the air/fuel ratio is adjusted to give the maximum knock. Thus it is not surprising that fuels behave rather differently in multi-cylinder engines, whether they are installed on dynamometers or in vehicles.

Clearly it is impractical to vary the compression ratio of a production engine. Instead the ignition timing is modified until the ignition timing is sufficiently advanced just to cause knock; this defines the knock limited spark advance (KLSA). Palmer and Smith (1985) report that a unity increase in RON will increase the KLSA by 1½° to 2° crank angle. Fuels can thus be tested in a steady-state test at full throttle, in which the KLSA is obtained as a function of speed for each fuel. The KLSA can then be compared with the manufacturer's ignition advance characteristics to establish the knock margin.

Transients are also of great importance. During an increase in load, the throttle is opened and the pressure in the inlet manifold rises. At a low throttle setting there will be a high proportion of fuel vaporised, as the partial pressure of the fuel represents a larger fraction of the manifold pressure. When the throttle opens, and the manifold pressure rises, the partial pressure of the fuel cannot rise, so that some previously vaporised fuel condenses. Furthermore, when the throttle is opened, the air flow increases almost instantaneously (within a cycle or so), but the larger fuel droplets and the fuel film on the manifold will lag behind the air flow. For these two reasons, extra fuel is supplied during a throttle opening transient, to ensure that the mixture supplied to the engine is flammable. During an increase in load it will be the more volatile components of the fuel (the front end) that will vaporise preferentially and enter the engine. The different fractions within a fuel will have different octane ratings, and this is illustrated by figure 3.13. The overall RON of the catalytic reformate fuel of figure 3.13 is 91, yet the RON of the fractions lies with the range of 58.5 to 114.

Such a fuel is unlikely to give an acceptable engine performance, as the fractions that boil in the temperature range of 45–105°C have a low RON. A useful concept here is the Delta octane number (ΔON). This is the difference between the knock rating of the whole gasoline, and the knock rating of the gasoline boiling below 100°C. The lower the ΔON, then the better the transient performance of the gasoline in avoiding knock.

However, it is the in-vehicle performance that is important, and this can be evaluated by an acceleration test. Suppose the vehicle is subject to a series of full throttle acceleration tests in a fixed gear (say direct drive) between specified speeds (for example, 50–100 km/h). In each test, the ignition timing can be advanced incrementally away from the vehicle manufacturer's setting until knock is detected.

Since a wide range of spark ignition engines are used in vehicles, fuel companies have to undertake extensive test programmes to ensure that their fuel will meet the market requirements. The test programmes have to cover a wide range of engines, and include supposedly identical engines in order to allow for variations in manufacture.

When methanol and ethanol are readily available, their properties make them attractive substitutes for petrol. Since alcohols burn more rapidly than petrol, the ignition timings in the tests for RON and MON are over-advanced, and this leads to an underestimate of the anti-knock

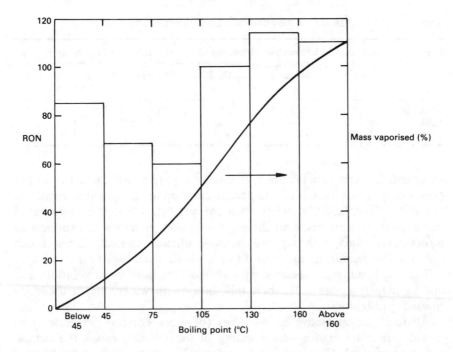

Figure 3.13   The RON and mass per cent of the fractions in a catalytic
                reformate fuel of 91 RON (data from Palmer and Smith (1985))

performance. However, methanol in particular is susceptible to pre-
ignition, and this is more likely to restrict the compression ratio (so as to
limit surface temperatures in the combustion chamber).

The evaporative and combustion properties of alcohols are given in table
3.7. Even though smaller percentages of the alcohols evaporate during
mixture preparation, the greater evaporative cooling effect is such that the
alcohols produce lower mixture temperatures, with consequential improvements to the volumetric efficiency of the engine. The lower air/fuel
ratios for alcohols mean that the chemical energy released per kg of
stoichiometric mixture burnt during combustion is greater than for petrol,
despite the lower specific enthalpies of combustion. Table 3.7 lists the
enthalpy of combustion on the basis of 1 kg of a stoichiometric mixture.
The improved volumetric efficiency and the higher combustion energy both
increase the output of the engine, and thus reduce the significance of
mechanical losses, thereby improving the overall efficiency. Goodger
(1975) reports the comparisons with hydrocarbon fuels made by Ricardo at
a fixed compression ratio, in which there was a 5 per cent improvement in
efficiency using ethanol, and a 10 per cent improvement when methanol

Table 3.7   The evaporative and combustion properties of alcohols

Fuel	Boiling point (°C)	Stoichiometric gravimetric air/fuel ratio	Enthalpy of evaporation (kJ/kg)	Enthalpy of combustion	
				(MJ/kg fuel)	(MJ/kg stoichiometric mixture)
Methanol	65	6.5	1170	22.2	3.03
Ethanol	78.5	9.0	850	29.7	2.97
Petrol	25–175	14.5	310	42.0	2.71

was used. In racing applications, methanol is particularly attractive as the power output increases with a richer mixture, up to an equivalence ratio of about 1.4 (substantially richer than for petrol). Also, the more rapid combustion of alcohols, and their greater ratio of moles of products to moles of reactants, both improve the cycle efficiency. Finally, alcohols can operate with leaner mixtures, and this leads to lower emissions.

The disadvantages associated with alcohols are their lower volatility and energy density (table 3.7), their miscibility with water, and a tendency towards pre-ignition.

Alcohols are attractive as additives (or more correctly extenders) to petrol, since the higher octane rating of the alcohols raises the octane rating of the fuel. Goodger (1975) reports that up to 25 per cent ethanol yields a linear increase in the octane rating (RON) of 8 with 92 octane fuel and an increase of 4 with 97 octane fuel.

In Europe, blends of petrol with 3 per cent alcohol have been commonly used to improve the octane rating, and West Germany has sponsored a programme to introduce 15 per cent of methanol into petrol.

Palmer (1986) reports on vehicle tests, which showed that all oxygenate blends gave a better anti-knock performance during low speed acceleration than hydrocarbon fuels of the same octane rating. Furthermore, there is a tendency for the anti-knock benefits of oxygenate fuels to improve in unleaded fuels. However, care is needed with ethanol blends to avoid possible problems of high-speed knock. Fuel consumption on a volumetric basis is almost constant as the percentage of oxygenate fuel in the blend increases. The lower calorific value of the oxygenates is in part compensated for by the increase in fuel density. None the less, Palmer (1986) shows that energy consumption per kilometre falls as the oxygenate content in the fuel increases.

Ethanol is entirely miscible with petrol, while methanol is only partially miscible. The miscibility of both alcohols in petrol reduces with the presence of water and lower temperatures. To avoid phase separation, as moisture becomes absorbed in the fuel, chemicals such as benzene, acetone or the higher alcohols can be added to improve the miscibility. Palmer (1986) also reports on materials compatibility, hot and cold weather driveability, altitude effects, and the exhaust emissions from oxygenate blends.

The major disadvantages of methanol and ethanol as alternative fuels concern their production and low volatility. Methanol can be produced from either coal or natural gas, but there is an associated energy cost and impact on the environment. Monaghan (1990) points out that when methanol is derived from natural gas, then the contribution to greenhouse gases (in terms of carbon dioxide equivalent) is less than from a petrol fuelled vehicle. However, for methanol derived from coal, there is a greater contribution of greenhouse gases.

The low volatility of methanol means that with conventional engine technology, priming agents are needed for cold starting below 10°C, Beckwith *et al.* (1986). Priming agents are hydrocarbons added to methanol, to improve the low temperature vapour pressure and the flammability. Beckwith *et al.* report tests on a range of priming agents for methanol, with concentrations of up to 18 per cent by volume. They point out that the priming agents all led to increased vapour pressure at high temperatures which might lead to vapour lock problems. They concluded that straight run gasoline (with cut points up to 150°C) gave the best compromise between cold and hot weather performance. A methanol blend known as M85 (meaning 85 per cent by volume methanol) is available in several countries.

3.7.3 Characteristics of diesel fuel

The most important characteristic of diesel fuel is the cetane number, as this indicates how readily the fuel self-ignites. Viscosity is also important, especially for the lower-grade fuels used in the larger engines; sometimes it is necessary to have heated fuel lines. Another problem with diesel fuels is that, at low temperatures, the high molecular weight components can precipitate to form a waxy deposit. This is defined in terms of the cold filter plugging point.

The properties of different fuel oils are specified in BS2869: 1988. The two fuel specifications quoted in table 3.8 are for high-quality automotive diesel fuel (A1), and general-purpose diesel fuel (A2).

The flashpoint is the temperature to which the liquid has to be heated for the vapour to form a combustible mixture with air at atmospheric pressure. Since the flashpoint of diesel fuel is at least 55°C, this makes it a safer fuel to store then either petrol or kerosene. The flashpoints of petrol and kerosene are about −40°C and 30°C respectively.

If an engine runs on a fuel with too low a cetane number, there will be diesel knock. Diesel knock is caused by too rapid combustion and is the result of a long ignition delay period, since during this period fuel is injected and mixes with air to form a combustible mixture. Ignition occurs

Table 3.8 Specifications for diesel fuels

Property		BS 2869	
		Class A1	Class A2
Viscosity, kinematic at	min.	1.5	1.5
37.8°C (centistokes)	max.	5.0	5.5
Cetane number	min.	50	45
Carbon residue, Ramsbottom per cent by mass on 10 per cent residue	max.	0.2	0.2
Distillation, recovery at 350°C, per cent by volume	min.	85	85.0
Flashpoint, closed Pensky Martins °C	min.	56	56
Water content, per cent by volume	max.	0.05	0.05
Sediment, per cent	max.	0.01	0.01
Ash, per cent by mass	max.	0.01	0.01
Sulphur	max.	0.3	0.5
Copper corrosion test	max.	1	1
Cold filter plugging point (°C) max.	Summer (16 Mar./30 Sept.)	−4	0
	Winter (1 Oct./15 Mar.)	−15	−9

only after the pressure and temperature have been above certain limits for sufficient time, and fuels with high cetane numbers are those that self-ignite readily.

As with octane numbers, a scale of 0–100 is constructed; originally a value of 0 was assigned, to α-methylnaphthalene ($C_{10}H_7CH_3$, a naphthenic compound with poor self-ignition qualities), and a value of 100 was assigned to n-cetane ($C_{16}H_{34}$, a straight-chain alkane with good self-ignition qualities). A 65 cetane fuel would have ignition delay performance equivalent to that of a blend of 65 per cent n-cetane and 35 per cent α-methylnaphthalene by volume. An isocetane, heptamethylnonane (HMN), is now used to define the bottom of the scale with a cetane number of 15.

The tests for determining cetane number in BS5580: 1978 refer to the *Annual Book of ASTM Standards, Part 47*. The tests are performed with an ASTM-CFR engine equipped with a special instrument to measure ignition delay. With standard operating conditions the compression ratio of the engine is adjusted to give a standard delay period with the fuel being tested. The process is repeated with reference fuel blends to find the compression ratios for the same delay period. When the compression ratio of the fuel being tested is bracketed by the reference fuels, the cetane number of the test fuel is found by interpolation.

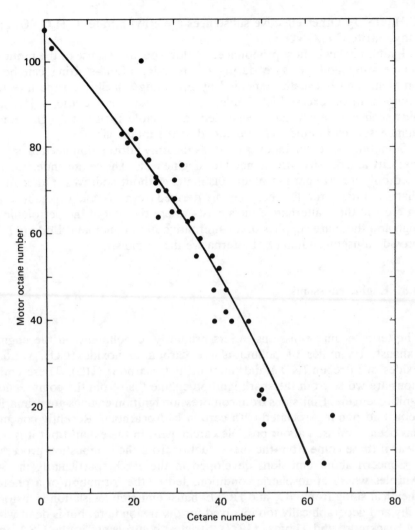

Figure 3.14   Relationship between cetane number and octane number for
pertroleum-derived fuels (adapted from Taylor (1985b))

As would be expected, fuels with high cetane numbers have low octane
numbers and vice versa. This relationship is shown in figure 3.14 using data
from Taylor (1985b); surprisingly there is a single line, thus showing
independence of fuel composition.

Additives in diesel fuel to improve the cetane number are referred to as
ignition accelerators. Their concentrations are greater than those of anti-
knock additives used in petrol. Typically an improvement of 6 on the
cetane scale is obtained by adding 1 per cent by volume of amyl nitrate,

$C_5H_{11}ONO_2$. Other effective substances are ethyl nitrate, $C_2H_5ONO_2$ and ethyl nitrite C_2H_5ONO.

Ignition delay is most pronounced at slow speeds because of the reduced temperature and pressure during compression. Cold-starting can be a problem, and is usually remedied by providing a facility on the injector pump to inject excess fuel. Under severe conditions, additional starting aids such as heaters may be needed, or volatile fuels with high cetane numbers, such as ether, can be added to the intake air.

Sometimes a cetane index is used, as the only information needed is fuel viscosity and density with no need for engine tests. The cetane index can be used only for straight petroleum distillates without additives. Other fuels that are suitable for diesel engines are derived from coal and vegetable oils. Interest in these alternative fuels exists where the cost of the petroleum is high and there are supplies of an alternative fuel. Onion and Bodo (1983) provide a useful summary of alternative diesel fuels.

3.8   Engine emissions

The term 'engine emissions' refers primarily to pollutants in the engine exhaust. Examples of pollutants are carbon monoxide (CO), various oxides of nitrogen (NO_x) and unburnt hydrocarbons (HC). These emissions are worse from the spark ignition engine than from the compression ignition engine. Emissions from compression ignition engines are primarily soot, and odour associated with certain hydrocarbons. Recently concern has been expressed about possible carcinogens in the exhaust but it is not clear if these come from the diesel fuel or from the combustion process.

Concern about emissions developed in the 1960s, particularly in Los Angeles where atmospheric conditions led to the formation of a photochemical smog from NO_x and HC. Exhaust emission legislation is historically and geographically too involved for discussion here, but is dealt with by Blackmore and Thomas (1977). Strictest controls are in the USA and Japan but European legislation is also building up.

The concentrations of CO and NO_x are greater than those predicted by equilibrium thermodynamics. The rate of the forward reaction is different from the backward reaction, and there is insufficient time for equilibrium to be attained. The chemical kinetics involved are complex and work is still proceeding to try and predict exhaust emissions (Mattavi and Amann (1980)).

Emissions of CO, NO_x and HC vary between different engines and are dependent on such variables as ignition timing, load, speed and, in particular, fuel/air ratio. Figure 3.15 shows typical variations of emissions with fuel/air ratio for a spark ignition engine.

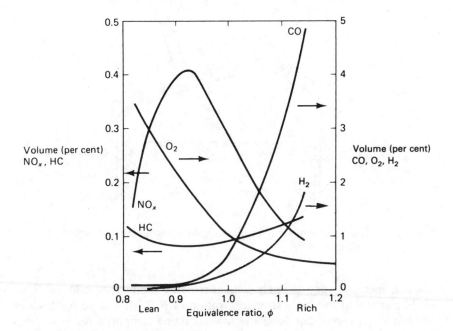

Figure 3.15   Spark ignition engine emissions for different fuel/air ratios
(courtesy of Johnson Matthey)

Carbon monoxide (CO) is most concentrated with fuel-rich mixtures, as there will be incomplete combustion. With lean mixtures, CO is always present owing to dissociation, but the concentration reduces with reducing combustion temperatures. Hydrocarbon (HC) emissions are reduced by excess air (fuel-lean mixtures) until the reduced flammability of the mixtures causes a net increase in HC emissions. These emissions originate from the flame quench layer — where the flame is extinguished by cold boundaries; regions like piston ring grooves can be particularly important. The outer edge of the quench can also contribute to the CO and aldehyde emissions.

The formation of NO_x is more complex since it is dependent on a series of reactions such as the Zeldovich mechanism:

$$O_2 \rightleftharpoons 2O$$
$$O + N_2 \rightleftharpoons NO + N$$
$$N + O_2 \rightleftharpoons NO + O$$

Some of the modifications to this reaction and the effects of different operating conditions are discussed by Benson and Whitehouse (1979). Chemical kinetics show that the formation of NO and other oxides of

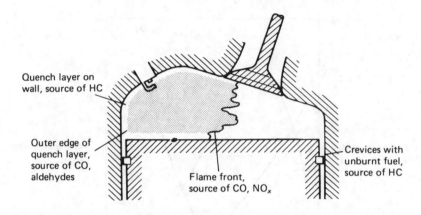

Figure 3.16   Source of emissions in spark ignition engine (from Mattavi and Amann (1980))

nitrogen increase very strongly with increasing flame temperature. This would imply that the highest concentration of NO_x should be for slightly rich mixtures, those that have the highest flame temperature. However, NO_x formation will also be influenced by the flame speed. Lower flame speeds with lean mixtures provide a longer time for NO_x to form. Similarly NO_x emissions increase with reduced engine speed. The sources of different emissions are shown in figure 3.16.

Emissions of HC and CO can be reduced by operating with lean mixtures; this has the disadvantage of reducing the engine power output. It is also difficult to ensure uniform mixture distribution to each cylinder in multi-cylinder engines. Alternatively, exhaust gas catalytic reactors or thermal reactors can complete the oxidation process; if necessary extra air can be admitted.

The ways of reducing NO_x emissions are more varied. If either the flame temperature or burn duration is reduced, the NO_x emissions will also be reduced. Retarding the ignition is very effective as this reduces the peak pressure and temperature, but it has an adverse effect on power output and economy. Another approach is to increase the concentration of residuals in the cylinder by exhaust gas recirculation (EGR). EGR lowers both flame temperature and speed, so giving useful reductions in NO_x. Between 5 and 10 per cent EGR is likely to halve NO_x emissions. However, EGR lowers the efficiency and reduces the lean combustion limit. Catalysts can be used to reduce the NO_x to oxygen and nitrogen but this is difficult to arrange if CO and HC are being oxidised. Such systems have complex arrangements and require very close to stoichiometric mixtures of fuels with no lead-based additives; they are discussed in chapter 4, section 4.3.

Compression ignition engines have fewer gaseous emissions than spark ignition engines, but compression ignition engines have greater particulate emissions. The equivalence ratio in a diesel engine is always less than unity (fuel lean), and this accounts for the low CO emissions, about 0.1 per cent by volume. Hydrocarbon emission (unburnt fuel) is also less, but rises towards the emission level of spark ignition engines as the engine load (bmep) rises.

The emissions of NO_x are about half those for spark ignition engines. This result might, at first, seem to contradict the pattern in spark ignition engines, for which NO_x emissions are worst for an equivalence ratio of about 0.9. In diffusion flames, fuel is diffusing towards the oxidant, and oxidant diffuses towards the fuel. The equivalence ratio varies continuously, from high values at the fuel droplet to values less than unity in the surrounding gases. The flame position can be defined for mathematical purposes as where the equivalence ratio is unity. However, the reaction zone will extend each side of the stoichiometric region to wherever the mixture is within the flammability limits. This will have an averaging effect on NO_x production. In addition, radiation from the reaction zone is significant, and NO_x production is strongly temperature-dependent. A common method to reduce NO_x emissions is to retard the injection timing, but this has adverse effects on fuel consumption and smoke emissions. Retarding the injection timing may be beneficial because this reduces the delay period and consequently the uncontrolled combustion period.

The most serious emission from compression ignition engines is smoke, with the characteristic grey or black of soot (carbon) particles. In this discussion, smoke does not include the bluish smoke that signifies lubricating oil is being burnt, or the white smoke that is characteristic of unburnt fuel. These types of smoke occur only with malfunctioning engines, both compression and spark ignition.

Smoke from compression ignition engines originates from carbon particles formed by cracking (splitting) of large hydrocarbon molecules on the fuel-rich side of the reaction zone. The carbon particles can grow by agglomeration until they reach the fuel-lean zone, where they can be oxidised. The final rate of soot release depends on the difference between the rate of formation and the rate of oxidation. The maximum fuel injected (and consequently power output) is limited so that the exhaust smoke is just visible. Smoke output can be reduced by advancing the injection timing or by injecting a finer fuel spray, the latter being obtained by higher injection pressures and finer nozzles. Smoke from a compression ignition engine implies a poorly calibrated injector pump or faulty injectors. Engine emissions are discussed further in chapter 4, section 4.3 for spark ignition engines, and chapter 5, section 5.6 for compression ignition engines.

3.9 Combustion modelling

3.9.1 Introduction

The combustion model is one of the key elements in any computer simulation of internal combustion engine cycles. In addition, all aspects of the engine operating cycle directly influence the combustion process. Heywood (1988) provides a very good introduction to the subject and he emphasises the interdependence and complication of the combustion and engine operation. The combustion occurs in a three-dimensional, time-dependent, turbulent flow, with a fuel containing a blend of hydrocarbons, and with poorly understood combustion chemistry. The combustion chamber varies in shape, and the heat transfer is difficult to predict.

There are three approaches to combustion modelling; in order of increasing complexity, they are:

 (i) Zero-dimensional models (or phenomenological models.) These use an empirical 'heat release' model, in which time is the only independent variable.
 (ii) Quasi-dimensional models. These use a separate submodel for turbulent combustion to derive a 'heat release' model.
(iii) Multi-dimensional models. These models solve numerically the equations for mass, momentum, energy and species conservation in one, two or three dimensions, in order to predict the flame propagation.

All models can be used for estimating engine efficiency, performance, and emissions. The zero-dimensional and quasi-dimensional models are readily incorporated into complete engine models, but there is no explicit link with combustion chamber geometry. Consequently, these models are useful for parametric studies associated with engine development. When combustion chamber geometry is important or subject to much change, multi-dimensional models have to be used. Since the computational demands are very high, multi-dimensional models are used for combustion chamber modelling rather than complete engine modelling.

The more complex models are still subject to much research and refinement, and rely on submodels for the turbulence effects and chemical kinetics. Review papers by Tabaczynski (1983) and by Greenhalgh (1983) illustrate the use of lasers in turbulence (laser doppler anemometry/ velocimetry — LDA, LDV) and the use of lasers in chemical species measurements (spectrographic techniques), respectively. These techniques can be applied to operating engines fitted with quartz windows for optical access. All models require experimental validation with engines, and combustion films can be invaluable for checking combustion models. The ways the in-cylinder flow can be measured and defined are discussed in chapter 8, along with how turbulent combustion can be modelled in spark ignition engines.

3.9.2 Zero-dimensional models

This approach to combustion modelling is best explained by reference to a particular model, the one described by Heywood *et al.* (1979), for spark ignition engines. This combustion model makes use of three zones, two of which are burnt gas:

  (i) unburnt gas
 (ii) burnt gas
(iii) burnt gas adjacent to the combustion chamber — a thermal boundary layer or quench layer.

This arrangement is shown in figure 3.16, in addition to the reaction zone or flame front separating the burnt and unburnt gases. The combustion does not occur instantaneously, and can be modelled by a Wiebe function (1967):

$$x(\theta) = 1 - \exp\{- a \, [(\theta - \theta_0)/\Delta\theta_b]^{\,m+1}\}$$

where $x(\theta)$ is the mass fraction burnt at crank angle θ
            θ_0 is the crank angle at the start of combustion
and $\Delta\theta_b$ is the duration of combustion.

a and m are constants that can be varied so that a computed p-V diagram can be matched to that of a particular engine. Typically

$$a = 5 \text{ and } m = 2$$

The effect of varying these parameters on the rate of combustion is shown in figure 3.17.

The heat transfer is predicted using the correlation developed by Woschni (1967); although this was developed for compression ignition engines, it is widely used for spark ignition engines. The correlation has a familiar form, in terms of Nusselt, Reynolds and Prandtl numbers:

$$Nu = a \, Re^b Pr^c$$

The constants a, b and c will depend on the engine geometry and speed but typical values are

$$a = 0.035, \, b = 0.8, \, c = 0.333$$

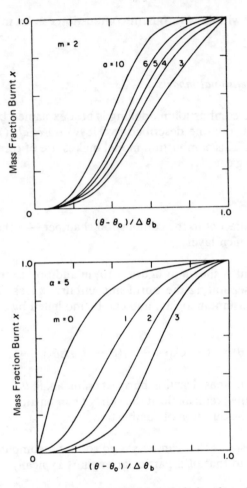

Figure 3.17   Wiebe functions (reprinted with permission from Heywood *et al.*
(1979), © Society of Automotive Engineers, Inc.)

As well as predicting the engine efficiency, this type of model is very useful in predicting engine emissions. The concentrations of carbon-oxygen-hydrogen species in the burnt gas are calculated using equilibrium thermodynamics. Nitric oxide emissions are more difficult to predict, since they cannot be described by equilibrium thermodynamics. The Zeldovich mechanism is used as a basis for calculating the nitrogen oxide production behind the flame front. This composition is then assumed to be 'frozen' when the gas comes into the thermal boundary layer. This approach is useful in assessing (for example) the effect of exhaust gas recirculation or ignition timing on both fuel economy and nitrogen oxide emissions. Some results from the Heywood model are shown in figure 3.18.

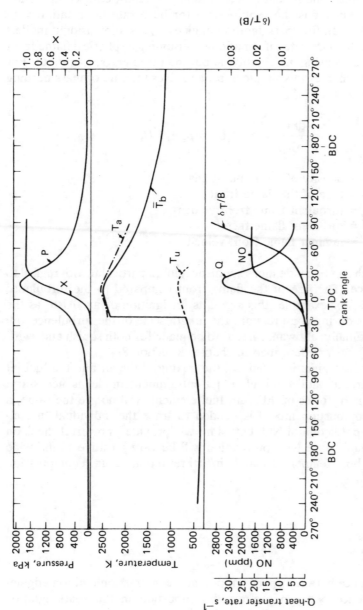

Figure 3.18 Profiles of variables predicted by the simulation throughout the four-stroke engine cycle for 5.7 litre displacement engine at base operating point. Plotted against crank angle are: mass fraction burnt x, unburnt mixture temperature T_u, mean burnt gas temperature $\bar{T}_b$, cylinder pressure p, temperature of burnt gas adiabatic core T_a, instantaneous heat transfer rate Q (normalised by the initial enthalpy of the fuel/air mixture within the cylinder), nitric oxide concentration NO, and thermal boundary layer thickness δ_T (normalised by the cylinder bore B) (reprinted with permission from Heywood *et al.* (1979). © Society of Automotive Engineers, Inc.)

3.9.3   Quasi-dimensional models

The simple three zone model is self-evidently unrealistic for compression ignition engines; nor is the requirement for burn rate information very convenient — even for spark ignition engines. Quasi-dimensional models try to predict the burn rate information by assuming a spherical flame front geometry, and by using information about the turbulence as an input. For spark ignition engines this simple approach gives the rate of mass burning (dm_b/dt) as

$$\frac{dm_b}{dt} = \rho_u A_f U_t = \rho_u A_f \text{ff } U_1$$

where ρ_u = density of the unburnt gas
    A_f = area of the flame front
    U_t = turbulent flame front velocity
    ff = turbulent flame factor
    U_1 = laminar flame front velocity.

This approach can be made more sophisticated, in particular with regard to the turbulence. The size of the flame front compared to the turbulence scale changes, for example; this accounts for ignition delay. Also, as the pressure rises during compression, the length scale of the turbulence will be reduced. Finally, allowance needs to be made for both squish and swirl. These techniques are described in chapter 8 section 8.4.

For compression ignition engines, the mixing of the air and the fuel jet are all important, and a turbulent jet-entrainment model is necessary. Since the time histories of different fuel elements will not be the same, a multi-zone combustion model is needed to trace the individual fuel elements. The prediction of NO_x emissions will provide a powerful check on any such model, since NO_x production will be very sensitive to the wide variations in both temperature and air/fuel ratio that occur in compression ignition engines.

3.10   Conclusions

A key difference between spark ignition and compression ignition engines is the difference between pre-mixed combustion in the spark ignition engine, and diffusion-controlled combustion. As a consequence, the two types of engine require different fuel properties. Spark ignition engines require volatile fuels that are resistant to self-ignition, while compression ignition engines require fuels that self-ignite readily. This also leads to the use of different fuel additives.

A proper understanding of normal and abnormal engine combustion follows from fuel and combustion chemistry. Engine emissions are also of great importance, but these are not explained entirely by predictions of dissociation from equilibrium thermodynamics. In particular, the production of nitrogen oxides involves a complex mechanism in which the different forward and reverse reaction rates are critical.

The effect of engine variables such as speed, load and ignition/injection timing on engine emissions and fuel economy is complex, and can only be explained qualitatively without the aid of engine models. In engine models, one of the critical parts is the combustion model and this in turn depends on being able to model the in-cylinder flows and reaction kinetics. Much work is still being devoted to an improved understanding of these aspects.

3.11 Examples

Example 3.1

A fuel has the following gravimetric composition;

hexane (C_6H_{14})	40 per cent
octane (C_8H_{18})	30 per cent
cyclohexane (C_6H_{12})	25 per cent
benzene (C_6H_6)	5 per cent

If the gravimetric air/fuel ratio is 17:1, determine the equivalence ratio. To calculate the equivalence ratio, first determine the stoichiometric air/fuel ratio. As the composition is given gravimetrically, this has to be converted to molar composition. For convenience, take 100 kg of fuel, with molar masses of 12 kg and 1 kg for carbon and hydrogen respectively.

Substance	Mass (m) kg	Molar mass (M) kg	No. of kmols (m/M)
C_6H_{14}	40	(6.12 + 14.1)	0.465
C_8H_{18}	30	(8.12 + 18.1)	0.263
C_6H_{12}	25	(6.12 + 12.1)	0.298
C_6H_6	5	(6.12 + 6.1)	0.064

Assuming that the combustion products contain only CO_2 and H_2O, the stoichiometric reactions in terms of kmols are found to be

$$C_6H_{14} + x \left(O_2 + \frac{79}{21} N_2^*\right) \rightarrow 6CO_2 + 7H_2O + yN_2^*$$

Variable x is found from the number of kilomoles of oxygen associated with the products, $x = 9\frac{1}{2}$ and by conservation of N_2^*, $y = (79/21)x$:

$$C_6H_{14} + 9\frac{1}{2} \left(O_2 + \frac{79}{21} N_2^*\right) \rightarrow 6CO_2 + 7H_2O + 35.74N_2^*$$

$$1 \text{ kmol of fuel: } 9\frac{1}{2} \left(1 + \frac{79}{21}\right) \text{ kmols of air} = 1 : 45.23$$

By inspection:

$$C_8H_{18} + 12\frac{1}{2} \left(O_2 + \frac{79}{21} N_2^*\right) \rightarrow 8CO_2 + 9H_2O + 47.0N_2^*$$

1 : 59.52 kmols

$$C_6H_{12} + 9 \left(O_2 + \frac{79}{21} N_2^*\right) \rightarrow 6CO_2 + 6H_2O + 33.86N_2^*$$

1 : 42.86 kmols

$$C_6H_6 + 7\frac{1}{2} \left(O_2 + \frac{79}{21} N_2^*\right) \rightarrow 6CO_2 + 3H_2O + 28.21N_2^*$$

1 : 35.71 kmols

Fuel	kmols of fuel	Stoichiometric molar air/fuel ratio	kmols of air
C_6H_{14}	0.465	45.23:1	21.03
C_8H_{18}	0.263	59.52:1	15.65
C_6H_{12}	0.298	42.86:1	12.77
C_6H_6	0.064	35.71:1	2.29
Total	100 kg fuel		51.74

$$100 \text{ kg fuel: } 51.74 \times 29 \text{ kg air}$$
$$1 : 15.00$$

$$\text{Equivalence ratio, } \phi = \frac{\text{(air/fuel ratio) stoichiometric}}{\text{(air/fuel ratio) actual}}$$

$$\phi = \frac{15}{17} = 0.882$$

Example 3.2

A fuel oil has a composition by weight of 0.865 carbon, 0.133 hydrogen and 0.002 incombustibles. Find the stoichiometric gravimetric air/fuel ratio.

When the fuel is burnt with excess air, the dry volumentric exhaust gas analysis is: CO_2 0.121, N_2^* 0.835, O_2 0.044. Determine the actual air/fuel ratio used and the wet volumetric exhaust gas analysis.

Consider 1 kg of fuel, and convert the gravimetric data to molar data. Molar mass of carbon = 12, hydrogen = 1.

$$\frac{0.865}{12} C + \frac{0.133}{1} H + Air \qquad \rightarrow Products$$

$$0.0721C + 0.133H + a\left(O_2 + \frac{79}{21} N_2^*\right) \rightarrow 0.0721CO_2 + 0.0665H_2O$$
$$+ 0.396N_2^*$$

For a stoichiometric reaction, $a = 0.0721 + \frac{1}{2}(0.0665) = 0.105$.

Thus 1 kg of fuel combines with $0.105\left(1 + \frac{79}{21}\right)$ kmol of air

$$or\ 0.105\left(1 + \frac{79}{21}\right) \times 29\ kg\ of\ air$$

and the stoichiometric gravimetric air/fuel ratio is 14.5: 1.
Note that the molar mass of the fuel is not known.

With excess air the equation can be rewritten with variable x. The incombustible material is assumed to occupy negligible volume in the products. A dry gas analysis assumes that any water vapour present in the combustion products has been removed.

$$0.0721C + 0.133H + 0.105x\ O_2 + 0.396x N_2^* \rightarrow 0.0721CO_2 + 0.0665H_2O$$
$$+ 0.396x\ N_2^* + 0.105(x - 1)O_2$$

The variable x is unity for stoichiometric reactions; if $x > 1$ there will be $0.105(x - 1)$ kmols of O_2 not taking part in the reaction with 1 kg fuel.

The dry volumetric gas analysis for each constituent in the products is in proportion to the number of moles of each dry constituent:

$$CO_2: 0.121 = \frac{0.0721}{0.0721 + 0.396x + 0.105(x - 1)}$$

$$0.0721/0.121 = 0.0721 + (0.396 + 0.105)x - 0.105;\ hence\ \underline{x = 1.255.}$$

$$O_2 : 0.044 = \frac{0.105(x - 1)}{0.0721 + 0.396x + 0.105(x - 1)}$$

$$0.105x - 0.044\,(0.396 + 0.105)x = 0.044\,(0.0721 - 0.105) + 0.105;$$

hence $x = 1.248$.

$$N_2^* : 0.835 = \frac{0.396x}{0.0721 + 0.396x + 0.105(x - 1)}$$

$$\left(0.396 + 0.105 - \frac{0.396}{0.835}\right) x = 0.105 - 0.0721; \text{ hence } x = 1.230.$$

As might be expected, each value of x is different. The carbon balance is most satisfactory since it is the largest measured constituent of the exhaust gases. The nitrogen balance is the least accurate, since it is found by difference ($0.835 = 1 - 0.044 - 0.121$). The redundancy in these equations can also be used to determine the gravimetric composition of the fuel.

Taking $x = 1.25$ the actual gravimetric air/fuel ratio is

$$14.5 \times 1.25: 1 = \underline{18.13: 1}$$

The combustion equation is actually

$$0.0721C + 0.0133H + 0.1313O_2 + 0.495N_2^* \rightarrow 0.0721CO_2 + 0.0665H_2O$$

$$+ 0.495N_2^* + 0.0263O_2$$

The wet volumetric analysis for each constituent in the products is in proportion to the number of moles of each constituent, including the water vapour

$$CO_2: \frac{0.0721}{0.0721 + 0.0665 + 0.495 + 0.0263} = \frac{0.0721}{0.6599} = \underline{0.1093}$$

$$H_2O: \frac{0.0665}{0.6599} = \underline{0.1008}$$

$$N_2^*: \frac{0.495}{0.6599} = \underline{0.7501}$$

$$O_2: \frac{0.0263}{0.6599} = \underline{0.0399}$$

Check that the sum equals unity (0.1093 + 0.1008 + 0.7501 + 0.0399 = 1.0001).

Example 3.3

Calculate the difference between the constant-pressure and constant-volume lower calorific values (LCV) for ethylene (C_2H_4) at 250°C and 1 bar. Calculate the constant-pressure calorific value at 2000 K. What is the constant-pressure higher calorific value (HCV) at 25°C? The ethylene is gaseous at 25°C, and the necessary data are in the tables by Rogers and Mayhew (1980b):

$$C_2H_4 \text{ (vap.)} + 3O_2 \rightarrow 2CO_2 + 2H_2O \text{ (vap.)}, \; (\Delta H_0)_{25°C}$$

$$= -1\,323\,170 \text{ kJ/kmol}$$

(i) Using equation (3.4):

$$(-\Delta H_0)_{25°C} - (-\Delta U_0)_{25°C} = (n_{GR} - n_{GP})R_0T$$

At 25°C the partial pressure of water vapour is 0.032 bar, so it is fairly accurate to assume that all the water vapour condenses; thus

n_{GR}, number of moles of gaseous reactants = 1 + 3

n_{PR}, number of moles of gaseous products = 2 + 0

$(n_{GR} - n_{GP})R_0T = (4 - 2) \times 8.3144 \times 298.15 = 4.958$ kJ/kmol

This difference in calorific values can be seen to be negligible. If the reaction occurred with excess air the result would be the same, as the atmospheric nitrogen and excess oxygen take no part in the reaction.

(ii) To calculate $(-\Delta H_0)_{2000K}$, use can be made of equation (3.3); this amounts to the same as the following approach:

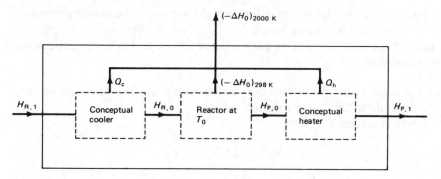

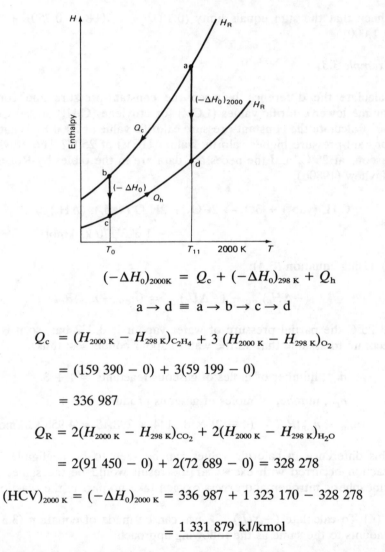

$$(-\Delta H_0)_{2000K} = Q_c + (-\Delta H_0)_{298\ K} + Q_h$$

$$a \to d \equiv a \to b \to c \to d$$

$$Q_c = (H_{2000\ K} - H_{298\ K})_{C_2H_4} + 3\,(H_{2000\ K} - H_{298\ K})_{O_2}$$

$$= (159\ 390 - 0) + 3(59\ 199 - 0)$$

$$= 336\ 987$$

$$Q_R = 2(H_{2000\ K} - H_{298\ K})_{CO_2} + 2(H_{2000\ K} - H_{298\ K})_{H_2O}$$

$$= 2(91\ 450 - 0) + 2(72\ 689 - 0) = 328\ 278$$

$$(HCV)_{2000\ K} = (-\Delta H_0)_{2000\ K} = 336\ 987 + 1\ 323\ 170 - 328\ 278$$

$$= \underline{1\ 331\ 879}\ kJ/kmol$$

Again this is not a significant difference. As before, if the reaction occurred with excess air the result would be the same, as the excess oxygen and atmospheric nitrogen take no part in the reaction, and have the same initial and final state.

(iii) As the enthalpy of reaction is for water in the vapour state, this corresponds to the lower calorific value.
   Using equation (3.5)

$$HCV - LCV = y\,h_{fg}$$

where y = mass of water vapour/kmol of fuel or

$$HCV - LCV = z\,h_{fg}$$

where z = no. of kmols of water vapour/kmol fuel.

$$H_{fg} = 43\,990 \text{ kJ/kmol of } H_2O \text{ at } 25°C$$

Thus

$$HCV = LCV + 2H_{fg}$$

$$= 1\,323\,170 + 2 \times 43\,990 = 1\,411\,150 \text{ kJ/kmol fuel}$$

Example 3.4

In a closed combustion vessel propane (C_3H_8) and air with an equivalence ratio of 1.11 initially at 25°C burn to produce products consisting solely of carbon dioxide (CO_2), carbon monoxide (CO), water (H_2O) and atmospheric nitrogen. If the heat rejected from the vessel is 770 MJ per kmol of fuel, show that the final temperature is about 1530°C. If the initial pressure is 1 bar, estimate the final pressure. Lower calorific value of propane $(-\Delta H_0)_a$, is 2 039 840 kJ/kmol of fuel at 25°C. As neither oxygen nor hydrogen is present in the combustion products, this implies that dissociation should be neglected. The stoichiometric reaction is

$$C_3H_8 + 5\left(O_2 + \frac{79}{21}N_2^*\right) \rightarrow 3CO_2 + 4H_2O + 5 \times \frac{79}{21}N_2^*$$

In this problem the equivalence ratio is 1.11, thus

$$C_3H_8 + 4.5\left(O_2 + \frac{79}{21}N_2^*\right) \rightarrow (3-x)\,CO_2 + x\,CO + 4H_2O + 16.93\,N_2^*$$

The temporary variable, x, can be eliminated by an atomic oxygen balance

$$4.5 \times 2 = (3-x) \times 2 + x + 4$$

Thus $x = 1$, and the actual combustion equation is

$$C_3H_8 + 4.5\left(O_2 + \frac{79}{21}N_2^*\right) \rightarrow 2CO_2 + CO + 4H_2O + 16.93N_2^*$$

Let the enthalpy of reaction be $(\Delta H_0)_b$ for this reaction.

If $(\Delta H_0)_c$ is the enthalphy of reaction for

$$CO + \tfrac{1}{2} O_2 \rightarrow CO_2$$

then $(\Delta H_0)_b = (\Delta H_0)_a - (\Delta H_0)_c$

Using data from Rogers and Mayhew (1980b)

$$(\Delta H_0)_b = -2\ 039\ 840 - (-282\ 990) = -1\ 756\ 850 \text{ kJ/kmol of fuel at } 25°C$$

Equation (3.4) converts constant-pressure to constant-volume calorific values:

where the number of kmols of gaseous reactants $n_{GR} = 1 + 4.5 \times \left(\dfrac{100}{21}\right)$

$$= 22.43$$

and the number of kmols of gaseous products $n_{GP} = 2 + 1 + 4 + 16.93$

$$= 23.93$$

$$(CV)_v = (CV)_p - (n_{GR} - n_{GP})R_0T$$

$$= 1\ 756\ 850 - \{(22.43 - 23.93) \times 8.3144 \times 298.15\}$$

$$= 1\ 760\ 568 \text{ kJ/kmol fuel at } 25°C$$

Assume the final temperature is 1800 K and determine the heat flow. Applying the 1st Law of Thermodynamics

$$(CV)_v = 2(U_{1800} - U_{298})_{CO_2} + 1(U_{1800} - U_{298})_{CO} + 4(U_{1800} - U_{298})_{H_2O}$$
$$+ 16.93(U_{1800} - U_{298})_{N_2^*} - Q$$

$$1\ 760\ 568 = 2(64\ 476 + 2479) + 1(34\ 556 + 2479) + 4(47\ 643 + 2479)$$
$$+ 16.93(34\ 016 + 2479) - Q$$

$$Q = -771\ 275 \text{ kJ/kmol fuel}$$

The negative sign indicates a heat flow from the vessel, and the numerical value is sufficiently close to 770 MJ.

To estimate the final pressure, apply the equation of state; use a basis of 1 kmol of fuel.

$$pV = n\ R_0T$$

Find V for the reactants

$$V = \frac{n_R R_0 T}{p} = \frac{(1 + 4.5 + 1693)R_0\ 298}{10^5}$$

for the products

$$p = \frac{n_R R_0 T}{V}$$

$$= \frac{(2 + 1 + 4 + 16.93)\ R_0\ 1800}{(1 + 4.5 + 16.93)\ R_0\ 298} \times 10^5$$

$$= \underline{6.53\ \text{bar}}$$

Example 3.5

Compute the partial pressures of a stoichiometric equilibrium mixture of CO, O_2, CO_2 at 3000 and 3500 K when the pressure is 1 bar. Show that for a stoichiometric mixture of CO and O_2 initially at 25°C and 1 bar, the adiabatic constant-pressure combustion temperature is about 3050 K.

Introduce a variable x into the combustion equation to account for dissociation

$$CO + \tfrac{1}{2} O_2 \rightarrow (1 - x)\ CO_2 + xCO + \frac{x}{2} O_2$$

The equilibrium constant K is defined by equation (3.7):

$$K = \frac{p'_{CO_2}}{(p'_{CO})(p'_{O_2})^{\frac{1}{2}}} \quad \text{where } p'_{CO_2} \text{ is the partial pressure of } CO_2 \text{ etc.}$$

The partial pressure of a species is proportional to the number of moles of that species. Thus

$$p'_{CO_2} = \frac{1 - x}{(1 - x) + x + \dfrac{x}{2}} p, \quad p'_{CO} = \frac{x}{1 + \dfrac{x}{2}} p, \quad p'_{O_2} = \frac{x/2}{1 + \dfrac{x}{2}} p$$

where p is the total system pressure. Thus

$$K = \frac{1 - x}{x(x/2)^{\frac{1}{2}}} \left(\frac{1 + (x/2)}{p} \right)^{\frac{1}{2}}$$

Again, data are obtained from Rogers and Mayhew (1980b). At 3000 K, $\log_{10} K = 0.485$ with pressure in atmospheres:

$$1 \text{ atmosphere} = 1.01325 \text{ bar}$$

Thus

$$10^{0.485} = \frac{1 - x}{x \sqrt{\left(\frac{x}{2}\right)}} \sqrt{\left(\frac{1 + x/2}{1.01325^{-1}}\right)}$$

$$10^{0.485} (1.01325)^{-\frac{1}{2}} x^{\frac{3}{2}} = \sqrt{(2)} (1 - x) \sqrt{(1 + (x/2))}$$

Square all terms

$$10^{0.97} (1.01325)^{-1} x^3 = (1 - x)^2 (2 + x)$$

$$8.211x^3 + 3x - 2 = 0$$

This cubic equation has to be solved iteratively. The Newton-Raphson method provides fast convergence; an alternative simpler method, which may not converge, is the following method

$$x_{n+1} = \sqrt[3]{\left(\frac{2 - 3x_n}{8.211}\right)} \quad \text{where } n \text{ is the iteration number}$$

Using a calculator without memory this readily gives

$$x = 0.436$$

At 3500 K use linear interpolation to give $\log_{10} K = -0.187$, leading to

$$x = 0.748$$

This implies greater dissociation at the higher temperature, a result predicted by Le Châtelier's Principle for an exothermic reaction.

The partial pressures are found by back-substitution, giving the results tabulated below:

T	p	x	p'_{CO_2}	p'_{O_2}	p'_{CO}	
3000 K	1	0.436	0.463	0.179	0.358	all pressures in bar
3500 K	1	0.748	0.183	0.272	0.544	

To find the adiabatic combustion temperature, assume a temperature and then find the heat transfer necessary for the temperature to be attained. For the reaction $CO + \frac{1}{2} O_2 \rightarrow CO_2$, $(\Delta H_0)_{298} = -282\,990$ kJ/kmol. Allowing for dissociation, the enthalpy for reaction is $(1 - x)(\Delta H_0)_{298}$.

It is convenient to use the following hypothetical scheme for the combustion calculation

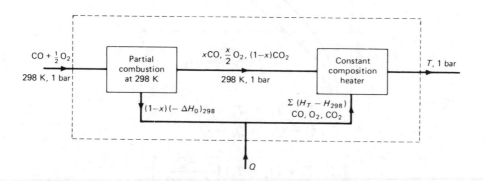

$$(1 - x)(-\Delta H_0)_{298} + Q = \sum_{CO, O_2, CO_2} (H_T - H_{298})$$

If temperature T is guessed correctly then $Q = 0$. Alternatively, make two estimates and interpolate.

1st guess: $T = 3000$ K, $x = 0.436$.

$$Q = \left\{(1 - x)(H_{3000})_{CO_2} + x(H_{3000})_{CO} + \frac{x}{2}(H_{3000})_{O_2}\right\} - (1 - x)(-\Delta H_0)_{298};$$

$$H_{298} = 0$$

$$= \{0.564 \times 152\,860 + 0.436 \times 93\,542 + 0.218 \times 98\,098\}$$
$$- 0.564 \times 282\,990$$

$$= -11\,223 \text{ kJ/kmol CO, that is, } T \text{ is too low.}$$

2nd guess: $T = 3500$ K, $x = 0.748$.

$$Q = \{0.252 \times 184\,130 + 0.748 \times 112\,230 + 0.374 \times 118\,315\} - 0.252$$
$$\times 282\,290$$

$$= 103\,462 \text{ kJ/kmol CO, that is, } T \text{ is much too high.}$$

To obtain a better estimate of T, interpolate between the 1st and 2nd guesses.

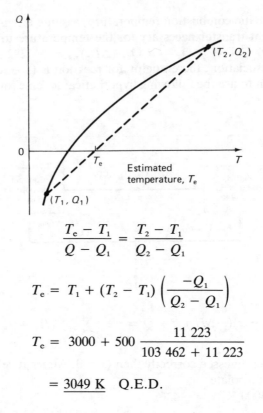

$$\frac{T_e - T_1}{Q - Q_1} = \frac{T_2 - T_1}{Q_2 - Q_1}$$

$$T_e = T_1 + (T_2 - T_1)\left(\frac{-Q_1}{Q_2 - Q_1}\right)$$

$$T_e = 3000 + 500 \frac{11\ 223}{103\ 462 + 11\ 223}$$

$$= \underline{3049\ K}\quad \text{Q.E.D.}$$

If a more accurate result is needed, recalculate Q when $T = 3050$ K. Assuming no dissociation, $T \approx 5000$ K.

Example 3.6

In fuel-rich combustion product mixtures, equilibrium between the species CO_2, H_2O, CO and H_2 is often assumed to be the sole determinant of the burned gas composition. If methane is burnt in air with an equivalence ratio of 1.25, determine the molar fractions of the products at 1800 K.

Comment on any other likely dissociation reactions, whether or not this is likely to be significant, and whether it is reasonable to assume equilibrium conditions.

Solution

First, establish stoichiometric combustion by introducing a temporary variable x:

$$CH_4 + x\left(O_2 + \frac{79}{21}N_2\right) \rightarrow CO_2 + 2H_2O + x\frac{79}{21}N_2$$

From the molecular oxygen balance: $x = 2$.

Thus, for the combustion of a methane/air mixture with an equivalence ratio of 1.25:

$$CH_4 + \frac{2}{1.25}\left(O_2 + \frac{79}{21}\,N_2\right) \rightarrow a\ CO_2 + b\ CO + c\ H_2O + d\ H_2 + 6.02\ N_2$$

There are now 4 unknown variables, and the atomic balance equations can be used to give 3 independent equations:

carbon balance	$a + b = 1$;	$b = 1 - a$
hydrogen balance	$2c + 2d = 4$;	$d = 2 - c$
oxygen balance	$2a + b + c = 3.2$;	

Substituting for b from the carbon balance:

$$2a + (1 - a) + c = 3.2; \quad a = 2.2 - c$$

Or if the carbon balance is used to eliminate a:

$$b = c - 1.2$$

The fourth equation has to come from the water gas equilibrium, as the nitrogen balance is not independent of the other atomic balances:

For $CO_2 + H_2 \rightleftharpoons CO + H_2O$ @ 1800 K, $\log_{10}K = 0.577$

Using equation (3.6)

$$\frac{\left(p'_{H_2O}\right)\left(p'_{CO}\right)}{\left(p'_{H_2}\right)\left(p'_{CO_2}\right)} = 3.776$$

As the stoichiometric coefficients are all unity, the system pressure is not relevant, nor is the total number of moles. Thus

$$b \times c = 3.776 \times a \times d$$

Substituting in terms of c

$$(c - 1.2) \times c = 3.776(2.2 - c)(2 - c)$$
$$c^2 - 1.2c = 16.61 - 15.86c + 3.776c^2$$
$$2.776c^2 - 14.66c + 16.61 = 0$$

and solving

$$c = \frac{14.66 \pm \sqrt{(14.66^2 - 4 \times 2.776 \times 16.61)}}{2 \times 2.776}$$

$$c = 3.635 \text{ or } c = 1.646$$

To establish which solution is valid, consider the result of the oxygen balance: as $a > 0$, then $c < 2.2$. Therefore, the required solution is $c = 1.646$. Back substitution then gives from:

the hydrogen balance $d = 0.354$
the carbon/oxygen balance $a = 2.2 - c = 0.554$
                            $b = c - 1.2 = 0.446$

(As a check $(b \times c)/(d \times a) = 3.74$, which is close enough)
The full combustion equation is thus

$$CH_4 + 1.6 \left(O_2 + \frac{79}{21} N_2\right) \rightarrow 0.554 \ CO_2 + 0.446 \ CO + 1.646 \ H_2O$$

$$+ \ 0.354 \ H_2 + 6.02 \ N_2$$

At 1800 K, the water will be in its vapour state and the molar fractions are:

$$\% CO_2 = a/\Sigma v_i = 0.554/9.02 = 6.14 \text{ per cent}$$

$$\% CO = 0.446/9.02 = 4.94 \text{ per cent}$$

$$\% H_2O = 18.25 \text{ per cent}$$

$$\% H_2 = 3.92 \text{ per cent}$$

$$\% N_2 = 66.7 \text{ per cent}$$

A possible dissociation reaction is:

$$CO + \tfrac{1}{2}O_2 \rightleftharpoons CO_2$$

At 1800 K:

$$\frac{p'_{CO_2}}{p'_{CO} \sqrt{p'_{O_2}}} = 4932 \text{ atm.}^{-\frac{1}{2}}$$

Assume $p = 1$ atm. p'_{O_2} is so small as not to affect p'_{CO} or p'_{CO_2}:

$$p'_{O_2} \approx \left(\frac{0.028}{4932 \times 1.916} \right)^2 = 8 \times 10^{-12} \text{ atm.}$$

However, equilibrium conditions at 1800 K are unlikely to pertain, as the composition will: (a) not be in equilibrium, (b) be frozen at its composition determined at a higher temperature before blowdown.

Example 3.7

A gas engine operating on methane (CH_4) at 1500 rpm, full throttle, generates the following emissions measured on a dry volumetric basis:

CO_2	10.4 per cent
CO	1.1 per cent
H_2	0.6 per cent
O_2	0.9 per cent
NO	600 ppm
HC	1100 ppm (as methane)

If the specific fuel consumption is 250 g/kWh, calculate the specific emissions of: carbon monoxide, nitrogen oxide and unburned hydrocarbons (that is, g/kWh). Why is this specific basis more relevant than a percentage basis?

Solution

Consider the combustion of sufficient fuel to produce 100 kmol of dry products. Introduce the following temporary variables

y kmol of CH_4 and x kmol of O_2 in the reactants
a kmol of H_2O and b kmol of N_2 in the products

$$yCH_4 + x\left(O_2 + \frac{79}{21} N_2\right) \rightarrow 1.1\,CO + 10.4\,CO_2 + 0.6\,H_2$$

$$+ 0.9\,O_2 + 0.11\,CH_4 + 0.06\,NO$$

$$+ a\,H_2O + b\,N_2$$

The quantity of N_2 is known, as the products add to 100 kmol. From the nitrogen balance:

$$b = 100 - (1.1 + 10.4 + 0.6 + 0.9 + 0.11 + 0.06) = 86.83$$

To find y, use the C balance:

$$y = 1.1 + 10.4 + 0.11 = 11.61$$

and find a from the H_2 balance:

$$2 \times 11.61 = 0.6 + 2 \times 0.11 + a$$
$$a = 22.4$$

Find x from the O_2 balance (or the N_2 balance — this will be used as a check):

$$x = \frac{1}{2} \times 1.1 + 10.4 + 0.9 + \frac{1}{2} \times 0.06 + \frac{1}{2} \times 22.4$$
$$x = 23.085$$

(Checking: $x \times \dfrac{79}{21} = 0.06/2 + 86.83$; $x = 23.089$, good enough.)

Thus 11.61 kmol of $CH_4 \rightarrow 1.1$ kmol of CO etc., or

$$11.61 \times (12 + 4 \times 1) \text{ kg of } CH_4 \rightarrow 1.1 \times (12 + 16) \text{ kg CO}$$

However 1 kWh is produced by 250 g of fuel. Specific emissions of CO

$$= \frac{n_{CO} \times M_{CO}}{n_{CH_4} \times M_{CH_4}} \times sfc = \frac{1.1 \times (12 + 16)}{11.61 \times (12 + 4 \times 1)} \times 250$$

$$= 41.5 \text{ g CO/kWh}$$

Specific emissions of NO

$$= \frac{n_{NO} \times M_{NO}}{n_{CH_4} \times M_{CH_4}} \times sfc = (0.06 \times 30) \times 1.3458$$

$$= 2.42 \text{ g NO/kWh}$$

Specific emissions of CH_4 (with subscript P denoting Products, and R denoting Reactants)

$$= \frac{(n_{CH_4})_P}{(n_{CH_4})_R} \times sfc$$

$$= 2.37 \text{ g } CH_4/kWh$$

A specific basis should be used for emissions, as the purpose of the engine is to produce work (measured as kWh); the volume flow rate will be a function of the engine efficiency and the engine equivalence ratio, so the volumetric composition of the exhaust does not give an indication of the amount of pollutants.

3.12 Problems

3.1 If a fuel mixture can be represented by the general formula C_xH_{2x}, show that the stoichiometric gravimetric air/fuel ratio is 14.8:1.

3.2 A fuel has the following molecular gravimetric composition

pentane (C_5H_{12})	10 per cent
heptane (C_7H_{16})	30 per cent
octane (C_8H_{18})	35 per cent
dodecane ($C_{12}H_{26}$)	15 per cent
benzene (C_6H_6)	10 per cent

Calculate the atomic gravimetric composition of the fuel and the gravimetric air/fuel ratio for an equivalence ratio of 1.1.

3.3 The dry exhaust gas analysis from an engine burning a hydrocarbon diesel fuel is as follows: CO_2 0.121, O_2 0.037, N_2^* 0.842. Determine the gravimetric composition of the fuel, the equivalence ratio of the fuel/air mixture, and the stoichiometric air/fuel ratio.

3.4 An engine with a compression ratio of 8.9:1 draws in air at a temperature of 20°C and pressure of 1 bar. Estimate the temperature and pressure at the end of the compression stroke. Why will the temperature and pressure be less than this in practice?

The gravimetric air/fuel ratio is 12:1, and the calorific value of the fuel is 44 MJ/kg. Assume that combustion occurs instantaneously at the end of the compression stroke. Estimate the temperature and pressure immediately after combustion.

3.5 Compute the partial pressures of a stoichiometric equilibrium mixture of CO, O_2, CO_2 at (i) 3000 K and (ii) 3500 K when the pressure is 10 bar. Compare the answers with example 3.5. Are the results in accordance with Le Châtelier's Principle?

3.6 In a test to determine the cetane number of a fuel, comparison was made with two reference fuels having cetane numbers of 50 and 55. In the test the compression ratio was varied to give the same ignition delay. The compression ratios for the reference fuels were 25.4 and 23.1 respectively. If the compression ratio for the test fuel was 24.9, determine its cetane number.

3.7 Contrast combustion in compression ignition engines and spark ignition engines. What are the main differences in fuel requirements?

3.8 What is the difference between 'knock' in compression ignition and spark ignition engines? How can 'knock' be eliminated in each case?

3.9 A Dual fuel engine operates by aspirating a mixture of air and fuel, which is then ignited by the spontaneous ignition of a small quantity of diesel fuel injected near the end of the compression stroke. A Dual fuel engine is operating with a gravimetric air fuel ratio of 20, and the effective atomic H:C composition of the fuel is 3.1:1.

(i)   Write down the stoichiometric and actual combustion equations, and calculate the volumetric wet gas composition.
(ii) If the calorific value of the fuel is 46 MJ/kg, what percentage of this could be recovered by cooling the exhaust gas from 627°C to 77°C?
(iii) If the fuel consists of methane (CH_4) and cetane ($C_{16}H_{34}$), what is the ratio of the fuel mass flowrates? What would be the advantages and disadvantages of cooling the exhaust below the dewpoint temperature?

In a Dual fuel engine, what determines the selection of the compression ratio? Explain under what circumstances emissions might be a problem.

3.10 Some combustion products at a pressure of 2 atm. are in thermal equilibrium at 1400 K, and a volumetric gas analysis yielded the following data:

CO_2	0.065
H_2O	0.092
CO	0.040

(i) Deduce the concentration of H_2 from the equilibrium equation.
(ii) If the fuel was octane, calculate the approximate percentage of

the fuel chemical energy not released as a result of the partial combustion.

(iii) Sketch how engine exhaust emissions vary with the air/fuel ratio, and discuss the shape of the nitrogen oxides and unburnt hydrocarbon curves.

3.11 The results of a dry gas analysis for a diesel engine exhaust are as follows:

Carbon dioxide, CO_2	10.1 per cent
Atmospheric nitrogen, N_2	82.6 per cent

Stating any assumption, calculate:

(i) the gravimetric air fuel ratio.
(ii) the atomic and gravimetric composition of the fuel.
(iii) the stoichiometric gravimetric air fuel ratio.
(iv) the actual equivalence ratio.

Why are the results derived from an exhaust gas analysis likely to be less accurate, when the mixture is rich or significantly weak of stoichiometric? What other diesel engine emissions are significant?

3.12 A four-stroke engine is running on methane with an equivalence ratio of 0.8. The air and fuel enter the engine at 25°C, the exhaust is at 527°C, and the heat rejected to the coolant is 340 MJ/kmol fuel. Write the equations for stoichiometric combustion and the actual combustion.

(i) By calculating the enthalpy of the exhaust flow, deduce the specific work output of the engine, and thus the overall efficiency.
(ii) If the engine has a swept volume of 5 litres and a volumetric efficiency of 72 per cent (based on the air flow), calculate the power output at a speed of 1500 rpm. The ambient pressure is 1 bar.

State clearly any assumptions that you make, and comment on any problems that would occur with a rich mixture.

3.13 The results of a dry gas analysis of an engine exhaust are as follows:

Carbon dioxide, CO_2	10.7 per cent
Carbon monoxide, CO	5.8 per cent
Atmospheric nitrogen, N_2^*	83.5 per cent

Stating any assumptions, calculate:

(i) the gravimetric air/fuel ratio
(ii) the gravimetric composition of the fuel
(iii) the stoichiometric gravimetric air/fuel ratio.

Discuss some of the means of reducing engine exhaust emission.

3.14 The following equation is a simple means for estimating the combustion temperature (T_c) of weak mixtures (that is, equivalence ratio, $\phi < 1.0$):

$$T_c = T_m + k\phi \qquad \phi < 1.0$$

where T_m is the mixture temperature prior to combustion, and k is a constant.

By considering the combustion of an air/octane (C_8H_{18}) mixture with an equivalence ratio of 0.9 at a temperature of 25°C, calculate a value for k (neglect dissociation).

For rich mixtures the following equation is applicable:

$$T_c = T_m + k - (\phi - 1.0) \times 1500 \qquad \phi \geqslant 1.0$$

Plot the combustion temperature rise for weak and rich mixtures, and explain why the slopes are different each side of stoichiometric. Sketch how these results would be modified by dissociation.

4 Spark Ignition Engines

4.1 Introduction

This chapter considers how the combustion process is initiated and constrained in spark ignition engines. The air/fuel mixture has to be close to stoichiometric (chemically correct) for satisfactory spark ignition and flame propagation. The equivalence ratio or mixture strength of the air/fuel mixture also affects pollutant emissions, as discussed in chapter 3, and influences the susceptibility to spontaneous self-ignition (that is, knock). A lean air/fuel mixture (equivalence ratio less than unity) will burn more slowly and will have a lower maximum temperature than a less lean mixture. Slower combustion will lead to lower peak pressures, and both this and the lower peak temperature will reduce the tendency for knock to occur. The air/fuel mixture also affects the engine efficiency and power output. At constant engine speed with fixed throttle, it can be seen how the brake specific fuel consumption (inverse of efficiency) and power output vary. This is shown in figure 4.1 for a typical spark ignition engine at full or wide open throttle (WOT). As this is a constant-speed test, power output is proportional to torque output, and this is most conveniently expressed as bmep since bmep is independent of engine size. Figure 4.2 is an alternative way of expressing the same data (because of their shape, the plots are often referred to as 'fish-hook' curves); additional part throttle data have also been included. At full throttle, the maximum for power output is fairly flat, so beyond a certain point a richer mixture significantly reduces efficiency without substantially increasing power output. The weakest mixture for maximum power (wmmp) will be an arbitrary point just on the lean side of the mixture for maximum power. This is also known as the mixture for Lean Best Torque (LBT). The minimum specific fuel consumption also occurs over a fairly wide range of mixture strengths. A simplified explanation for this is as follows. With dissociation occurring, maximum power will be with a rich mixture when as much as possible of the oxygen is consumed; this implies unburnt fuel and reduced efficiency. Conversely,

121

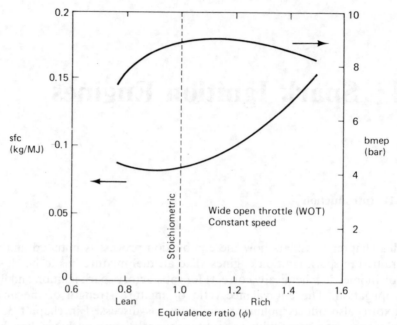

Figure 4.1   Response of specific fuel consumption and power output to changes in air/fuel ratio

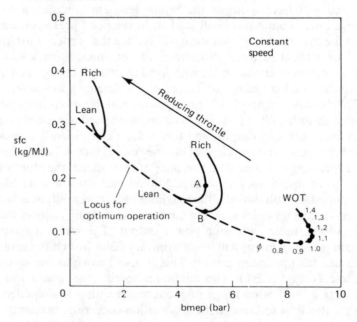

Figure 4.2   Specific fuel consumption plotted against power output for varying air/fuel ratios at different throttle settings

for maximum economy as much of the fuel should be burnt as possible, implying a weak mixture with excess oxygen present. In addition, it was shown in chapter 2 that the weaker the air/fuel mixture the higher the ideal cycle efficiency. When the air/fuel mixture becomes too weak the combustion becomes incomplete and the efficiency again falls.

The air/fuel mixture can be prepared either by a carburettor or by fuel injection. In a carburettor air flows through a venturi, and the pressure drop created causes fuel to flow through an orifice, the jet. There are two main types of carburettor — fixed jet and variable jet. Fixed jet carburettors have a fixed venturi, but a series of jets to allow for different engine-operating conditions from idle to full throttle. Variable jet carburettors have an accurately profiled needle in the jet. The needle position is controlled by a piston which also varies the venturi size. The pressure drop is approximately constant in a variable jet carburettor, while pressure drop varies in a fixed jet carburettor.

The alternative to carburettors is fuel injection. Early fuel injection systems were controlled mechanically, but the usual form of control is now electronic. Fuel is not normally injected directly into the cylinder during the compression stroke. This would require high-pressure injection equipment, and it would reduce the time for preparation of an homogeneous mixture. Also, the injectors would have to withstand the high temperature and pressures during combustion and be resistant to the build-up of combustion deposits. With low-pressure fuel injection systems, the fuel is usually injected close to the inlet valve of each cylinder. Alternatively a single injector can be used to inject fuel at the entrance to the inlet manifold.

The ignition timing also has to be controlled accurately, and a typical response for power output (which will also equate for efficiency, since the amount of fuel being burned is unchanged) is shown in figure 4.3. If ignition is too late, then although the work done by the piston during the compression stroke is reduced, so is the work done on the piston during the expansion stroke, since all pressures during the cycle will be reduced. Furthermore, there is a risk that combustion will be incomplete before the exhaust valve opens at the end of the expansion stroke, and this may overheat the exhaust valve. Conversely, if ignition is too early, there will be too much pressure rise before the end of the compression stroke (tdc) and power will be reduced. Thus, the increase in work during the compression stroke is greater than the increase in work done on the piston during the expansion stroke. Also, with early ignition the peak pressure and temperature may be sufficient to cause knock. Ignition timing is optimised for maximum power; as the maximum is fairly flat the ignition timing is usually arranged to occur on the late side of the maximum. This is shown by the MBT ignition timing (minimum advance for best torque) in figure 4.3. The definition of MBT timing is somewhat arbitrary. It may

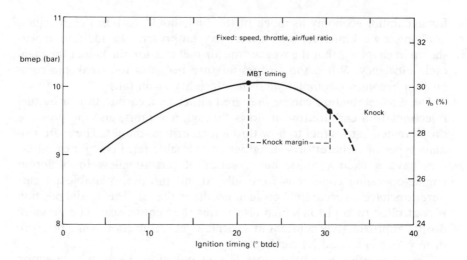

Figure 4.3   The effect of ignition timing on the output and efficiency of a spark ignition engine

correspond to the timing that gives a 1 per cent fall in the peak torque. Alternatively, it might be when the operator of an engine test first detects a fall in torque, as the ignition timing is being retarded. Figure 4.3 shows that MBT ignition timing increases the knock margin. However, for some engines operating at full throttle (particularly at low speed operation) knock will be encountered before the MBT ignition timing. In which case the ignition will be retarded from MBT to preserve a knock margin — this is to allow for manufacturing tolerances and engine ageing making some engines more susceptible to knock. At higher engine speeds, the lower volumetric efficiency (reducing peak pressures in the cylinder) and the less time available for the knock mechanism to occur, mean that knock is less likely to occur. Since engines rarely operate with full throttle at low speeds, it is common for a compression ratio/octane requirement combination to be adopted, that leads knock being encountered before MBT ignition timing at low speeds. Retarding the ignition timing in the low speed, full throttle part of the engine operating envelope, results in a local reduction in efficiency and output. However, everywhere else in the engine operating envelope, there is an efficiency and output benefit from the higher compression ratio.

At part throttle operation, the cylinder pressure and temperature are reduced and flame propagation is slower; thus ignition is arranged to occur earlier at part load settings. Ignition timing can be controlled either electronically or mechanically.

The different types of combustion chamber and their characteristics are discussed in the next section.

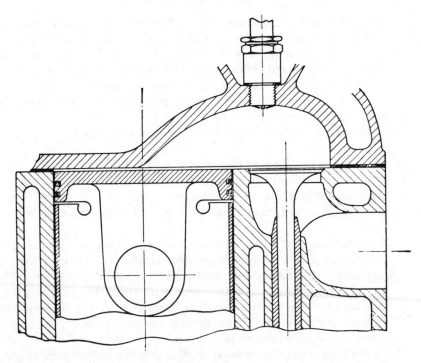

Figure 4.4   Ricardo turbulent head for side valve engines (reproduced with permission from Ricardo and Hempson (1968))

4.2 Combustion chambers

4.2.1 Conventional combustion chambers

Initially the cylinder head was little more than a cover for the cylinder. The simplest configuration was the side valve engine, figure 4.4, with the inlet and exhaust valves together at one side of the cylinder. The most successful combustion chamber for the side valve engine was the Ricardo turbulent head, as shown in figure 4.4. This design was the result of extensive experimental studies aimed at improving combustion. The maximum compression ratio that was reasonable with this geometry was limited to about 6:1, but this was not a restriction since the octane rating of fuels in the 1920s and 1930s was only about 60–70.

In the Ricardo turbulent head design the clearance between part of the cylinder head and piston at the end of the compression stroke is very small. This forms an area of 'squish', from which gas is ejected into the main volume. The turbulence that this jet generates ensures rapid combustion.

If too large a squish area is used the combustion becomes too rapid and noisy. This design also reduces the susceptibility to knock, since the gas furthest from the sparking plug is in the squish area. The end gas in the squish area is less prone to knock since it will be cooler because of the close proximity of the cylinder head and piston. Excessive turbulence also causes excessive heat transfer to the combustion chamber walls, and should be avoided for this reason also.

The main considerations in combustion chamber design are:

(i)   the distance travelled by the flame front should be minimised
(ii) the exhaust valve(s) and spark plug(s) should be close together
(iii) there should be sufficient turbulence
(iv) the end gas should be in a cool part of the combustion chamber.

(i)   By minimising the distance between the spark plug and the end gas, combustion will be as rapid as possible. This has two effects. Firstly, it produces high engine speeds and thus higher power output. Secondly, the rapid combustion reduces the time in which the chain reactions that lead to knock can occur. This implies that, for geometrically similar engines, those with the smallest diameter cylinders will be able to use the highest compression ratios.
(ii) The exhaust valve should be as close as possible to the sparking plug. The exhaust valve is very hot (possibly incandescent) so it should be as far from the end gas as possible to avoid inducing knock or pre-ignition.
(iii) There should be sufficient turbulence to promote rapid combustion. However, too much turbulence leads to excessive heat transfer from the chamber contents and also to too rapid combustion, which is noisy. The turbulence can be generated by squish areas or shrouded inlet valves.
(iv) The small clearance between the cylinder head and piston in the squish area forms a cool region. Since the inlet valve is cooled during the induction stroke, this too can be positioned in the end gas region.

For good fuel economy all the fuel should be burnt and the quench areas where the flame is extinguished should be minimised. The combustion chamber should have a low surface-to-volume ratio to minimise heat transfer. The optimum swept volume consistent with satisfactory operating speeds is about 500 cm^3 per cylinder. For high-performance engines, smaller cylinders will enable more rapid combustion, so permitting higher operating speeds and consequently greater power output. For a given geometry, reducing the swept volume per cylinder from 500 cm^3 to 200 cm^3 might increase the maximum engine speed from about 6000 rpm to 8000 rpm.

The ratio of cylinder diameter to piston stroke is also very important.

When the stroke is larger than the diameter, the engine is said to be 'under-square'. In Britain the car taxation system originally favoured under-square engines and this hindered the development of higher-performance over-square engines. In over-square engines the cylinder diameter is larger than the piston stroke, and this permits larger valves for a given swept volume. This improves the induction and exhaust processes, particularly at high engine speeds. In addition the short stroke reduces the maximum piston speed at a given engine speed, so permitting higher engine speeds. The disadvantage with over-square engines is that the combustion chamber has a poor surface-to-volume ratio, so leading to increased heat transfer. More recently there has been a return to under-square engines, as these have combustion chambers with a better surface-to-volume ratio, and so lead to better fuel economy. It will be seen later in chapter 5, section 5.2 that in general, the maximum power output of an engine is proportional to the piston area, while the maximum torque output is proportional to the swept volume.

Currently most engines have a compression ratio of about 9:1, for which a side valve geometry would be unsuitable. Overhead valve (ohv) engines have a better combustion chamber for these higher compression ratios. If the camshaft is carried in the cylinder block the valves are operated by push rods and rocker arms. A more direct alternative is to mount the camshaft in the cylinder head (ohc — overhead camshaft). The camshaft can be positioned directly over the valves, or to one side with valves operated by rocker arms. These alternatives are discussed more fully in chapter 6.

Figure 4.5a–d shows four fairly typical combustion chamber configurations; where only one valve is shown, the other is directly behind. Very often it will be production and economic considerations rather than thermodynamic considerations that determine the type of combustion chamber used. If combustion chambers have to be machined it will be cheapest to have a flat cylinder head and machined pistons. If the finish as cast is adequate, then the combustion chamber can be placed in the cylinder head economically.

Figure 4.5a shows a wedge combustion chamber; this is a simple chamber that produces good results. The valve drive train is easy to install, but the inlet and exhaust manifold have to be on the same side of the cylinder head. The hemispherical head, figure 4.5b has been popular for a long time in high-performance engines since it permits larger valves to be used than those with a flat cylinder head. The arrangement is inevitably expensive, with perhaps twin overhead camshafts. With the inlet and exhaust valves at opposite sides of the cylinder, it allows crossflow from inlet to exhaust. Crossflow occurs at the end of the exhaust stroke and the beginning of the induction stroke when both valves are open; it is benficial since it reduces the exhaust gas residuals. More recently 'pent-roof' heads with four valves

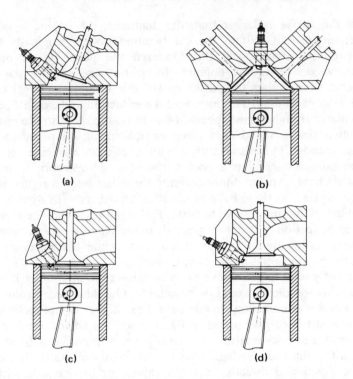

Figure 4.5   Combustion chambers for spark ignition engines. (a) Wedge
chamber; (b) hemispherical head; (c) bowl in piston chamber;
(d) bath-tub head

per cylinder have become popular; these have a shape similar to that of a
house roof. The use of four valves gives an even greater valve area than
does the use of two valves in a hemispherical head. A much cheaper
alternative, which also has good performance, is the bowl in piston (Heron
head) combustion chamber, figure 4.5c. This arrangement was used by
Jaguar for their V12 engine and during development it was only marginally
inferior to a hemispherical head engine with twin overhead camshafts
(Mundy (1972)). The bath-tub combustion chamber, figure 4.5d, has a very
compact combustion chamber that might be expected to give economical
performance; it can also be used in a crossflow engine. All these combustion chambers have:

   (i)   short maximum flame travel
   (ii) the spark plug close to the exhaust valve
   (iii) a squish area to generate turbulence
   (iv) well-cooled end gas.

The fuel economy of the spark ignition engine is particularly poor at part load; this is shown in figure 4.2. Although operating an engine on a very lean mixture can cause a reduction in efficiency, this reduction is less than if the power was controlled by throttling, with its ensuing losses. Too often, engine manufacturers are concerned with performance (maximum power and fuel economy) at or close to full throttle, although in automotive applications it is unusual to use maximum power except transiently.

In chapter 3 it was stated that the maximum compression ratio for an engine is usually dictated by the incipience of knock. If the problem of knock could be avoided, either by special fuels or special combustion chambers, there would still be an upper useful limit for compression ratio. As compression ratio is raised, there is a reduction in the rate at which the ideal cycle efficiency improves, see figure 4.6. Since the mechanical efficiency will be reduced by raising the compression ratio (owing to higher pressure loadings), the overall efficiency will be a maximum for some finite compression ratio, see figure 4.6.

Some of the extensive work by Caris and Nelson (1958) is summarised in figure 4.7. This work shows an optimum compression ratio of 16:1 for maximum economy, and 17:1 for maximum power. The reduction in efficiency is also due to poor combustion chamber shape — at high compression ratios there will be a poor surface-to-volume ratio. Figures for

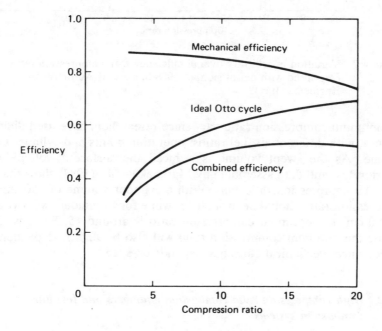

Figure 4.6   Variation in efficiency with compression ratio

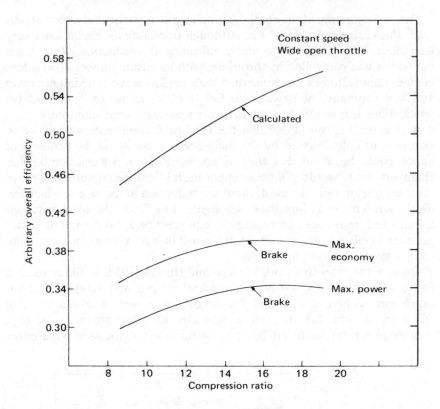

Figure 4.7   Variation in arbitrary overall efficiency with compression ratio
(reprinted with permission, © 1958 Society of Automotive
Engineers, Inc.)

the optimum compression ratio vary since researchers have used different
engines, with different air/fuel ratios, operating points and cylinder swept
volume. As the swept volume is reduced the surface-to-volume ratio
deteriorates, and data assembled by Muranaka *et al.* (1987) show that the
optimum compression ratio falls. With a cylinder volume of 250 cm³ the
compression ratio should be around 11, while for a cylinder swept volume
of 500 cm³ the optimum compression ratio is around 15. For any given
engine the optimum compression ratio will also be slightly dependent on
speed, since mechanical efficiency depends on speed.

4.2.2   *High compression ratio combustion chambers and fast burn combustion systems*

An approach that permits the use of high compression ratios with ordinary
fuels is the high turbulence, lean-burn, compact combustion chamber

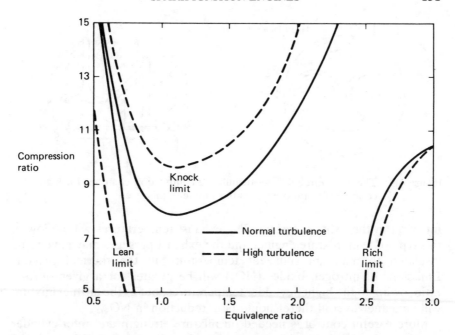

Figure 4.8   Effect of turbulence on increasing the operating envelope of spark
ignition engines (adapted from Ford (1982))

engine. The concepts behind these engines and a summary of the different
types are given by Ford (1982).

Increasing the turbulence allows leaner mixtures to be burnt, and these
are less prone to knock, owing to the reduced combustion temperatures.
Increasing the turbulence also reduces the susceptibility to knock since
normal combustion occurs more rapidly. These results are summarised in
figure 4.8. A compact combustion chamber is needed to reduce heat
transfer from the gas. The chamber is concentrated around the exhaust
valve, in close proximity to the sparking plug; this is to enable the mixture
around the exhaust valve to be burnt soon after ignition, otherwise the hot
exhaust valve would make the combustion chamber prone to knock. The
first design of this type was the May Fireball (May (1979)), with a flat
piston and the combustion chamber in the cylinder head, figure 4.9.
Subsequently another design has been developed with the combustion
chamber in the piston, and a flat cylinder head. In an engine with the
compression ratio raised from 9.7:1 to 14.6:1, the gain in efficiency was up
to 15 per cent at full throttle, with larger gains at part throttle.

There are also disadvantages associated with these combustion
chambers. Emissions of carbon monoxide should be reduced, but hydrocarbon emissions (unburnt fuel) will be increased because of the large
squish areas and poor surface-to-volume ratios. Hydrocarbon emissions
can be removed by oxidation in a thermal reactor, with secondary air

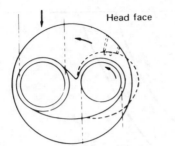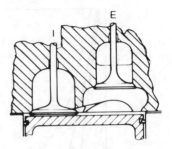

Figure 4.9   The May Fireball high-turbulence combustion chamber for high
compression ratio engines burning lean mixtures

injection in the exhaust system. The exhaust temperatures will be low in
this type of lean mixture engine, and the exhaust passages may have to be
insulated to maintain sufficient temperature in the thermal reactor.
Emissions of nitrogen oxides (NO_x) will be greater for a given air/fuel
ratio, owing to the higher cylinder temperatures, but as the air/fuel mixture
will be leaner overall there should be a reduction in NO_x.

More careful control is needed on mixture strength and inter-cylinder
distribution, in order to stay between the lean limit for misfiring and the
limit for knock. More accurate control is also needed on ignition timing.
During manufacture greater care is necessary, since tolerances that are
acceptable for compression ratios of 9:1 would be unacceptable at 14:1; in
particular combustion chambers need more accurate manufacture. Combustion deposits also have a more significant effect with high compression
ratios since they occupy a greater proportion of the clearance volume.

Three more combustion systems that have a fast burn characteristic are
the Ricardo High Ratio Compact chamber (HRCC), the Nissan NAPS-Z
(or ZAPS) and the four-valve pent-roof combustion chamber. The Ricardo
HRCC combustion system is similar in concept to the May Fireball (Figure
4.9), but has a straight passage from the inlet valve to the combustion
chamber. The four-valve pent-roof system and the Nissan NAPS-Z combustion system are shown in figure 4.10.

These three different combustion systems have been the subject of
extensive tests conducted by Ricardo, and reported by Collins and Stokes
(1983). The main characteristic of the four-valve pent-roof combustion
chamber is the large flow area provided by the valves. Consequently there
is a high volumetric efficiency, even at high speeds, and this produces an
almost constant bmep from mid speed upwards. The inlet tracts tend to be
almost horizontal, and to converge slightly. During the induction process,
barrel swirl (rotation about an axis parallel to the crankshaft) is produced
in the cylinder. The reduction in volume during compression firstly causes
an increase in the swirl ratio through the conservation of the moment of

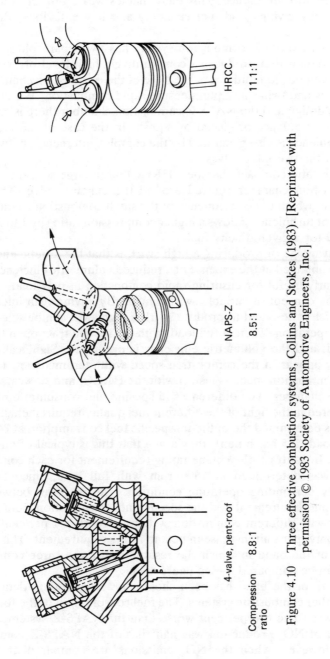

Figure 4.10 Three effective combustion systems, Collins and Stokes (1983). [Reprinted with permission © 1983 Society of Automotive Engineers, Inc.]

HRCC

11.1:1

NAPS-Z

8.5:1

4-valve, pent-roof

8.8:1

Compression
ratio

momentum. Subsequently, the further reduction in volume causes the swirl to break up into turbulence. This then enables weak air/fuel ratios to be burnt, thereby giving good fuel economy and low emissions, Benjamin (1988).

The Nissan NAPS-Z combustion system has twin spark plugs, and an induction system that produces a comparatively high level of axial swirl. While the combustion initiates at the edge of the combustion chamber, the swirling flow and twin spark plugs ensure rapid combustion. With both the four-valve design and the NAPS-Z combustion chamber there is comparatively little turbulence produced by squish. In the case of the four-valve head, turbulence is also generated by the complex interaction between the flows from the two inlet valves.

The high ratio compact chamber (HRCC) has a large squish area, with the combustion chamber centred around the exhaust valve. The rapid combustion, which is a consequence of the small combustion chamber and high level of turbulence, allows a higher compression ratio (by 1 to 2 ratios) to be used for a given quality fuel.

A disadvantage of producing a high swirl, is that the kinetic energy for the flow is obtained at the expense of a reduced volumetric efficiency. Swirl is particularly useful for ensuring rapid combustion at part load, and this leads to the concept of variable swirl control. By having twin inlet tracts, one of which is designed to produce swirl, a high swirl can be selected for part load operation. Then at full load, with the second tract open the swirl is reduced, and the volumetric efficiency is optimised. Significant differences only appear in the combustion speed with lean mixtures, in which case the combustion speed is fastest with the HRCC, and slowest with the four-valve chamber. The differences in specific fuel consumption need to be considered in the light of the different fuel quality requirements. Collins and Stokes determined the optimum specific fuel consumption at 2400 rpm and part load (2.5 bar bmep); they argue that this is typical of mid-speed operation. In contrast, the octane rating requirement for each combustion chamber was determined at 1800 rpm with full load, since this is a particularly demanding operating condition. The trade-off between the octane rating requirement and the specific fuel consumption from a range of engines with different combustion systems suggests a 1 per cent gain in fuel economy, per unit increase in octane rating requirement. This defines the slope of the band in which the results from these three combustion systems can be compared — see figure 4.11.

The width of the band has been chosen to accommodate data from a range of other combustion systems. The fuel consumption of the four-valve chamber was some 6–8 per cent worse than the NAPS-Z system, but the peak level of NO_x production was half that of the NAPS-Z combustion system. However, when the NO_x emissions are compared at the full

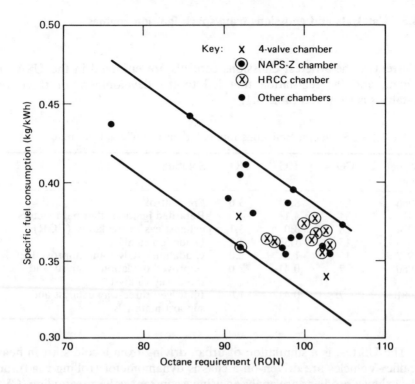

Figure 4.11 Trade-off between fuel economy and octane requirement for different combustion chambers, adapted from Collins and Stokes (1983). [Reprinted with permission © 1983 Society of Automotive Engineers, Inc.]

throttle maximum economy setting, there is little difference between the three combustion chamber designs. Thus a combustion system needs to be selected not only in terms of its efficiency and output for a given quality fuel, but also its level of emissions in the light of its likely operating regime.

In addition to the low fuel consumption of the HRCC-type system, it also allows a leaner mixture to be burned; the equivalence ratio can be as low as 0.6, compared with 0.7 for the four-valve or NAPS-Z systems. If an engine is operated solely with a lean mixture and a high level of turbulence, then high compression ratios can be obtained with conventional fuels. The attraction of such an engine is the potential improvement in fuel economy and, more significantly, the potential for reduced emissions of CO and NO_x. However, these compact combustion chambers are prone to knock and pre-ignition, and are often limited to applications with automatic gearboxes.

4.3   Catalysts and emissions from spark ignition engines

Currently, the strictest emissions controls are enforced in the USA and Japan, and the legislation that led to the development of three way catalysts is shown in table 4.1.

Table 4.1   US Federal Emissions Limits (grams of pollutant per mile)

Model year	CO	HC	NO_x	Solution
1966	87	8.8	3.6	Pre-control
1970	34	4.1	4.0	⎫ Retarded ignition, thermal reactors,
1974	28	3.0	3.1	⎭ exhaust gas recirculation (EGR)
1975	15	1.5	3.1	Oxidation catalyst
1977	15	1.5	2.0	Oxidation catalyst and improved EGR
1980	7	0.41	2.0	Improved oxidation catalysts and three-way catalysts
1981	7	0.41	1.0	Improved three-way catalyst and support materials

The US test is a simulation of urban driving from a cold start in heavy traffic. Vehicles are driven on a chassis dynamometer (rolling road), and the exhaust products are analysed using a constant-volume sampling (CVS) technique in which the exhaust is collected in plastic bags. The gas is then analysed for carbon monoxide (CO), unburnt hydrocarbons (HC) and nitrogen oxides (NO_x), using standard procedures. In 1970, three events — the passing of the American Clean Air Act, the introduction of lead-free petrol, and the adoption of cold test cycles for engine emissions — led to the development of catalyst systems.

Catalysts in the process industries usually work under carefully controlled steady-state conditions, but this is obviously not the case for engines — especially after a cold start. While catalyst systems were being developed, engine emissions were controlled by retarding the ignition and using exhaust gas recirculation (both to control NO_x) and a thermal reactor to complete oxidation of the fuel. These methods of NO_x control led to poor fuel economy and poor driveability (that is, poor transient engine response). Furthermore, the methods used to reduce NO_x emissions tend to increase CO and HC emissions and vice versa — see figure 3.15. The use of EGR and retarding the ignition also reduce the power output and fuel economy of engines.

Catalysts (Anon (1984a)) were able to overcome these disadvantages and meet the 1975 emissions requirements. The operating regimes of the different catalyst systems are shown in figure 4.12. With rich-mixture

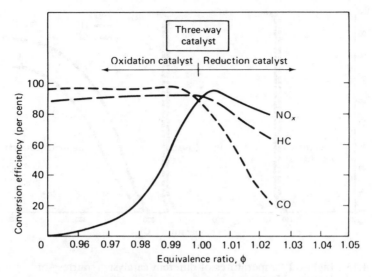

Figure 4.12   Conversion efficiencies of catalyst systems (courtesy of Johnson Matthey)

running, the catalyst promotes the reduction of NO_x by reactions involving HC and CO:

$$4HC + 10NO \rightarrow 4CO_2 + 2H_2O + 5N_2$$

and

$$2CO + 2NO \rightarrow 2CO_2 + N_2$$

Since there is insufficient oxygen for complete combustion, some HC and CO will remain. With lean-mixture conditions the catalyst promotes the complete oxidation of HC and CO:

$$4HC + 5O_2 \rightarrow 4CO_2 + 2H_2O$$

$$2CO + O_2 \rightarrow 2CO_2$$

With the excess oxygen, any NO_x present would not be reduced.

Oxidation catalyst systems were the first to be introduced, but NO_x emissions still had to be controlled by exhaust gas recirculation. Excess oxygen was added to the exhaust (by an air pump), to ensure that the catalyst could always oxidise the CO and HC. The requirements of the catalyst system were:

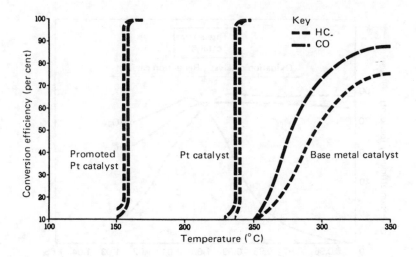

Figure 4.13   Light-off temperatures of different catalysts (courtesy of
Johnson Matthey)

(1) High conversion of CO and HC at low operating temperatures.
(2) Durability — performance to be maintained after 80 000 km (50 000 miles).
(3) A low light-off temperature.

Light-off temperature is demonstrated by figure 4.13. The light-off temperature of platinum catalysts is reduced by adding rhodium, which is said to be a 'promoter'.

Dual catalyst systems control NO_x emissions without resort to exhaust gas recirculation or retarded ignition timings. A feedback system incorporating an exhaust oxygen sensor is used with a carburettor or fuel injection system to control the air/fuel ratio. The first catalyst is a reduction catalyst, and by maintaining a rich mixture the NO_x is reduced. Before entering the second catalyst, air is injected into the exhaust to enable oxidation of the CO and HC to take place in the oxidation catalyst.

Conventional reduction catalysts are liable to reduce the NO_x, but produce significant quantities of ammonia (NH_3). This would then be oxidised in the second catalyst to produce NO_x. However, by using a platinum/rhodium system the selectivity of the reduction catalyst is improved, and a negligible quantity of ammonia is produced.

Three-way catalyst systems control CO, HC and NO_x emissions as a result of developments to the platinum/rhodium catalysts. As shown by figure 4.13 very close control is needed on the air/fuel ratio. This is normally achieved by electronic fuel injection, with a lambda sensor to provide feedback by measuring the oxygen concentration in the exhaust. A typical air/fuel ratio perturbation for such a system is ±0.25 (or $\pm0.02\phi$).

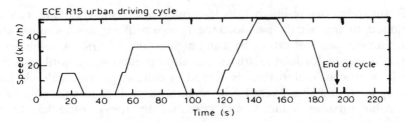

Figure 4.14   The ECE R15 urban driving cycle; the cycle is repeated four times and is preceded by a warm-up idle time of 40 s (total test duration is 820 s)

Table 4.2   European emissions legislation

Engine size (litres)	CO (g/test)	HC + NO$_x$ (g/test)	New models	New vehicles
<1.4	19	5	1.7.92	1.12.92
1.4–2.0	30	8	1.10.91	1.10.93
>2.0	25	6.5	1.10.88	1.10.89

In Europe, the emissions legislation is defined by the ECE15 European urban driving cycle which is shown here in figure 4.14. The emissions legislation agreed in June 1989 is listed in table 4.2.

This legislation makes some allowance for larger engines being installed in larger vehicles and therefore generating more emissions. However, for vehicles with engines over 2.0 litres swept volume, the implication is that these are expensive vehicles which can better absorb the cost of meeting stricter emissions limits. It is generally accepted that the only way to meet these emissions levels is through the use of three-way catalysts.

During the development of emissions legislation in Europe there was much research into lean-burn engines, as this technology offered an alternative to three-way catalysts. Lean-burn engines employed combustion systems such as those described in section 4.2.2. Compact, highly turbulent combustion allowed mixtures that were sufficiently weaker than stoichiometric to enable the NO$_x$ emissions to be reduced. Lean-burn engines offered the potential for lower capital costs and more efficient operation.

When a three-way catalyst is used, it requires: firstly, an engine management system capable of very accurate air/fuel ratio control, and secondly, a catalyst. Both these requirements add considerably to the cost of an engine. Since a three-way catalyst always has to operate with a stoichiometric air/fuel ratio, then at part load, this means that the maximum economy cannot be achieved (see figure 4.1). The fuel consumption penalty (and also the increase in carbon dioxide emissions) associated with stoichiometric operation is around 10 per cent.

Fortunately, the research into lean-burn combustion systems can be exploited, to improve the part-load fuel economy of engines operating with stoichiometric air/fuel ratios, by using high levels of EGR. A combustion system designed for lean mixtures can also operate satisfactorily when a stoichiometric air/fuel mixture is diluted by exhaust gas residuals. At part load up to around 30 per cent EGR can be used; this reduces the volume of flammable mixture induced, and consequently the throttle has to be opened slightly to restore the power output. With a more open throttle the depression across the throttle plate is reduced, and the pumping work (or pmep) is lower. Nakajima *et al.* (1979) show that for a bmep of 3.24 bar with stoichiometric operation at a speed of 1400 rpm, they were able to reduce the fuel consumption by about 5 per cent through the use of 20 per cent EGR on an engine with a fast burn combustion system.

4.4   Cycle-by-cycle variations in combustion

Cycle-by-cycle variation of the combustion (or cyclic dispersion) in spark ignition engines was mentioned in section 3.5.1. It is illustrated here by figure 4.15, the pressure–time record for five successive cycles. Clearly not all cycles can be optimum, and Soltau (1960) suggested that if cyclic dispersion could be eliminated, there would be a 10 per cent increase in the power output for the same fuel consumption with weak mixtures. Similar conclusions were drawn by Lyon (1986), who indicated that a 6 per cent improvement in fuel economy could be achieved if all cycles burned at the optimum rate. Perhaps surprisingly, the total elimination of cyclic dispersion may not be desirable, because of engine management systems that retard the ignition when knock is detected. If there was no cyclic dispersion, then either none or all of the cycles would knock. It would be acceptable to the engine and driver for only a few cycles to knock, and the ignition control system can then introduce the necessary ignition retard. If all the cycles were to knock it is likely to lead to runaway knock, in which case retarding the ignition would have no effect.

Cyclic dispersion occurs because the turbulence within the cylinder varies from cycle to cycle, the air/fuel mixture is not homogeneous (there may even be droplets of fuel present) and the exhaust gas residuals will not be fully mixed with the unburned charge. It is widely accepted that the early flame development can have a profound effect on the subsequent combustion. Firstly, the formation of the flame kernel will depend on: the local air/fuel ratio, the mixture motion, and the exhaust gas residuals, in the spark plug gap at the time of ignition.

De Soete (1983) conducted a parametric study of the phenomena con-

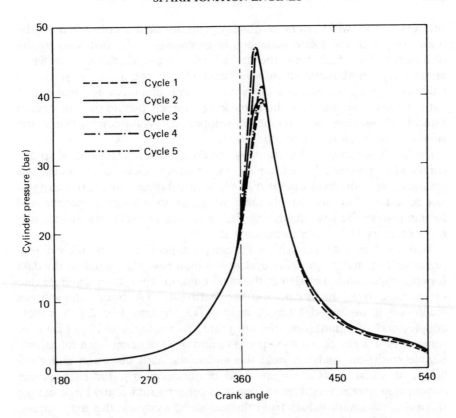

Figure 4.15   Pressure–time diagrams for five successive cycles in a Ricardo E6
engine (compression ratio 8:1, stoichiometric air/iso-octane
mixture, 1000 rpm, 8.56 bar bmep) (reprinted from Stone, C. R.
and Green-Armytage (1987) by permission of the Council of the
Institution of Mechanical Engineers)

trolling the initial behaviour of spark-ignited flames. He confirmed that
combustion starts as self-ignition, occurring in the volume of very hot gases
(spark kernel) behind the expanding, spark-created, shock wave. He also
observed that spark-ignited flames pass through a non-steady propagation
period before reaching a steady speed. This transient period is relatively
important, compared with the total time available for combustion, in an
engine cycle. In the early stages of flame growth, the flame is small
compared with the turbulence length scales, and the flame can be convected away from the spark plug. If the flame nucleus is moved into the
thermal-boundary layer surrounding the combustion chamber, then it will
burn slowly. Furthermore, at a given flame radius, the greater contact with
the wall will reduce the flame front area.
    Conversely, if the flame is moved away from the combustion chamber

surfaces, then it will burn more quickly. The turbulence only enhances the burn rate, once the flame surface is large enough to be distorted by the turbulence, by which time the turbulence no longer moves the flame around the combustion chamber. Lyon (1986) conducted a statistical analysis of groups of pressure distributions. This demonstrated that events early in the development of the flame kernel largely dictate the subsequent rate of combustion and pressure development. When combustion starts slowly, then it tends to continue slowly.

Cyclic dispersion is increased by anything that tends to slow-up the combustion process, for example: lean mixture operation, exhaust gas residuals, and low load operation (in part attributable to greater exhaust gas residuals, but also attributable to lower in-cylinder pressures and temperatures). Before illustrating the variations in cyclic dispersion, it is first necessary to be able to measure it.

With modern data acquisition systems, it is possible to log the cylinder pressure from many successive cycles. It is then possible to analyse the data for each cycle, and to evaluate: the maximum pressure, the maximum rate of pressure rise, the imep, and the burn rate. (A burn rate analysis technique is discussed later in chapter 13, section 13.5.2.) A widely adopted way of summarising the burn rate, is to note the 0–10 per cent (or sometimes 0–1 per cent), 0–50 per cent and 0–90 per cent burn durations. Simple statistical analyses yield the mean, standard deviation and coefficient of variation (CoV = standard deviation/mean), for each of the combustion parameters. It is necessary to collect a sufficiently large sample to ensure stationary values from the statistical analyses; this may require data from up to 1000 cycles when there is a high level of cyclic dispersion.

Table 4.3 presents some cyclic dispersion data from a gas engine. Table 4.3 shows that each performance parameter has a different value of its coefficient of variation, this means that the way cyclic dispersion has been measured should always be defined. The peak pressure is often used, since it is easy to measure. However, the imep is probably the parameter with the most relevance to the overall engine performance. At sufficiently high levels of cyclic dispersion the driver of a vehicle will become aware of fluctuations in the engine output, and this is clearly linked to the imep, because of the integrating effect of the engine flywheel. Ultimately, cyclic dispersion leads into misfire, but a driver will detect poor driveability much earlier than this. For good driveability, the coefficient of variation in the imep should be no greater than 5–10 per cent. Not only is there no direct link between the coefficients of variation of the peak pressure and the imep, they can also respond to variables in opposite ways.

Figure 4.16 shows the variation in the mean imep, and the coefficients of variation of imep and peak pressure, when the ignition timing is varied. The imep is fairly insensitive to the changes in ignition timing (MBT ignition timing 17° btdc), but it can be seen that the maximum imep

Table 4.3   Statistical summary of the combustion performance from a fast burn
            gas engine combustion system operating at 1500 rpm, full
            throttle with an equivalence ratio of 1.12 and MBT ignition timing

Peak pressure	mean (bar)	59.5
	standard deviation (bar)	2.7
	coefficient of variation (per cent)	4.5
imep	mean (bar)	7.40
	standard deviation (bar)	0.14
	coefficient of variation (per cent)	1.9
0–10 per cent burn duration	mean (°ca)	10.3
	standard deviation (°ca)	1.2
	coefficient of variation (per cent)	11.6
0–50 per cent burn duration	mean (°ca)	18.4
	standard deviation (°ca)	1.54
	coefficient (per cent)	8.4
0–90 per cent burn duration	mean (°ca)	32.7
	standard deviation (°ca)	3.07
	coefficient of variation (per cent)	9.4

corresponds quite closely to a minimum in the coefficient of variation of the
imep, although the coefficient of variation of the peak pressure increases as
the ignition timing is moved closer to tdc, until a maximum (7.6 per cent)
occurs at about 4° btdc. In general, the coefficient of variation of imep is a
minimum in the region of MBT ignition timing. This means that any
uncertainty in locating MBT ignition timing has a lesser effect on the
coefficient of variation of imep than the coefficient of variation of the
maximum cylinder pressure. The dependence of the coefficients of variation on the ignition timing does not appear to be widely appreciated, but
it has been studied in detail by Brown, A. G. (1991). When coefficients of
variation are quoted it is important to identify both the ignition timing and
the sensitivity of the coefficient of variation to the ignition timing.

4.5   Ignition systems

4.5.1   Ignition system overview

Most engines have a single sparking plug per cylinder, a notable exception
being in aircraft where the complete ignition system is duplicated to
improve reliability. The spark is usually provided by a battery and coil,

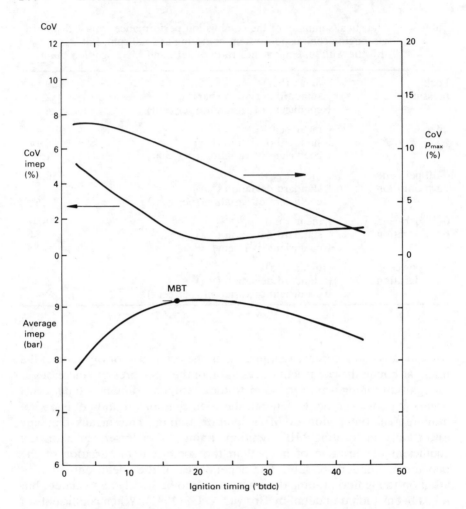

Figure 4.16   The influence of ignition timing on the imep, and cycle-by-cycle variation in combustion. Ricardo E6 operating at 1500 rpm with full throttle

though for some applications a magneto is better.

For satisfactory performance, the central electrode of the sparking plug should operate in the temperature range 350–700°C; if the electrode is too hot, pre-ignition will occur. On the other hand, if the temperature is too low carbon deposits will build up on the central insulator, so causing electrical breakdown. The heat flows from the central electrode through the ceramic insulator; the shape of this determines the operating temperature of the central electrode. A cool-running engine requires a 'hot' or

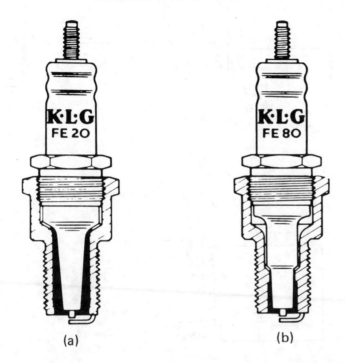

(a)   (b)

Figure 4.17   Sparking plugs. (a) Hot running; (b) cool running (from Campbell
              (1978))

'soft' sparking plug with a long heat flow path in the central electrode,
figure 4.17a. A hot-running engine, such as a high-performance engine or a
high compression ratio engine, requires a 'cool' or 'hard' sparking plug.
The much shorter heat flow path for a 'cool' sparking plug is shown in
figure 4.17b. The spark plug requires a voltage of 5–15 kV to spark; the
larger the electrode gap and the higher the cylinder pressure the greater
the required voltage.

   Both a conventional coil ignition system and a magneto ignition system
are shown in figure 4.18. The coil is in effect a transformer with a primary
or LT (Low Tension) winding of about 200 turns, and a secondary or HT
(High Tension) winding of about 20 000 turns of fine wire, all wrapped
round an iron core. The voltage V induced in the HT winding is

$$V = M \frac{\mathrm{d}I}{\mathrm{d}t} \tag{4.1}$$

where I is the current flowing in the LT winding
         M is the mutual inductance $= k \sqrt{(L_1 L_2)}$
         L_1, L_2 are the inductances of the LT and HT windings,
              respectively (proportional to the number of turns squared)

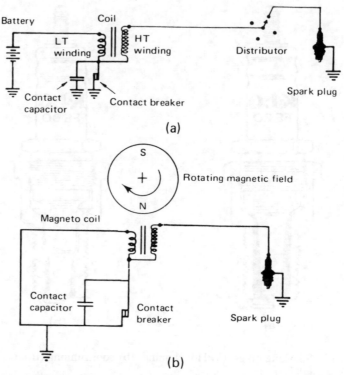

Figure 4.18   Mechanically operated ignition systems. (a) Conventional coil
ignition system; (b) magneto ignition system (adapted from
Campbell (1978))

and k is a coupling coefficient (less than unity)
or $V = k$ (turns ratio of windings) (Low Tension Voltage).

When the contact breaker closes to complete the circuit a voltage will be
induced in the HT windings, but it will be small since dI/dt is limited by the
inductance and resistance of the LT winding. Equation (4.2) defines the
current flow in the LT winding:

$$I = \left(\frac{V_s}{R}\right)\left[1 - \exp\left(-\frac{Rt}{L_1}\right)\right] \qquad (4.2)$$

where V_s = supply voltage
         R = resistance of the LT winding
         t = time after application of V_s.

When the contact breaker opens dI/dt is much greater and sufficient
voltage is generated in the HT windings to jump the gaps between elec-

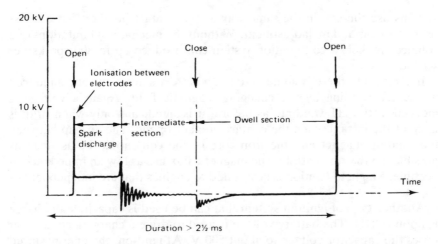

Figure 4.19   HT output from ignition coil (adapted from Campbell (1978))

trodes. A higher voltage (200-300 V) is generated in the LT windings and this energy is stored in the capacitor. Without the capacitor there would be severe arcing at the contact breaker. Once the spark has ended, the capacitor discharges.

The energy input to the LT side (E_p) of the coil is the integration of the instantaneous current (I) and the supply voltage (V_s) over the period the coil is switched on:

$$E_p = \int_0^{t'} IV_s \, dt \qquad\qquad (4.3)$$

where t' is the time at which the coil is switched off, the coil-on-time.

The energy stored in the coil (E_s) is defined as

$$E_s = 0.5 \, L_1 I_p^2 \qquad\qquad (4.4)$$

where I_p is the LT current at the time when the coil is switched off.

The HT output is shown in figure 4.19. Initially 9 kV is needed to ionise the gas sufficiently before the spark jumps with a voltage drop of 2 kV. As engine speeds increase, the dwell period becomes shorter and the spark energy will be reduced. Such a system can produce up to 400 sparks per second; beyond this the spark energy becomes too low and this leads to misfiring. For higher spark rates, twin coil/contact breaker systems or alternatively electronic systems can be used. The current through the contact breaker can be reduced by using it to switch the base of a transistor that controls the current to the LT winding; this prolongs the life of the contact breaker. The contact breaker can be replaced by making use of opto-electronic, inductive or magnetic switching. However, all these systems use the coil in the same way as the contact breaker, but are less prone to wear and maladjustment. Without the mechanical limitations of a contact breaker, a coil ignition system can produce up to 800 sparks per second.

In a magneto there is no need to use a battery since a current is induced in the LT winding by the changing magnetic field. Again a voltage is induced in the HT winding; as before, it is significant only when dI/dt is large at the instant when the contact breaker opens. The air gap between the rotating magnet and the iron core of the coil should be as small as possible, so that the path for the magnetic flux has as low an impedance as possible. Magneto ignition is best suited to engines that are independent of a battery.

Another type of ignition system that can be used is capacitive discharge ignition (CDI). The battery voltage is used to drive a charging circuit that raises the capacitor voltage to about 500 V. At ignition, the energy stored in the capacitor is discharged through an ignition transformer (that is, a coil with primary and secondary windings), the circuit being controlled by a thyristor. The discharge from the capacitor is such, that a short duration (about 0.1 ms) spark is generated; the rapid discharge makes this ignition system less susceptible to spark plug fouling.

Ignition timing is usually expressed as degrees before top dead centre (°btdc), that is, before the end of the compression stroke. The ignition timing should be varied for different speeds and loads. However, for small engines, particularly those with magneto ignition, the ignition timing is fixed.

Whether the ignition is by battery and coil (positive or negative earth) or magneto, the HT windings are usually arranged to make the central electrode of the spark plug negative. The electron flow across the electrode gap comes from the negative electrode (the cathode), and the electrons flow more readily from a hot electrode. Since the central electrode is not in direct contact with the cylinder head, this is the hotter electrode. By arranging for the hotter electrode to be the cathode the breakdown voltage is reduced.

In chapter 3, section 3.5, it was explained how turbulent flame propagation occupies an approximately constant fraction of the engine cycle, since at higher speeds the increased turbulence gives a nearly corresponding increase in flame propagation rate. However, the initial period of flame growth occupies an approximately constant time ($\frac{1}{2}$ ms) and this corresponds to increased crank angles at increased speeds. The ignition advance is often provided by spring-controlled centrifugal flyweights. Very often two springs of different stiffness are used to provide two stages of advance rate.

Ignition timing has to be advanced at part throttle settings since the reduced pressure and temperature in the cylinder cause slower combustion. The part throttle condition is defined by the pressure drop between atmosphere and the inlet manifold, the so-called engine 'vacuum'. An

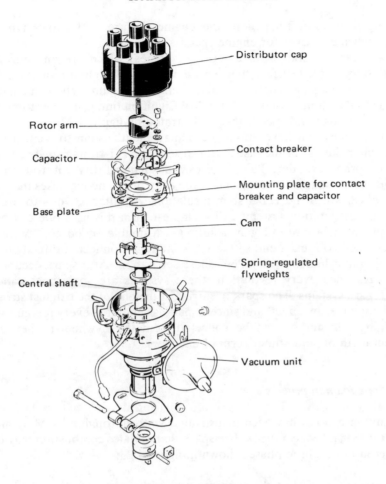

Figure 4.20 Automotive distributor (courtesy of Lucas Electrical Ltd)

exploded view of a typical automotive distributor is shown in figure 4.20. The central shaft is driven at half engine speed (for four-stroke cycles) and the rotor arm directs the HT voltage to the appropriate sparking plug via the distributor cap. For a four-cylinder engine there is a four-lobed cam that operates the contact breaker. The contact breaker and capacitor are mounted on a plate that can rotate a limited amount around the cams relative to the base plate. The position of this plate is controlled by the vacuum unit, a spring-controlled diaphragm that is connected to the inlet manifold. The cams are on a hollow shaft that can rotate around the main shaft. The relative angular position of the two shafts is controlled by the spring-regulated flyweights. The ignition timing is set by rotating the complete distributor relative to the engine. Figure 4.21 shows typical ignition advance curves for engine speed and vacuum.

These characteristics form a series of planes, which are an approximation to the curved surface that represents the MBT ignition timing on a plot of advance against engine speed and engine vacuum. When an engine management system is used, then the MBT ignition timing can be stored as a function of load and speed; this is illustrated by figure 4.22.

Many engines that have an engine management system to control the spark, still make use of the distributor to send the HT from a single coil to the appropriate cylinder. This requires a mechanical drive, introduces a loss (due to the spark gap at the end of the rotor arm) and increases the risk of a breakdown in the HT system insulation. An alternative is to use a distributorless ignition system. In its simplest form this uses one coil per spark plug, but a more elegant solution is a double-ended coil. With a double-ended coil, each end of the HT winding is connected directly to a spark plug in cylinders with a 360° phase separation. As a spark occurs at both plugs once every revolution, these systems are sometimes called wasted spark systems. The spark occurring at the end of the exhaust stroke should not have any effect, and since a spark is generated every revolution, the timing information can be collected from the flywheel (either the flywheel teeth or some other form of encoder).

4.5.2   The ignition process

The ignition process has been investigated very thoroughly by Maly and Vogel (1978) and Maly (1984). The spark that initiates combustion may be considered in the three phases shown in figure 4.23:

1. *Pre-breakdown*. Before the discharge occurs, the mixture in the cylinder is a perfect insulator. As the spark pulse occurs, the potential difference across the plug gap increases rapidly (typically 10–100 kV/ms). This causes electrons in the gap to accelerate towards the anode. With a sufficiently high electric field, the accelerated electrons may ionise the molecules they collide with. This leads to the second phase — avalanche breakdown.
2. *Breakdown*. Once enough electrons are produced by the pre-breakdown phase, an overexponential increase in the discharge current occurs. This can produce currents of the order of 100 A within a few nanoseconds. This is concurrent with a rapid decrease in the potential difference and electric field across the plug gap (typically to 100 V and 1 kV/cm respectively). Maly suggests that the minimum energy required to initiate breakdown at ambient conditions is about 0.3 mJ. The breakdown causes a very rapid temperature and pressure increase.

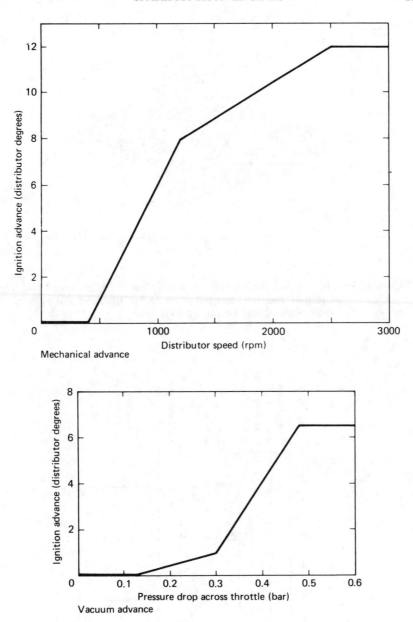

Figure 4.21   Typical ignition advance curves

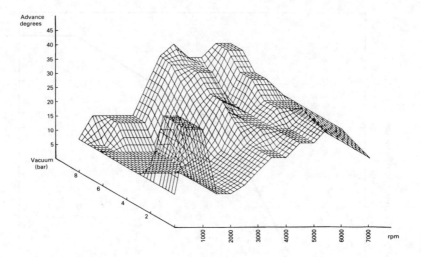

Figure 4.22   The ignition timing map (as a function of engine speed and inlet manifold pressure) as used in an engine management system (from Forlani and Ferrati (1987))

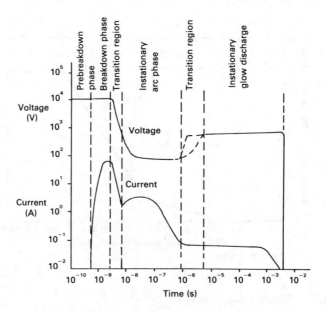

Figure 4.23   The current and voltage as a function of time during a spark discharge (adapted from Maly (1984))

Temperatures of 60 000 K give rise to pressures of several hundred bars. These high pressures cause an intense shock wave as the spark channel expands at supersonic speed. Expansion of the spark channel allows the conversion of potential energy to thermal energy, and facilitates cooling of the plasma. Prolonged high currents lead to thermionic emission from hot spots on the electrodes and the breakdown phase ends as the arc phase begins.

3. *Arc discharge.* The characteristics of the arc discharge phase are controlled by the external impedances of the ignition circuit. Typically, the burning voltage is about 100 V and the current is greater than 100 mA, and is dependent on external impedances. The arc discharge is sustained by electrons emitted from the cathode hot spots. This process causes erosion of the electrodes, with the erosion rate increasing with the plug gap. Depending on the conditions, the efficiency of the energy-transfer process from the arc discharge to the thermal energy of the mixture is typically between 10 and 50 per cent.

   With currents of less than 100 mA, this phase becomes a glow discharge, which is distinguished from an arc discharge by the cold cathode. Electrons are liberated by ion impact, a less efficient process than thermionic emission. Even though arc discharges are inherently more efficient, glow discharges are more common in practice, because of the high electrode erosion rates associated with arc discharges.

The determination of the optimum spark type and duration has resulted in disagreement between researchers. Some work concludes that longer arc durations improve combustion system, while other work indicates that short-duration (10–20 ns) high-current arcs (such as occur with capacitative discharge ignition systems) can be beneficial. These arguments have been reviewed by Stone, C. R. and Steele (1989), who also report on tests in which the spark energy and the spark plug gap were varied. The spark energy was measured in a special calorimeter, and it was controlled by varying the coil-on-time and the spark plug gap. The engine performance was characterised by the bsfc and cyclic dispersion of tests at: 1200 rpm with a bmep of 3.2 bar and an air/fuel ratio of 17. It was found that spark plug gap was a stronger determinant of engine performance than spark energy, and there was little to be gained by using spark plug gaps above 0.75 mm. However, for small spark plug gaps there were advantages in increasing the spark energy.

The apparent conflict between claims for long-duration and short-duration sparks can be reconciled. The short-duration spark has a better thermal conversion efficiency, and can overcome in-cylinder variations by reliable ignition and accelerated flame kernel development. In contrast, the long-duration discharge is successful, since it provides a time window long enough to mask the effects of in-cylinder variations. Similarly, a large spark plug gap is beneficial, since it increases the likelihood of there existing a favourable combination of turbulence and mixture between the electrodes.

4.6 Mixture preparation

4.6.1 Introduction

The air/fuel mixture can be prepared by either a carburettor or a fuel injection system. In both cases fuel will be present in the inlet manifold as: vapour, liquid droplets and a liquid film. Although emissions legislation is now reducing the scope for using carburettors, their use is still widespread. There are two main types of carburettor:

 (i) fixed jet (or fixed venturi) described in section 4.6.3, or
(ii) variable jet (or variable venturi) described in section 4.6.2.

There are also two types of fuel injection system used on spark ignition engines, multi-point injection and single-point injection; both of these systems are described in section 4.6.4. The multi-point injection system employs injectors usually mounted close to the inlet port(s) of each cylinder. The single-point injection system can look very much like a carburettor, and as with the carburettor, the throttle plate and inlet manifold play an important part in mixture preparation. Even with multi-point fuel injection systems, a liquid fuel film will develop on the walls of the inlet manifold.

The carburettor (or fuel injection system) and manifold have to perform satisfactorily in both steady-state and transient conditions. When an engine is started, extra fuel floods into the inlet manifold. Under these conditions the engine starts on a very rich mixture and the inlet manifold acts as a surface carburettor; often there will be small ribs to control the flow of liquid fuel.

In a simple branched manifold with a carburettor or single-point injection system, the intersections will often have sharp corners, although for good gas flow rounded intersections would be better. The reason is that the sharp corners help to break up the liquid film flowing on the manifold walls, see figure 4.24. In automotive applications the manifold is sometimes inclined relative to the vehicle. This is so that the fuel distribution becomes optimum when the vehicle is ascending a gradient. When carburettors are used, a way of improving the aerodynamic performance of the inlet manifold is to use multiple carburettor installations. The problem

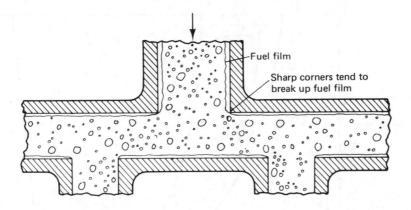

Fuel film

Sharp corners tend to
break up fuel film

Figure 4.24   Flow of air, fuel vapour, droplets and liquid film in the inlet
manifold

then becomes one of balancing the carburettors — that is, ensuring that the
flow is equal through all carburettors, and that each carburettor is producing the same mixture strength. A cheaper alternative to twin carburettors is
the twin choke carburettor. The term 'choke' is slightly misleading, as in
this context it means the venturi. The saving in a twin choke carburettor is
because there is only one float chamber.

Referring back to figure 4.2, points A and B represent the same power
output but it is obviously more economical to operate the engine with a
wider throttle opening and leaner mixture. When emissions legislation
permits, the engine should normally receive a lean mixture and at full
throttle a rich mixture. This ensures economical operation, yet maximum
power at full throttle. If a lean mixture were used at full throttle, this would
reduce the power output and possibly overheat the exhaust valve because
of the slower combustion. When the engine is idling or operating at low
load the low pressure in the inlet manifold increases the exhaust gas
residuals in the cylinder, and consequently the carburettor has to provide a
rich mixture. The way that the optimum air/fuel ratio changes for maximum power and maximum economy with varying power output for a
particular engine at constant speed is shown in figure 4.25. The variations
in the lean limit and rich limit are also shown.

When the throttle is opened, extra fuel is needed for several reasons.
The air flow into the engine increases more rapidly than the fuel flow, since
some fuel is in the form of droplets and some is present as a film on the
manifold walls. Secondly, for maximum power a rich mixture is needed.
Finally, when the throttle is opened the vaporised fuel will tend to condense. When the throttle opens the pressure in the manifold increases and
the partial pressure of the fuel vapour will increase (the partial pressure of

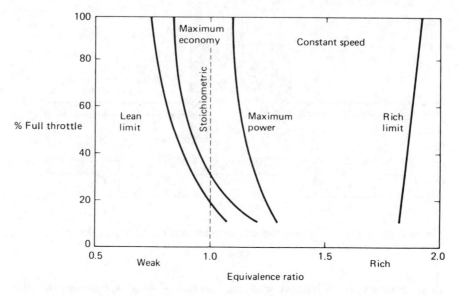

Figure 4.25   Variation in air/fuel ratio requirement with load at constant speed

fuel vapour depends on the air/fuel ratio). If the partial pressure of the fuel rises above its saturation pressure then fuel will condense, and extra fuel is injected to compensate.

Long inlet manifolds will be particularly bad in these respects because of the large volume in the manifold and the length that fuel film and droplets have to travel. In engines with horizontally opposed cylinders it is very difficult to arrange satisfactory carburation from a single carburettor.

When the throttle is suddenly closed, the reduced manifold pressure causes the fuel film to evaporate. This can provide an over-rich mixture, and so lead to emissions of unburnt hydrocarbons. The problem is overcome by a spring-loaded over-run valve on the throttle value plate that by-passes air into the manifold. Sometimes heated manifolds are used to reduce the liquid film and droplets. The manifold can be heated by the engine coolant, or by conduction from the exhaust manifold. The disadvantage of a heated inlet manifold is the ensuing reduction in volumetric efficiency.

The volumetric efficiency penalty is reduced if the engine coolant is used to heat the inlet manifold, but it should be appreciated that the coolant warms-up more slowly than the exhaust manifold. However, supplementary electrical heaters can be used during warm-up. These heaters often use PTC (positive temperature coefficient) materials, so as to give automatic temperature control. The heaters employ extended surfaces (usually spines), and are located in the manifold under the carburettor or a single-point fuel injector.

Despite the careful attention paid to manifold design, it is quite usual for carburettors to give ±5 per cent variation in mixture strength between cylinders, even for steady-state operation.

4.6.2   Variable jet carburettor

A cross-section of a variable jet or variable venturi carburettor is shown in figure 4.26. The fuel is supplied to the jet ① from an integral float chamber. This has a float-operated valve that maintains a fuel level just below the level of the jet. The pressure downstream of the piston ② is in constant communication with the suction disc (the upper part of the piston) through the passage ③. If the throttle ④ is opened, the air flow through the venturi ⑤ increases. This decreases the pressure downstream of the venturi and causes the piston ② to rise. The piston will rise until the pressure on the piston is balanced by its weight and the force from the light spring ⑥. The position of the tapered needle ⑦ in the jet or orifice ① varies with piston position, thus controlling the air/fuel mixture. The

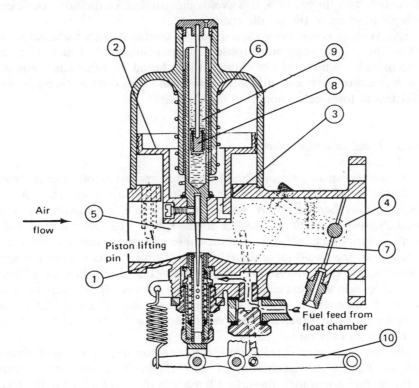

Figure 4.26   Variable jet or variable venturi carburettor (from Judge (1970))

damper ⑧ in the oil ⑨ stops the piston ① oscillating when there is a change in load. A valve in the damper causes a stronger damping action when the piston rises than when it falls. When the throttle is opened the piston movement is delayed by the damper, and this causes fuel enrichment of the mixture. For an incompressible fluid the flow through an orifice or venturi is proportional to the square root of the pressure drop. Thus if both air and fuel were incompressible, the air/fuel mixture would be unchanged with increasing flow and a fixed piston. However, as air is compressible its pressure drop will be greater than that predicted by incompressible flow and this will cause extra fuel to flow.

For starting, extra fuel is provided by a lever ⑩ that lowers the jet. A linkage and cam also operate the throttle valve to raise the idling speed. Several modifications are possible to improve the carburettor performance. The position of the jet can be controlled by a bi-metallic strip to allow for the change in fuel properties with temperature. Fuel flow will vary with the eccentricity of the needle in the jet, with the largest flow occurring when the needle touches the side of the jet. Rather than maintain exact concentricity, the needle is lightly sprung so as always to be in contact with the jet wall; this avoids the problem of the flow coefficient being a function of the needle eccentricity.

This type of carburettor should not be confused with carburettors in which there is no separate throttle and the piston and needle are lifted directly. This simple type of carburettor is found on some small engines (such as motorcycles and outboard motors) and does not have facilities like enrichment for acceleration.

4.6.3 Fixed jet carburettor

The cross-section of a simple fixed jet or fixed venturi carburettor is shown in figure 4.27. This carburettor can only sense air flow rate without distinguishing between fully open throttle at a slow engine speed or partially closed throttle at a higher engine speed. The fuel outlet is at the smallest cross-sectional area so that the maximum velocity promotes break-up of the liquid jet and mixing with the air; the minimum pressure also promotes fuel evaporation. The fuel outlet is a few millimetres above the fuel level in the float chamber so that fuel does not spill or syphon from the float chamber. If the air flow were reversible, there would be no pressure drop; in practice the pressure drop might be 0.05 bar at the maximum air flow rate.

The fuel/air ratio change in response to a change in air flow rate is shown in figure 4.28 for this carburettor. No fuel will flow until the pressure drop in the venturi overcomes the surface tension at the fuel outlet and the head difference from the float chamber. As the air flow increases to its maximum, the air/fuel mixture becomes richer. The maximum air flow is when

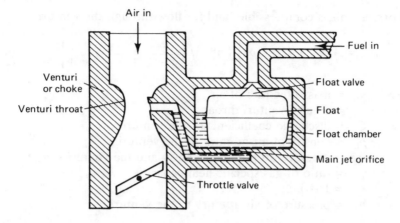

Figure 4.27 Simple fixed jet carburettor

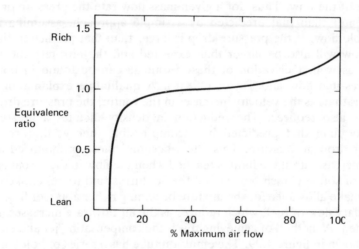

Figure 4.28 Mixture characteristics of a simple carburettor

the velocity at the venturi throat is supersonic. The reason for the change
in air/fuel ratio is as follows.

Fuel can be treated as incompressible, and for flow through an orifice

$$\dot{m}_f = A_o C_o \sqrt{(2\rho_f \, \Delta p)} \qquad (4.5)$$

where $\dot{m}_f$ = mass flow rate of fuel
        A_o = orifice area
        C_o = coefficient of discharge for the orifice
        ρ_f = density of fuel
        Δp = pressure difference across the orifice.

In contrast, air is compressible, and for flow through the venturi

$$\dot{m}_a = A_t \, C_v \, \sqrt{(2\rho_a \Delta p)} \left[(r)^{1/\gamma} \sqrt{\left\{ \frac{\gamma}{\gamma - 1} \frac{1 - r^{(\gamma-1)/\gamma}}{1 - r} \right\}} \right] \qquad (4.6)$$

where $\dot{m}_a$ = mass flow rate of air
$\quad\quad\quad A_t$ = area of venturi throat
$\quad\quad\quad C_v$ = discharge coefficient for the venturi
$\quad\quad\quad \rho_a$ = density of air at entry to the venturi
$\quad\quad\quad \Delta p$ = pressure drop between entry and the venturi throat
$\quad\quad\quad \gamma$ = ratio of gas specific heat capacities
$\quad\quad\quad r$ = $1 - p/p_a$
$\quad\quad\quad p_a$ = pressure of air at entry to the venturi.

Since r is always less than unity the square bracket term in equation (4.6) will always be less than unity. This term accounts for the compressible nature of the flow. Thus, for a given mass flow rate the pressure drop will be greater than that predicted by a simple approach, assuming incompressible flow. If the pressure drop is larger than that predicted, then the fuel flow will also be larger than expected and the air/fuel ratio will be richer as well. Derivation of these formulae can be found in books on compressible flow and Taylor (1985b). A qualitative explanation of the effect is that, as the velocity increases in the venturi the pressure drops and density also reduces. The reduction in density dictates a greater flow velocity than that predicted by incompressible theory, thus causing a greater drop in pressure. This effect becomes more pronounced as flow rates increase, until the limit is reached when the flow in the throat is at the speed of sound (Mach No. 1) and the venturi is said to be 'choked'.

To make allowance for the mixture becoming richer at larger flow rates a secondary flow of fuel, which reduces as the air flow rate increases, should be added. A method of achieving this is the compensating jet and emulsion tube shown in figure 4.29. The emulsion tube has a series of holes along its length, and an air bleed to the centre. At low flow rates the emulsion tube will be full of fuel. As the flow rate increases the fuel level will fall in the emulsion tube, since air is drawn in through the bleed in addition to the fuel through the compensating jet. The fuel level will be lower inside the emulsion tube than outside it, owing to the pressure drop associated with the air flowing through the emulsion tube holes. As air emerges from the emulsion tube it will evaporate the fuel and form a two-phase flow or emulsion. This secondary flow will assist the break-up of the main flow. The cumulative effects of the main and secondary flows are shown in figure 4.30.

A rich mixture for full throttle operation can be provided by a variety of means, by either sensing throttle position or manifold pressure. The mixture can be enriched by an extra jet (the 'power' jet) or the air supply to

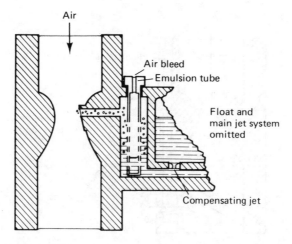

Figure 4.29   Emulsion tube and compensating jet in a fixed jet carburettor

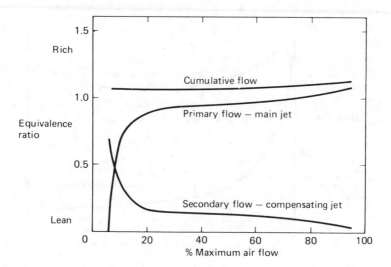

Figure 4.30   Cumulative effect of main jet and compensating jet

the emulsion system can be reduced. Alternatively an air bleed controlled by manifold pressure can be used to dilute a normally rich mixture. This might be a spring-loaded valve that closes when the manifold pressure approaches atmospheric pressure at full throttle.

At low air flow rates no fuel flows, so an additional system is required for idling and slow running. Under these conditions the pressure drop in the venturi is too small and advantage is taken of the pressure drop and venturi

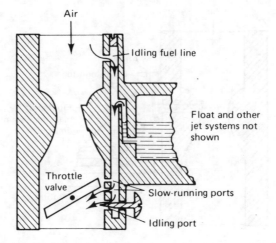

Figure 4.31   Slow-running and idling arrangement in a fixed jet carburettor

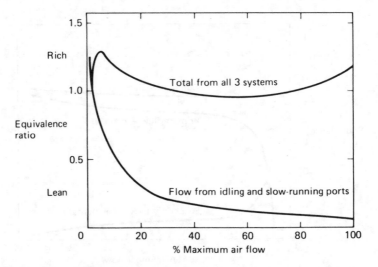

Figure 4.32   Contribution to the air/fuel mixture from the idling and
slow-running ports

effect at the throttle valve. A typical arrangement is shown in figure 4.31.
Fuel is drawn into the idling fuel line by the low-pressure region around the
throttle valve. A series of ports are used to provide a smooth progression
to the main jet system, and the idling mixture is adjusted by a tapered
screw. When this is added to the result of the other jet systems shown in
figure 4.30 the results will be as shown in figure 4.32.

With fixed jet carburettors there is no automatic mixture enrichment as
the throttle is opened, instead a separate accelerator pump is linked to the

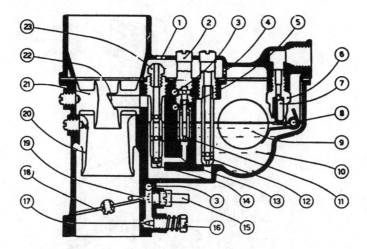

Figure 4.33   Fixed jet or fixed venturi carburettor. ① Air intake; ② idling jet holder; ③ idling mixture tube; ④ air intake to the bowl; ⑤ air intake for idling mixture; ⑥ needle valve seat; ⑦ needle valve; ⑧ float fulcrum pivot; ⑨ float; ⑩ carburettor bowl; ⑪ idling jet; ⑫ main jet; ⑬ emulsioning holes; ⑭ emulsioning tube; ⑮ tube for connecting automatic spark advance; ⑯ idling mixture adjusting screw; ⑰ idling hole to the throttle chamber; ⑱ throttle butterfly; ⑲ progression hole; ⑳ choke tube; ㉑ auxiliary venturi; ㉒ discharge tube; ㉓ emulsioning tube air bleed screw

throttle. For starting, a rich mixture is provided by a choke or strangler valve at entry to the carburettor. When this is closed the whole carburettor is below atmospheric pressure and fuel is drawn from the float chamber directly, and the manifold acts as a surface carburettor. The choke valve is spring loaded so that once the engine fires the choke valve is partially opened. The choke is usually linked to a cam that opens the main throttle to raise the idling speed. A complete fixed jet carburettor is shown in figure 4.33; by changing the jet size a carburettor can be adapted for a range of engines.

4.6.4   Fuel injection

The original purpose of fuel injection was to obtain the maximum power output from an engine. The pressure drop in a carburettor impairs the volumetric efficiency of an engine and reduces its power output. The problems of balancing multiple carburettors and obtaining even distribution in the inlet manifold can also be avoided with fuel injection. Early

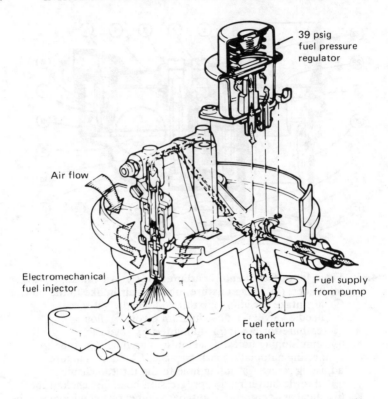

Figure 4.34   Ford single-point fuel injection system (from Ford (1982))

fuel injection systems were mechanical, and complex two-dimensional cams have now been superseded by electronic systems.

Normal practice is to have low-pressure injection operating in the inlet manifold. If the injection were direct into the cylinders, there would be problems of charge stratification. The injectors would also have to withstand the high temperatures in the cylinder and be resistant to the build-up of combustion deposits.

As explained in section 4.6.1, there are two types of fuel injection system:

 (i) single-point fuel injection shown by figure 4.34
(ii) multi-point fuel injection, for which an injector is shown in figure 4.35.

The single-point fuel injection system (figure 4.34) is a cheaper alternative to multi-point fuel injection, it can lead to about a 10 per cent lower power output than a multi-point injection system, and this helps the motor industry to maintain product differentiation. Multi-point fuel injection has

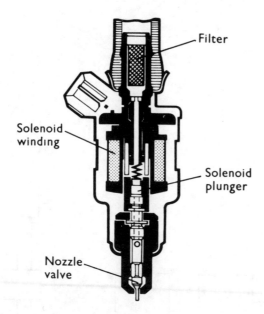

Figure 4.35   Solenoid-operated fuel injection valve (courtesy of Lucas
                Electrical Ltd)

the potential for a higher power output, since the manifold can be designed
for optimum air flow, and perhaps also include some induction tuning
features. The fuelling level is controlled by the fuel supply pressure, and
the duration of the injection pulses. As with carburettors, the single-point
fuel injection system leads to fuel transport delays in the inlet manifold.

The control system for a fuel injection system is shown in figure 4.36 for
a multi-point fuel injection system, but the same principles apply to the
control of the single-point injection system.

The fuel flow rate through an injector is controlled by the differential
pressure across the injector. In the case of a single-point injection system,
the injector sprays fuel into a region at atmospheric pressure, so a constant
gauge pressure is maintained by the fuel pressure regulator. In contrast,
the fuel injectors for a multi-point injection system inject into the reduced
pressure of the inlet manifold. Thus, to make the fuel flow rate a function
only of injection duration, the fuel pressure regulator senses the inlet
manifold pressure, so as to maintain a constant differential pressure
(typically 2 bar) across the injector. The control of the injectors will be
discussed later in section 4.7.

In multi-point injection systems the pulse duration is typically in the
range of 2–8 ms. The ratio of maximum to minimum fuel flow rate in a
spark ignition engine can be 50 or so. This is a consequence of the speed

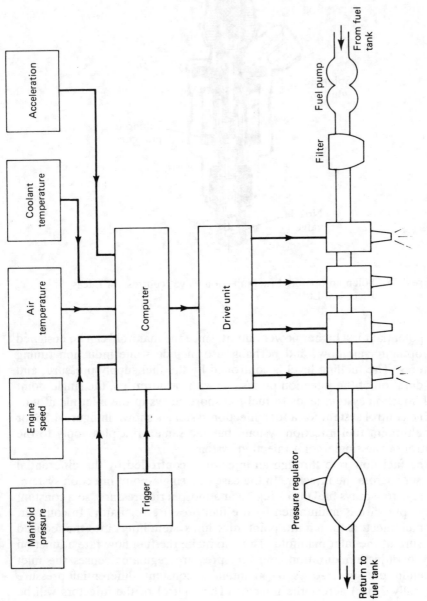

Figure 4.36   Electronic fuel injection system

range and the use of throttling. At maximum power, the injectors will be open almost continuously, while at light load, then the injection duration will be an order of magnitude less. The range of pulse durations is much lower since a finite time is required to open the injector (about 2 ms).

In order to simplify the injector drive electronics, it has been common practice to fire the injectors in two groups. (If all the injectors fired together then there would be a very uneven fuel demand.) Clearly, when the fuel injection pulse starts is not going to affect the maximum power performance, since the injection is more-or-less continuous under these conditions. However, for part load operation, it has been found that if injection occurs when the inlet valve is closed, then this leads to lower emissions of NO_x and unburnt hydrocarbons. The explanation for this is that fuel sprayed on to a closed inlet valve is vaporised by the hot valve (and in doing so helps to cool the valve), and there is also more time for heat transfer to the fuel from the inlet port. Thus a more homogeneous mixture is formed when the injection pulse is phased with regard to inlet valve events, these systems are known variously as: timed, sequential or timed sequential injection.

Although fuel transport delays are reduced with multi-point injection systems they are not eliminated, and multi-point fuel injection systems can also be subject to other complications during load increase transients. In particular, when the throttle is opened rapidly, the inlet manifold pressure will rise more quickly than the fuel supply pressure (because of the lag inherent in the pressure regulator). This means that during a load increase transient the differential pressure across the injector falls. A similar problem can occur as a result of pressure pulsations in the fuel supply rail and in the inlet manifold. These pressure pulsations are an inevitable consequence of the unsteady flow, and they mean that the instantaneous pressure difference across the fuel injectors will vary, and will not correspond to the mean pressure differential being controlled by the pressure regulator.

4.6.5 *Mixture preparation*

It might be assumed that multi-point fuel injection systems give the best mixture preparation and most uniform fuel distribution; this is not necessarily the case. At very light loads the fuel injectors will only open for a very short period of time, and under these conditions differences in the response time become significant. This is because the duration of the spray is small compared with the activation time of the injector. An exhaust gas oxygen sensor (described in section 4.7) can control the overall air/fuel ratio, but a single sensor cannot compensate for inter-cylinder variations. However, the exhaust gas oxygen sensor can allow for uniform ageing effects, such as the formation of deposits in injectors. The formation of gum deposits in injectors was a problem mostly associated with the USA in the late 1980s, and was linked with high levels of unsaturated hydrocarbons (notably alkyls). These problems led to detail design changes in the fuel injector and improved detergent additives in the fuel.

The above arguments have shown that the mixture mal-distribution will be worst from a multi-point fuel injection system at light loads. In contrast, under light load conditions, the mixture mal-distribution will be least from a carburettor or single-point fuel injection system. The mixture mal-distribution arises through unequal flows of liquid fuel (from the wall film and droplets) entering the cylinders. At light loads, these problems are ameliorated for two reasons.

Firstly, at light loads the low manifold pressure ensures that a higher proportion of the fuel evaporates, and enters the cylinder as vapour. Secondly, when the pressure ratio across the throttle plate (or butterfly valve) exceeds 1.9, then the flow will become sonic in the throat between the throttle plate and the throttle body. The fuel is sprayed towards the throttle plate and impinges on it, to form a liquid film. The fuel film travels to the edge of the throttle plate, and the very high air velocity in this region causes the liquid film to be broken into very small droplets. The small droplets evaporate more readily, and are also more likely to move with the air flow — as opposed to being deposited on the walls of the inlet manifold.

The spray quality of different mixture preparation systems has been systematically investigated by Fraidl (1987). Fraidl illustrated the change in droplet size distribution from a carburettor as the load increased (the droplet size distribution moved towards the larger droplet sizes), and the spatial distribution along the axis of the throttle plate spindle. Fraidl also provides a comparison between the droplet size distributions for: a carburettor, single-point injector and multi-point injector at full load and part load, (figure 4.37).

At full load, figure 4.37 shows that the multi-point fuel injection system gives the distribution with the smallest droplet sizes. Under these conditions the throttle plate is fully open, and so cannot contribute any improvements to the atomisation of the liquid fuel. At part load, the situation is reversed, and the ordinary multi-point fuel injection system has the droplet size distribution with the largest droplets. However, the fuel atomisation can be improved by the use of air-containment, in which an annular flow of high velocity air encircles the fuel spray. A liquid spray is broken into droplets by the shear forces between the liquid and the surrounding gas. The higher the relative velocity between the liquid and gas, then the greater the shear force, and the smaller the size of the droplets that are formed. A high relative velocity between the liquid and gas can be provided by one of two ways. Firstly, a high velocity jet (v_j) can be

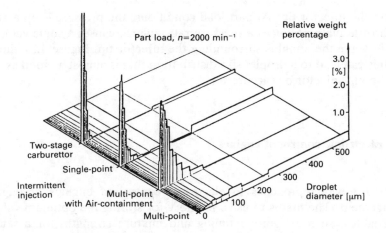

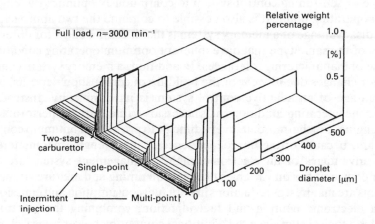

Figure 4.37   Representative droplet size spectra of mixture preparation systems at part load and full load, from Fraidl (1987) with permission of EAEC

produced by using a high differential pressure (Δp_i) across the injector; to a first order

$$v_j \propto \sqrt{(\Delta p_i)} \qquad (4.7)$$

Secondly, it is possible to increase the velocity of the air surrounding the injector. The 'air-containment' technique exploits the high pressure ratio that occurs across the throttle plate at light loads. Air from upstream of the throttle plate by-passes the inlet manifold and enters an annular cavity around the injector tip. At part load conditions, the pressure drop across the throttle plate produces a pressure ratio, which causes a sonic velocity flow to leave the annulus surrounding the injector tip. Figure 4.37 shows that this can lead to a droplet size distribution that is almost as good as the single-point injector or carburettor.

4.7   Electronic control of engines

There are two approaches to electronic control of engines or engine management. The first is to use a memory for storing the optimum values of variables, such as ignition timing and mixture strength, for a set of discrete engine-operating conditions. The second approach is to use an adaptive or self-tuning control system to continuously optimise the engine at each operating point. It is also possible to combine the two approaches.

The disadvantage of a memory system is that it cannot allow for different engines of the same type that have different optimum operating conditions because of manufacturing tolerances. In addition, a memory system cannot allow for changes due to wear or the build-up of combustion deposits. The disadvantage of an adaptive control system is its complexity. Instead of defining the operating conditions, it is necessary to measure the performance of the engine. Furthermore, it is very difficult to provide an optimum control algorithm, because of the interdependence of many engine parameters.

The advantages of an electronic engine management system are the greater control it has on variables like ignition timing and mixture strength. The gains are manifest as reductions in both fuel consumption and emissions.

With electronic ignition and fuel injection, combining the electronic control is a logical step since the additional computing power is very cheap. In vehicular applications the natural extension will be control of transmission ratios in order to optimise the overall fuel economy.

With memory systems the engine-operating conditions are derived from engine maps. Figure 4.38 is an example of a typical engine map, derived from experimental results. The map shows contours for fuel economy and manifold pressure. Additional contours could be added for emissions, ignition timing and mixture strength, but these have been omitted to avoid confusion.

When an engine is tested the power output, emissions, manifold depression, optimum ignition timing and air/fuel mixture will all be recorded for each throttle setting and speed. The results are plotted against engine speed and bmep, since bmep is a measure of engine output, independent of its size. In a microprocessor-controlled system the optimum operating

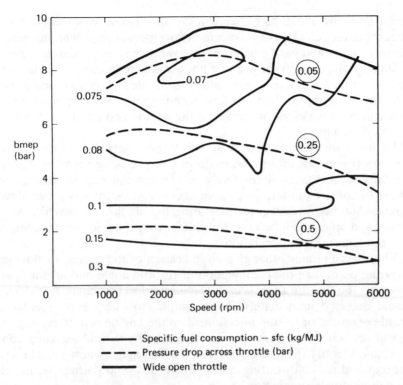

Figure 4.38   Engine map for fuel economy and manifold pressure (reprinted
with permission, © 1982 Society of Automotive Engineers, Inc.)

conditions will be stored in ROM (Read Only Memory) for each operating
point. Since it is difficult to measure the output of an engine, except on a
test bed, the operating point is identified by other parameters, such as
engine speed and manifold pressure.

In an engine management system, some of the parameters that can be
measured have been shown in figure 4.38. This information is then used by
the engine management system to control: the ignition timing, the exhaust
gas recirculation valve, and the fuel injection equipment. Since the engine
will have been calibrated to operate with a particular schedule of air/fuel
ratio, it is very important to know the air flow rate. This leads to two types
of electronic fuel injection control, those which measure the air flow rate
and those that deduce the air flow — the so-called speed–density systems.
The speed–density systems measure the manifold pressure and air temperature, and then from the engine speed/manifold pressure relationship
(stored in memory), the engine management system can deduce the air
flow rate. This approach is less direct (and less accurate) than measuring

the air flow into the engine. Nor can the speed–density system allow for exhaust gas recirculation. Two common flow measuring techniques are: (a) the use of a pivoted vane connected to a variable resistor, that is deflected by the air flow; and (b) the use of a hot wire anemometer. Unfortunately, the hot wire anemometer is insensitive to the flow direction, and as wide open throttle is approached, the flow becomes more strongly pulsating and there can be a backflow of air out of the engine (especially with a tuned induction system).

If a hot wire anemometer is used it is necessary to deduce the flow direction (by means of a pressure drop, or by looking for when the flow rate falls to zero), or alternatively to rely on the engine management system to correct the hot wire anemometer results for when backflow is present. Another advantage of measuring the air flow, is that the speed–density system can still be used for either cross-checking or deducing the level of exhaust gas recirculation.

Other measurements that give an indication of the engine air flow (and operating point) are the combination of throttle angle and engine speed. However, the throttle angle is usually measured so that changes in demand can be detected immediately. For example, to inject extra fuel during throttle opening, or to stop injection when the engine is decelerating. The air and coolant temperature sensors also identify the cold-starting conditions, and identify the operating point during engine warm-up. The extra fuel required for cold-starting can be quantified, and perhaps be injected through an auxiliary injector.

The type of memory-based systems described so far can make no allowances for change or differences in engine performance, nor can they allow for fuel changes. Suppose a fuel with a different density and octane rating was used, then the air/fuel ratio would be changed, and combustion knock might be encountered at certain full throttle conditions. Thus it is common practice to employ knock detection, and measurement of the exhaust gas oxygen level to deduce the air/fuel ratio.

If a knock detector is fitted, this can retard the ignition at the onset of knock, thereby preventing damage to the engine. Combustion knock causes the engine structure to vibrate, and an accelerometer can be used as a knock sensor. Forlani and Ferranti (1987) report that the signal is typically filtered with a pass band of 6–10 kHz, and examined in a window from tdc to 70° after tdc. Since the signal is examined for a particular time window, then the knocking cylinder can be identified and the ignition timing can be retarded selectively. The knock sensor provides a safety margin, which would otherwise be obtained by having a lower compression ratio or permanently retarded ignition. A very striking example of this is provided by Meyer et al. (1984), in which an engine fitted with a knock sensor is run with fuels of a lower octane rating than it had been designed for. The results are shown in figure 4.39, and it can be seen that there is no significant change in the fuel consumption, for a range of driving condi-

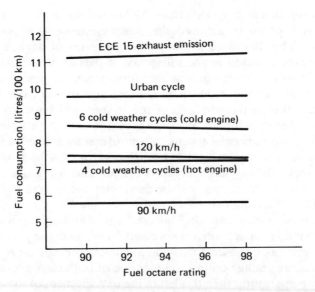

Figure 4.39 The effect on fuel consumption of reduced fuel quality on an engine designed for 98 octane fuel, but fitted with a knock sensor controlled electronic ignition (adapted from Meyer *et al.* (1984))

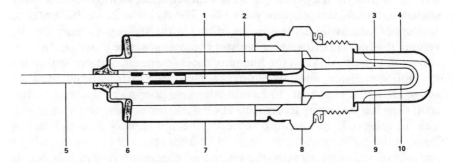

Lambda sensor
1 Contact element. 2 Protective ceramic element. 3 Sensor ceramic. 4 Protective tube (exhaust end). 5 Electrical connection. 6 Disc spring. 7 Protective sleeve (atmosphere end). 8 Housing (−). 9 Electrode (−). 10 Electrode (+).

Figure 4.40 Cross-section drawing of an exhaust oxygen sensor, reproduced by permission of Robert Bosch Ltd

tions. Indeed, only a slight deterioration was found in the full load fuel economy.

An exhaust gas oxygen sensor is shown in figure 4.40. When a three-way catalyst is to be used, then it is essential to use a feedback system incorporating such a sensor, to maintain an air/fuel ratio that is within about 1 per cent of stoichiometric (as discussed in section 4.3). The oxygen or lambda sensor has been described by Wiedenmann *et al.* (1984). One electrode is exposed to air, and the other electrode is exposed to the exhaust gas. The difference in the partial pressures of oxygen leads to a flow of electrons related to the difference in partial pressures. Since the platinum electrode also acts as a catalyst for the exhaust gases, then for rich or stoichiometric air/fuel ratios there is a very high output from the lambda sensor, since the partial pressure of the oxygen will be many orders of magnitude lower than for air. Since the sensor is used in a way that decides whether the mixture is rich or weak, a control system is needed that makes the air/fuel ratio perturbate around stoichiometric. As the lambda sensor will only work when it has reached a temperature of about 300°C, this feedback control system can only be used after the engine has started to warm-up (20–30 s).

This type of sensor can also be used for lean-burn engine control. However, for this application it is necessary to evaluate the partial pressure of the oxygen (as opposed to deciding whether or not there is oxygen present). As the voltage output is a function of both the oxygen level and the sensor temperature, then this led to the development of the electrically heated oxygen sensor described by Wiedenmann *et al.* (1984).

A closed loop control system needs to be used when a carbon canister is used for the control of evaporative emissions. The major source of evaporative emissions from a passenger car is a consequence of the fuel tank being subject to diurnal temperature variations. The fuel tank has to be vented to atmosphere, to avoid a pressure build-up, as the fuel tank warms-up. By venting the fuel tank through a canister containing active charcoal, the fuel vapour is absorbed. When the fuel tank cools down, air is drawn in through the carbon canister, and this removes some of the hydrocarbons from the active charcoal. However, to ensure adequate purging of the active charcoal, then it is necessary to draw additional air through the active charcoal. This is achieved, by drawing some of the air flowing into the engine through the carbon canister. Clearly, the air/fuel ratio of this stream will be unpredictable, but a closed-loop engine management system will enable the engine to always adopt the intended air/fuel ratio.

It is also expected that the EC will adopt measures to control vapour emissions during fuel tank filling; these vapour emissions represent about a quarter of the evaporative emissions.

In an adaptive control system the operating point would be found by optimising the fuel economy or emissions. With the multitude of control loops the hierarchy has to be carefully defined, otherwise one loop might be working directly against another. The computational requirements are not a problem since they can be readily met by current microprocessors. Only this approach offers fully flexible systems.

An example of an adaptive engine control system is provided by Wakeman *et al.* (1987) and Holmes *et al.* (1988). Perturbations are applied to the ignition timing, and the slope of the ignition timing/torque curve is inferred from the response of the engine speed. This way the control system can find the MBT ignition timing for a particular operating condition, and then store this as a correction value from the ignition timing map. A similar system can be used for controlling EGR to minimise NO_x emissions. Holmes *et al.* also describe an approach to knock control in which the timing retard from the ignition map is stored as a function of operating point. The system is arranged so that when there is a change in fuel quality the ignition timing will converge to the new optimum timing schedule. The fuel economy benefits of an adaptive ignition system are shown by Wakeman *et al.* Four nominally identical vehicles were tested on the ECE 15 urban driving cycle, and it was found that the adapted ignition timing map was different for each vehicle (and different from the manufacturer's calibration). Fuel savings of up to 9 per cent were achieved.

An important method of improving part load fuel economy is cylinder disablement. In its simplest form this consists of not supplying a cylinder with fuel — a technique used on gas engines at the turn of the century. When applied to fuel-injected multi-cylinder engines a group of cylinders can be disabled, such as a bank of three cylinders in a V6. Alternatively, a varying disablement can be used; for example, every third injection pulse might be omitted in a four-cylinder engine. Slightly greater gains would occur if the appropriate inlet valves were not opened since this would save the pumping losses. This approach could also be used with single-point injection systems or carburettors. However, this requires electrically controlled inlet valves.

In vehicular applications a very significant amount of use occurs in short journeys in which the engine does not reach its optimum temperature — the average British journey is about 10 miles. To improve the fuel economy of carburetted engines under these conditions the choke can be controlled by a stepper motor which responds fast enough to prevent the engine stalling at slower than normal idling speeds.

The questions about electronic engine management concern not just economics but also reliability and whether or not an engine is fail safe.

4.8 Conclusions

Spark ignition engines can operate only within a fairly narrow range of mixture strengths, typically within a gravimetric air/fuel ratio range of 10–18:1 (stoichiometric 14.8:1). The mixture strength for maximum power is about 10 per cent rich of stoichiometric, while the mixture strength for maximum economy is about 10 per cent weak of stoichiometric. This can be explained qualitatively by saying that the rich mixture ensures optimum utilisation of the oxygen (but too rich a mixture will lead to unburnt fuel, which will lower the combustion temperatures and pressures). Conversely the lean mixture ensures optimum combustion of the fuel (but too weak a mixture leads to increasingly significant mechanical losses and ultimately to misfiring). The power output and economy at constant speed for a range of throttle settings are conveniently shown by the so-called 'fish-hook' curves, figure 4.2. From these it is self-evident that at part throttle it is always more economical to run an engine with a weak mixture as opposed to a slightly more closed throttle with a richer mixture.

The main combustion chamber requirements are: compactness, sufficient turbulence, minimised quench areas, and short flame travel from the spark plug to the exhaust valve. Needless to say there is no unique solution, a fact demonstrated by the number of different combustion chamber designs. However, the high-turbulence, high compression ratio, lean-burn combustion chamber should be treated as a separate class. This type of combustion chamber is exemplified by the May Fireball combustion chamber. The high turbulence enables lean mixtures to be burnt rapidly, and the compact combustion chamber (to minimise heat transfer) is located around the exhaust valve. All these attributes contribute to knock-free operation, despite the high compression ratios; the compression ratio can be raised from the usual 9:1 to 15:1, so giving a 15 per cent improvement in fuel economy. However, such combustion chambers place a greater demand on manufacturing tolerances, mixture preparation and distribution, and ignition timing.

Mixture preparation is either by carburettor or by fuel injection. There are two main types of carburettor — fixed or variable venturi (or jet) — and both types can be used for single or multi-carburettor installations. Fuel injection can provide a much closer control on mixture preparation and injection is invariably at low pressure into the induction passage. When a single-point injection system is used the problems associated with maldistribution in the inlet manifold still occur, but are ameliorated by the better atomisation of the fuel. It is very difficult (maybe even impossible) to design an inlet manifold that gives both good volumetric efficiency and uniform mixture distribution.

With the decreasing cost and increasing power of micro-electronics it is logical to have complete electronic control of ignition timing and mixture preparation. High compression ratio engines need the sophisticated control to avoid both knock and misfiring. The low-emission engines need careful engine control to ensure operation on a stoichiometric mixture, so that the three-way exhaust catalyst can function properly.

4.9 Example

A variable jet carburettor is designed for a pressure drop (Δp) of 0.02 bar and an air/fuel ratio of 15:1. If the throttle is suddenly opened and the pressure drop quadruples, calculate the percentage increases in air flow and fuel, and the new air/fuel ratio. Neglect surface tension and the difference in height between the jet and the liquid level in the float chamber; atmospheric pressure is 1 bar, and $\gamma = 1.4$.

Immediately after the throttle opens there will be no change in either orifice, as the piston and needle will not have moved.

Rewriting equations (4.5) and (4.6) gives

$$\dot{m}_f = k_f \sqrt{(\Delta p)}$$

$$\dot{m}a = k_a \sqrt{(\Delta p)} \left[(r)^{1/\gamma} \sqrt{ \left\{ \frac{\gamma}{\gamma - 1} \frac{1 - r}{\gamma - 1}^{(\gamma - 1/\gamma)} \right\} } \right]$$

where $r = 1 - \Delta p/p$; k_f, k_a are constants.

If Δp quadruples, the increase in $\dot{m}_f$ is

$$\frac{\sqrt{(4\,\Delta p)} - \sqrt{(\Delta p)}}{\sqrt{(\Delta p)}} \times 100 = 100 \text{ per cent}$$

Using suffix 1 to denote conditions before the step change and suffix 2 to denote conditions after

$$r_1 = 1 - \frac{0.02}{1} = 0.98 \; ; r_2 = 1 - \frac{0.08}{1} = 0.92$$

$$\dot{m}_{a1} = k_a \sqrt{(0.02)} \, 0.098^{\,(1/1.4)} \sqrt{\left(\frac{1.4}{1.4 - 1} \frac{1 - 0.98}{1 - 0.98}^{(1.4-1)/1.4} \right)}$$

$$= k_a \sqrt{(0.02)} \, 0.98^{\,0.714} \sqrt{\left(3.5 \frac{1 - 0.98}{1 - 0.98}^{0.286} \right)} = k_a \times 0.140$$

$$\dot{m}_{a2} = k_a \sqrt{(0.08)} \, 0.92^{\,0.714} \sqrt{\left(3.5 \frac{1 - 0.92}{1 - 0.92}^{0.286} \right)} = k_a \times 0.271$$

$$\% \text{ increase in } \dot{m}_a = \frac{0.271 - 0.140}{0.140} 100 = 93.6 \text{ per cent}$$

$$\text{New air fuel ratio} = \frac{15 \times 0.936}{1.00} : 1 = 14.04 : 1$$

4.10 Problems

4.1 Why does the optimum ignition timing change with engine-operating conditions? What are the advantages of electronic ignition with an electronic control system?

4.2 Explain the principal differences between fixed jet and variable jet carburettors. Why does the mixture strength become richer with increasing flow rate in a simple carburettor?

4.3 What are the air/fuel requirements for a spark ignition engine at different operating conditions? How are these needs met by a fixed jet carburettor?

4.4 List the advantages and disadvantages of electronic fuel injection.

4.5 Contrast high-turbulence, high compression ratio combustion chambers with those designed for lower compression ratios.

4.6 Two spark ignition petrol engines having the same swept volume and compression ratio are running at the same speed with wide open throttles. One engine operates on the two-stroke cycle and the other on the four-stroke cycle. State with reasons:

(i) which has the greater power output.
(ii) which has the higher efficiency.

4.7 The Rover M16 spark ignition engine has a swept volume of 2.0 litres, and operates on a 4-stroke cycle. When installed in the Rover 800, the operating point for a vehicle speed of 120 km/h corresponds to 3669 rpm and a torque of 71.85 N m, for which the specific fuel consumption is 298 g/kWh.

Calculate the bmep at this operating point, the arbitrary overall efficiency and the fuel consumption (litres/100 km). If the gravimetric air/fuel ratio is 20:1, calculate the volumetric efficiency of the engine, and comment on the value.

The calorific value of the fuel is 43 MJ/kg, and its density is 795 kg/m^3. Ambient conditions are 27°C and 1.05 bar.

Explain how both lean-burn engines and engines fitted with three-way catalyst systems obtain low exhaust emissions. What are the advantages and disadvantages of lean-burn operation?

4.8 A spark ignition engine is to be fuelled by methanol (CH_3OH). Write down the equation for stoichiometric combustion with air, and calculate the stoichiometric gravimetric air/fuel ratio.

Compare the charge cooling effects of fuel evaporation in the following two cases:

(i) A stoichiometric air/methanol mixture in which 25 per cent of the methanol evaporates,

(ii) 60 per cent of a 14.5:1 gravimetric air/petrol mixture evaporates.

The enthalpy of evaporation for methanol is 1170 kJ/kg and that of petrol is 310 kJ/kg. State clearly any assumptions made.

If in a lean-burn engine the gravimetric air/fuel ratio is 8:1, rewrite the combustion equation, and determine the volumetric composition of the products that should be found from a dry gas analysis. What is the equivalence ratio for the air/fuel ratio? Why might methanol be used as a fuel for spark ignition engines?

4.9 The Rover K16 engine is a four-stroke spark ignition engine, with a swept volume of 1.397 litres and a compression ratio of 9.5. The maximum torque occurs at a speed of 4000 rpm, at which point the power output is 52 kW; the maximum power of the engine is 70 kW at 6250 rpm. Suppose the minimum brake specific fuel consumption is 261.7 g/kWh, using 95 RON lead-free fuel with a calorific value of 43 MJ/kg.

Calculate the corresponding Otto cycle efficiency, the maximum brake efficiency and the maximum brake mean effective pressure. Give reasons why the brake efficiency is less than the Otto cycle efficiency. Show the Otto cycle on the p–V state diagram, and contrast this with an engine indicator diagram — identify the principal features.

Explain why, when the load is reduced, the part-load efficiency of a diesel engine falls less rapidly than the part-load efficiency of a spark ignition engine.

4.10 Since the primary winding of an ignition coil has a finite resistance, energy is dissipated while the magnetic field is being established after switching on the coil. If the primary winding has an inductance of 5.5 mH, and a resistance of 1.9 Ohms, and the supply voltage is 11.6 V, show that:

(i) the energy supplied after 2 ms would be 39 mJ, increasing to 246 mJ after 6 ms,

(ii) the theoretical efficiency of energy storage is about 65 per cent for a 2 ms coil-on-time, falling to 32 per cent for a 6 ms coil-on-time.

What are the other sources of loss in an ignition coil?

Define MBT ignition timing, and describe how and why it varies with engine speed and load. Under what circumstances are ignition timings other than MBT used?

5 Compression Ignition Engines

5.1 Introduction

Satisfactory operation of compression ignition engines depends on proper control of the air motion and fuel injection. The ideal combustion system should have a high output (bmep), high efficiency, rapid combustion, a clean exhaust and be silent. To some extent these are conflicting requirements; for instance, engine output is directly limited by smoke levels. There are two main classes of combustion chamber: those with direct injection (DI) into the main chamber, figure 5.1, and those with indirect injection (ID), figure 5.5, into some form of divided chamber. The fuel injection system cannot be designed in isolation since satisfactory combustion depends on adequate mixing of the fuel and air. Direct injection engines have inherently less air motion than indirect injection engines and, to compensate, high injection pressures (up to 1000 bar) are used with multiple-hole nozzles. Even so, the speed range is more restricted than for indirect injection engines. Injection requirements for indirect injection engines are less demanding; single-hole injectors with pressures of about 300 bar can be used.

There are two types of injector pump for multi-cylinder engines, either in-line or rotary. The rotary pumps are cheaper, but the limited injection pressure makes them more suited to indirect injection engines.

The minimum useful cylinder volume for a compression ignition engine is about 400 cm^3, otherwise the surface-to-volume ratio becomes disadvantageous for the normal compression ratios. The combustion process is also slower than in spark ignition engines and the combined effect is that maximum speeds of compression ignition engines are much less than those of spark ignition engines. Since speed cannot be raised, the output of compression ignition engines is most effectively increased by turbocharging. The additional benefits of turbocharging are improvements in fuel economy, and a reduction in the weight per unit output.

The compression ratio of turbocharged engines has to be reduced, in order to restrict the peak cylinder pressure; the compression ratio is typically in the range 12–24:1. The actual value is usually determined by the cold starting requirements, and the compression ratio is often higher than optimum for either economy or power. Another compromise is the fuel injection pattern. For good cold starting the fuel should be injected into the air, although very often it is directed against a combustion chamber wall to improve combustion control.

There are many different combustion chambers designed for different sizes of engine and different speeds, though inevitably there are many similarities. Very often it will be the application that governs the type of engine adopted. For automotive applications a good power-to-weight ratio is needed and some sacrifice to economy is accepted by using a high-speed engine. For marine or large industrial applications size and weight will matter less, and a large slow-running engine can be used with excellent fuel economy.

All combustion chambers should be designed to minimise heat transfer. This does not of itself significantly improve the engine performance, but it will reduce ignition delay. Also, in a turbocharged engine a higher exhaust temperature will enable more work to be extracted by the exhaust turbine. The so-called adiabatic engine, which minimises (not eliminates) heat losses, uses ceramic materials, and will have higher exhaust temperatures.

Another improvement in efficiency is claimed for the injection of water/fuel emulsions; for example, see Katsoulakos (1983). By using an emulsion containing up to 10 per cent water, improvements in economy of 5–8 per cent are reported. Improvements are not universal, and it has been suggested that they occur only in engines in which the air/fuel mixing has not been optimised. For a given quantity of fuel, a fuel emulsion will have greater momentum and this could lead to better air/fuel mixing. An additional mechanism is that when the small drops of water in the fuel droplets evaporate, they do so explosively and break up the fuel droplet. However, the preparation of a fuel emulsion is expensive and it can lead to problems in the fuel injection equipment. If a fuel emulsion made with untreated water is stored, bacterial growth occurs. Fuel emulsions should reduce NO_x emissions since the evaporation and subsequent dissociation of water reduce the peak temperature.

As in spark ignition engines, NO_x emissions can be reduced by exhaust gas recirculation since this lowers the mean cylinder temperature. Alternatively NO_x emissions can be reduced by retarded injection. However this has an adverse effect on output, economy and emissions of unburnt hydrocarbons and smoke.

Section 5.6 discusses the emissions from Diesel engines and summarises the relevant emissions legislation. The ways of predicting the ignition delay, and the combustion rate are discussed in chapter 10, section 10.2, where modelling techniques are discussed.

Before discussing Diesel and indirect Diesel engines, it is instructive to compare their part-load efficiency with that of a spark ignition engine; this had been done in figure 5.1. The difference in efficiency of direct and indirect injection Diesel engines is discussed later in section 5.3. When comparing the sfc of compression ignition and spark ignition engines, it must be remembered that on a specific basis the energy content of diesel fuel is about 4.5 per cent lower. Figure 5.1 shows that the Diesel engines have a higher maximum efficiency than the spark ignition engine for three reasons:

(a) The compression ratio is higher.
(b) During the initial part of compression, only air is present.
(c) The air/fuel mixture is always weak of stoichiometric.

Furthermore, the Diesel engine is, in general, designed to operate at lower speeds, and consequently the frictional losses are smaller.

In a Diesel engine the air/fuel ratio is always weak of stoichiometric, in order to achieve complete combustion. This is a consequence of the very limited time in which the mixture can be prepared. The fuel is injected into the combustion chamber towards the end of the compression stroke, and around each droplet the vapour will mix with air, to form a flammable mixture. Thus the power can be regulated by varying the quantity of fuel injected, with no need to throttle the air supply.

As shown in figure 5.1, the part-load specific fuel consumption of a Diesel engine rises less rapidly than for a spark ignition engine. A fundamental difference between spark ignition and Diesel engines is the manner in which the load is regulated. A spark ignition engine always requires an air/fuel mixture that is close to stoichiometric. Consequently, power regulation is obtained by reducing the air flow as well as the fuel flow. However, throttling of the air increases the pumping work that is dissipated during the gas exchange processes.

Also, since the output of a Diesel engine is regulated by reducing the amount of fuel injected, the air/fuel ratio weakens and the efficiency will improve. Finally, as the load is reduced, the combustion duration decreases, and the cycle efficiency improves. To summarise, the fall in part load efficiency is moderated by:

(a) The absence of throttling.
(b) The weaker air/fuel mixtures.
(c) The shorter duration combustion.

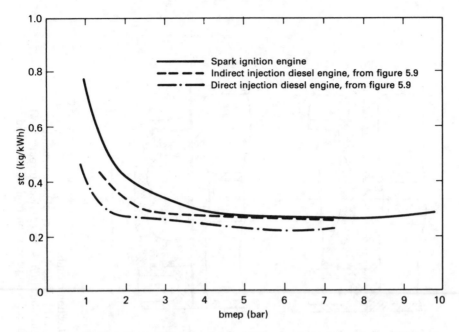

Figure 5.1   Comparison of the part load efficiency of spark ignition and Diesel engines at 2000 rpm (Stone, R. (1989a))

Low heat loss (so-called Adiabatic) Diesel engines are discussed later in chapter 9, section 9.3, as the benefits are most significant for turbocharged engines.

5.2   Direct injection (DI) systems

Some typical direct injection combustion chambers given by Howarth (1966) are shown in figure 5.2. Despite the variety of shapes, all the combustion chambers are claimed to give equally good performance in terms of fuel economy, power and emissions, when properly developed. This suggests that the shape is less critical than careful design of the air motion and fuel injection. The most important air motion in direct injection Diesel engines is swirl, the ordered rotation of air about the cylinder axis. Swirl can be induced by shrouded or masked inlet valves and by design of the inlet passage — see figure 3.2.

$$\text{swirl (ratio)} = \frac{\text{swirl speed (rpm)}}{\text{engine speed (rpm)}} \qquad (5.1)$$

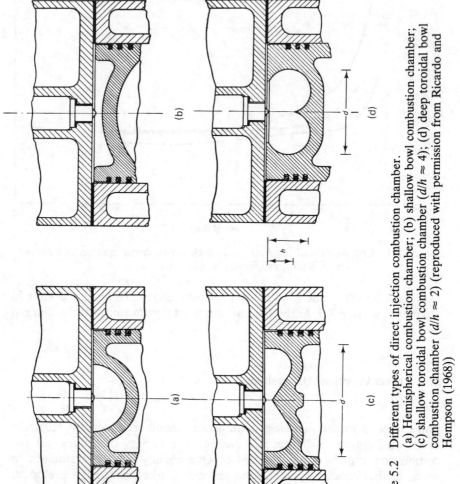

Figure 5.2   Different types of direct injection combustion chamber.
(a) Hemispherical combustion chamber; (b) shallow bowl combustion chamber;
(c) shallow toroidal bowl combustion chamber ($d/h \approx 4$); (d) deep toroidal bowl
combustion chamber ($d/h \approx 2$) (reproduced with permission from Ricardo and
Hempson (1968))

The swirl speed will vary during the induction and compression strokes and an averaged value is used.

The methods by which the swirl is measured, how the swirl ratio is defined, and how the averaging is undertaken all vary widely. Heywood (1988) summarises some of the approaches, but a more comprehensive discussion is provided by Stone, R and Ladommatos (1992). For steady-flow tests, the angular velocity of the swirl will vary across the bore, but an averaged value is given directly by a paddle wheel (or vane) anemometer. Alternatively, a flow straightener can be used, and the moment of momentum of the flow corresponds to the torque reaction on the flow straightener.

If there is a uniform axial velocity in the cylinder, then

$$M = \frac{1}{8} \, \dot{m}\omega B^2 \tag{5.2}$$

where M = moment of momentum flux (N m)
   m = mass flow rate (kg/s)
   ω = swirl angular velocity (rad/s)
   B = cylinder bore (m).

If hot wire anemometry or laser doppler anemometry are used to define the flow field in the cylinder, and a single value is required to quantify the swirl, then

$$M = \int_0^{B/2} \int_0^{2\pi} \rho \, v_t \, (r, \, \theta) \, v_a \, (r, \, \theta) r^2 \, dr d\theta \tag{5.3}$$

where ρ = air density
   v_t = tangential velocity
   v_a = axial velocity
   r = radial coordinate
   θ = angular coordinate.

Because of the different ways in which swirl can be measured and defined, published swirl data should be treated on a comparative basis, unless the measurement method has been fully defined. The means by which swirl is generated in a helical inlet port is illustrated by figure 5.3.

Ricardo and Hempson (1968) report extensive results from an engine in which the swirl could be varied. The results for a constant fuel flow rate are shown in figure 5.4. For these conditions the optimum swirl ratio for both optimum economy and power output is about 10.5. However, the maximum pressure and rate of pressure rise were both high, which caused noisy and rough running. When the fuel flow rate was reduced the optimum swirl

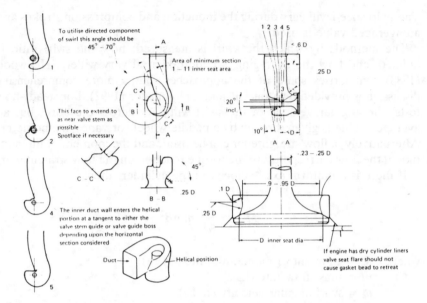

Figure 5.3   Geometric details of a helical inlet port, Beard (1984). Reproduced by kind permission of Butterworth–Heinemann Ltd, Oxford

ratio was reduced, but this incurred a slight penalty in performance and fuel economy. Tests at varying speeds showed that the optimum relationship of fuel injection rate to swirl was nearly constant for a wide range of engine speeds. Constant speed tests showed that the interdependence of swirl and fuel injection rate became less critical at part load. Increasing swirl inevitably increases the convective heat transfer coefficient from the gas to the cylinder walls. This is shown by a reduction in exhaust temperature and an increase in heat transfer to the coolant.

Care is needed in the method of swirl generation in order to avoid too great a reduction in volumetric efficiency, since this would lead to a corresponding reduction in power output.

In two-stroke engines there are conflicting requirements between swirl and the scavenging process. The incoming air will form a vortex close to the cylinder wall, so trapping exhaust products in the centre.

The combustion chamber is invariably in the piston, so that a flat cylinder head can be used. This gives the largest possible area for valves. Very often the combustion chamber will have a raised central portion in the piston, on the grounds that the air motion is minimal in this region. In engines with a bore of less than about 150 mm, it is usual for the fuel to impinge on the piston during injection. This breaks up the jet into droplets, and a fuel film forms on the piston to help control the combustion rate.

The role of squish, the inward air motion as the piston reaches the end

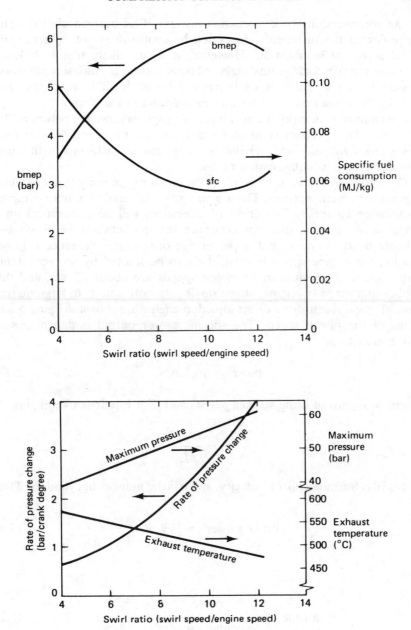

Figure 5.4 Variation of engine performance with swirl ratio at constant fuelling rate (adapted from Ricardo and Hempson (1968))

of the compression stroke, remains unclear. Combustion photography suggests that the turbulence generated by squish does not influence the initial stages of combustion. However, it seems likely that turbulence increases the speed of the final stages of combustion. In coaxial combustion chambers, such as those shown in figure 5.2, squish will increase the swirl rate, by the conservation of angular momentum in the charge.

The compression ratio of direct injection engines is usually between 12:1 and 16:1. The compact combustion chamber in the piston reduces heat losses from the air, and reliable starting can be achieved with these comparatively low compression ratios.

The stroke-to-bore ratio is likely to be greater than unity (under-square engine) for several reasons. The longer stroke will lead to a more compact combustion chamber. The effect of tolerances will be less critical on a longer stroke engine, since the clearance volume between the piston and cylinder head will be a smaller percentage of the total clearance volume. Finally, the engine speed is most likely to be limited by the acceptable piston speed. Maximum mean piston speeds are about 12 m/s, and this applies to a range of engines, from small automotive units to large marine units. Typical results for a direct injection engine are shown in figure 5.5 in terms of the piston speed. The specific power output is dependent on piston area since

$$\text{power} = \bar{p}_b \, \text{LAN}' \tag{2.17}$$

where $N' = $ no. of firing strokes per second. For a four-stroke engine

$$N' = \frac{n \, \bar{v}_p}{4L}$$

where $\bar{v}_p$ is the mean piston velocity, and n is the number of cylinders. Thus

$$\text{brake power} = \bar{p}_b \bar{v}_p \, n \, \frac{A}{4} \tag{5.4}$$

But

$$\text{torque} = \frac{\text{power}}{\text{angular velocity}} = \frac{\bar{p}_b \cdot V}{4 \, \pi} \tag{5.5}$$

The fuel injector is usually close to the centre-line of the combustion chamber and can be either normal or angled to the cylinder head. To provide good mixing of the fuel and air during the injection period, all the

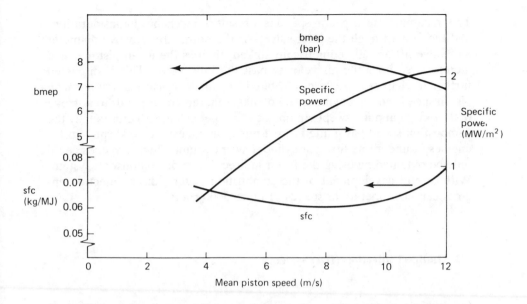

Figure 5.5   Typical performance for direct injection engine (reproduced from Howarth (1966), courtesy of M. H. Howarth, Atlantic Research Associates)

air should move past a jet of fuel. By using multi-hole injectors the amount of swirl can be reduced. However, multi-hole injectors require higher injection pressures for the same flow rate and jet penetration.

Traditionally, low swirl or quiescent combustion chambers (such as those shown in figures 5.2b and 5.2c) tend to have been used in the larger diesel engines (say above 200 mm bore). With a high enough injection pressure, 1500 bar, a better emissions/fuel consumption trade-off can be obtained, and there is now a trend towards quiescent combustion systems for small direct injection Diesel engines (Frankl *et al.* (1989)).

In the past the maximum speed of small direct injection engines has been limited to about 3000 rpm by combustion speed. High-speed direct injection engines have been developed by meticulous attention during the development of the air/fuel mixing; such an engine is discussed in chapter 14.

When combustion speed is not the limiting factor, there is none the less a limitation on the maximum mean piston speed ($\bar{v}_p$), that limits the engine speed (N):

$$N = \frac{\bar{v}_p \times 60}{2 \times L} \qquad (5.6)$$

The maximum mean piston speed is a result of mechanical considerations and the flow through the inlet valve. As the bore, stroke, valve diameter and valve lift are all geometrically linked, then as the mean piston speed increases, so does the air velocity past the inlet valve. This is discussed further in chapter 6, section 6.3, but 12 m/s is a typical maximum mean piston speed. For an engine speed of 3000 rpm this corresponds to a stroke of 0.12 m, or about a swept volume of 1 litre per cylinder. Traditionally the combustion speed has limited direct injection engines to 3000 rpm, so for engines of under one litre per cylinder swept volume, higher speeds could only be obtained by using the faster indirect injection combustion system. With careful development of the combustion system, direct injection engines can now operate at speeds of up to 5000 rpm.

5.3   Indirect injection (IDI) systems

Indirect injection systems have a divided combustion chamber, with some form of pre-chamber in which the fuel is injected, and a main chamber with the piston and valves. The purpose of a divided combustion chamber is to speed up the combustion process, in order to increase the engine output by increasing engine speed. There are two principal classes of this combustion system; pre-combustion chamber and swirl chamber. Pre-combustion chambers rely on turbulence to increase combustion speed and swirl chambers (which strictly are also pre-combustion chambers) rely on an ordered air motion to raise combustion speed. Howarth (1966) illustrates a range of combustion chambers of both types, see figure 5.6.

Pre-combustion chambers are not widely used, but a notable exception is Mercedes-Benz. The development of a pre-combustion chamber to meet emissions legislation is described by Fortnagel (1990), who discusses the influence of the pre-chamber geometry and injection parameters on emissions and fuel consumption.

Both types of combustion chamber use heat-resistant inserts with a low thermal conductivity. The insert is quickly heated up by the combustion process, and then helps to reduce ignition delay. These combustion chambers are much less demanding on the fuel injection equipment. The fuel is injected and impinges on the combustion chamber insert, the jet breaks up and the fuel evaporates. During initial combustion the burning air/fuel mixture is ejected into the main chamber, so generating a lot of turbulence. This ensures rapid combustion in the main chamber without having to provide an ordered air motion during the induction stroke. Since these systems are very effective at mixing air and fuel, a large fraction of the air can be utilised, so giving a high bmep with low emissions of smoke.

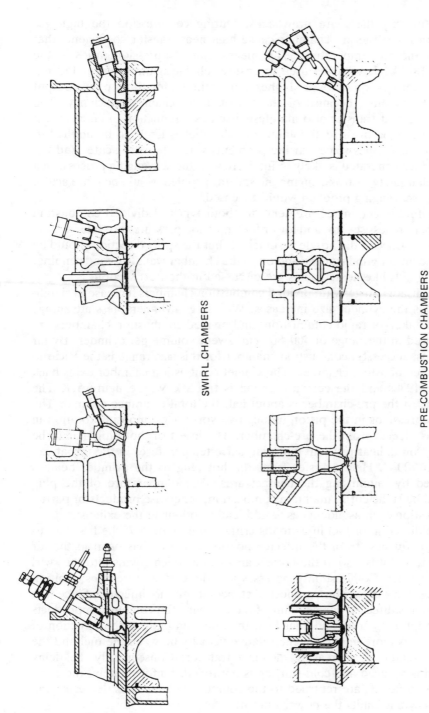

SWIRL CHAMBERS

PRE-COMBUSTION CHAMBERS

Figure 5.6 Different types of pre-combustion and swirl combustion chambers (reproduced from Howarth (1966), courtesy of M. H. Howarth Atlantic Research Associates)

Unfortunately there are drawbacks. During compression the high gas velocities into the pre-chamber cause high heat transfer coefficients that reduce the air temperature. This means that compression ratios in the range 18–24:1 have to be used to ensure reliable ignition when starting. These compression ratios are higher than optimum for either power output or fuel economy, because of the fall-off in mechanical efficiency. The increased heat transfer also manifests itself as a reduction in efficiency.

Neither type of divided combustion chamber is likely to be applied to two-stroke engines since starting problems would be very acute, and the turbulence generated is likely to interfere with the scavenging process. In a turbocharged two-stroke engine the starting problem would be more acute, but the scavenging problem would be eased.

The fuel injection requirements for both types of divided combustion chamber are less severe, and lower fuel injection pressures are satisfactory. A single orifice in the nozzle is sufficient, but the spray direction should be into the air for good starting and on to the chamber walls for good running. Starting aids like heater plugs are discussed in the next section.

The disadvantages with divided combustion chambers increase in significance as the cylinder size increases. With large cylinders, less advantage can be taken of rapid combustion, and divided combustion chambers are only used in the range of 400–800 cm³ swept volume per cylinder. By far the most successful combustion chamber for this size range is the Ricardo Comet combustion chamber. The Comet combustion chamber dates back to the 1930s and the current version is the Mk V, see figure 5.7. The volume of the pre-chamber is about half the total clearance volume. The two depressions in the piston induce two vortices of opposing rotation in the gas ejected from the pre-chamber. The insert or 'hot plug' has to be made from a heat-resistant material, since temperatures in the throat can rise to 700°C. Heat transfer from the hot plug to the cylinder head is reduced by minimising the contact area. The temperature of the plug should be sufficient to maintain combustion, otherwise products of partial combustion such as aldehydes would lead to odour in the exhaust.

The direction of fuel injection is critical, see figure 5.7; the first fuel to ignite is furthest from the injector nozzle. This fuel has been in the air longest, and it is also in the hottest air — that which comes into the swirl chamber last. Combustion progresses and the temperature rises; ignition spreads back to within a short distance from the injector. Since the injection is directed downstream of the air swirl, the combustion products are swept away from and ahead of, the injection path. If the direction of injection is more upstream, the relative velocity between the fuel and the air is greater, so increasing the heat transfer. Consequently the delay period is reduced and cold starting is improved. Unfortunately, the combustion products are returned to the combustion zone; this decreases the efficiency and limits the power output.

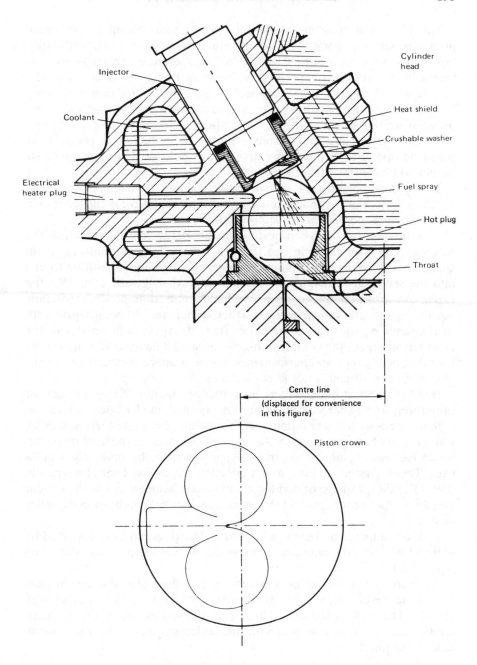

Figure 5.7   Ricardo Comet Mk V combustion chamber (reproduced with permission from Ricardo and Hempson (1968))

The high heat transfer coefficients in the swirl chamber can cause problems with injectors; if the temperature rises above 140°C carbonisation of the fuel can occur. The injector temperature can be limited by increasing the cooling or, preferably, by using a heat shield to reduce the heat flow to the injector. Typical performance figures are shown in figure 5.8, which includes part load results at constant speed. As with all compression ignition engines, the power output is limited by the fuelling rate that causes just visible (jv) exhaust smoke. The reduction in economy at part load operation is much less than for a spark ignition engine with its output controlled by throttling.

The Comet combustion system is well suited to engines with twin overhead valves per cylinder. If a four-valve arrangement is chosen then a pre-chamber is perhaps more appropriate.

A final type of divided combustion chamber is the air cell, of which the MAN air cell is an example, see figure 5.9. Fuel is injected into the main chamber and ignites. As combustion proceeds, fuel and air will be forced into the secondary chamber or air cell, so producing turbulence. As the expansion stroke continues the air, fuel and combustion products will flow out of the air cell, so generating further turbulence. In comparison with swirl chambers, starting will be easier since the spray is directed into the main chamber. As the combustion-generated turbulence and swirl will be less, the speed range and performance will be more restricted than in swirl chambers; the air cell is not in common use.

Divided combustion chambers have reduced ignition delay, greater air utilisation, and faster combustion; this permits small engines to run at higher speeds with larger outputs. Alternatively, for a given ignition delay lower quality fuels can be used. As engine size increases the limit on piston speed reduces engine speed, and the ignition delay becomes less significant. Thus, in large engines, direct injection can be used with low-quality fuels. The disadvantage of divided combustion chambers is a 5–15 per cent penalty in fuel economy, and the more complicated combustion chamber design.

A direct comparison between DI and IDI engines has been reported by Hahn (1986), and the differences in specific fuel consumption are shown by figure 5.10.

As energy costs rise the greater economy of direct injection engines has led to the development of small direct injection engines to run at high speeds. The fuel economy of high compression ratio, leanburn spark ignition engines is comparable with that of indirect injection compression ignition engines.

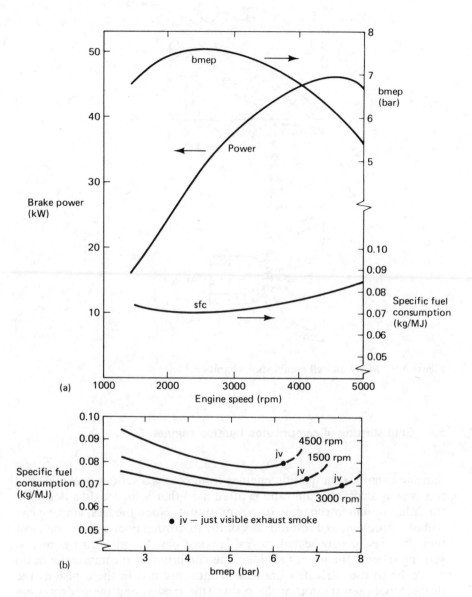

Figure 5.8 Typical performance for a 2 litre engine with Comet combustion chamber. (a) Full load; (b) part load at different speeds (adapted from Howarth (1966))

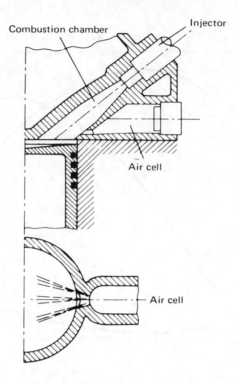

Figure 5.9   MAN air cell combustion chamber

5.4   Cold starting of compression ignition engines

Starting compression ignition engines from cold is a serious problem. For this reason a compression ratio is often used that is higher than desirable for either optimum economy or power output. None the less, starting can still be a problem, as a result of any of the following: poor-quality fuel, low temperatures, poorly seated valves, leakage past the piston rings or low starting speed. One way of avoiding the compromise in compression ratio would be to use variable compression ratio pistons. In these pistons the distance between the top of the piston (the crown) and the gudgeon pin (little end) can be varied hydraulically. So far these pistons have not been widely used. Ignition in a compression ignition engine relies on both a high temperature and a high pressure. The fuel has to evaporate and reach its self-ignition temperature with sufficient margin to reduce ignition delay to an acceptable level. The high pressure brings air and fuel molecules into more intimate contact, so improving the heat transfer. With too long an

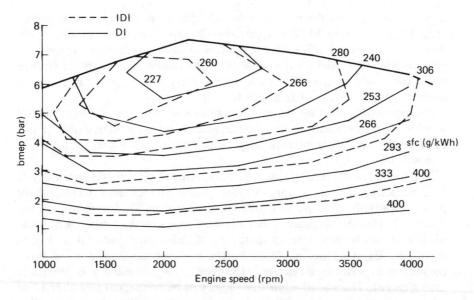

Figure 5.10 Comparison of specific fuel consumption (g/kWh) maps for four-cylinder 2.5 litre naturally aspirated DI and IDI engines (adapted from Hahn (1986))

ignition delay the expansion stroke will have started before ignition is established.

While at low turning speeds there is greater time for ignition to be established, the peak pressures and temperatures are much reduced. The peak pressure is reduced by greater leakage past the valves and piston rings. The peak temperatures are reduced by the greater time for heat transfer. The situation is worst in divided chamber engines, since the air passes through a cold throat at a high velocity before the fuel is injected. The high velocity causes a small pressure drop, but a large temperature drop owing to the high heat transfer coefficient in the throat. The starting speed is not simply the mean engine speed; of greatest significance is the speed as a piston reaches the end of its compression stroke. For this reason it is usual to fit a large flywheel; starting systems cannot usually turn engines at the optimum starting speed. Sometimes a decompression lever is fitted, which opens the valves until the engine is being turned at a particular starting speed.

Various aids for starting can be fitted to the fuel injection system: excess fuel injection, late injection timing, and extra nozzles in the injector. Excess fuel injection is beneficial for several reasons: its bulk raises the compression ratio, any unburnt fuel helps to seal the piston rings and valves, and extra fuel increases the probability of combustion starting. It is essential to have an interlock to prevent the excess fuel injection being used during normal operation; although this would increase the power output the smoke would be unacceptable. Retarded injection timing means that fuel is injected when the temperature and pressure are higher. In systems where the fuel spray impinges on the combustion chamber surface it is particularly beneficial to have an auxiliary nozzle in the injector, in order to direct a spray of fuel into the air.

An alternative starting aid is to introduce with the air a volatile liquid which self-ignites readily. Ether (diethyl ether) is very effective since it also burns over a very wide mixture range; self-igniton occurs with compression ratios as low as 3.8:1.

The final type of starting aids is heaters. Air can be heated electrically or by a burner prior to induction. A more usual arrangement is to use heater plugs, especially in divided combustion chambers — for example, see figure 5.7. Heater plugs are either exposed loops of thick resistance wire, or finer multi-turned wire insulated by a refractory material and then sheathed. Exposed heater plugs had a low resistance and were often connected in series to a battery. However, a single failure was obviously inconvenient. The sheathed heater plugs can be connected to either 12 or 24 volt supplies through earth and are more robust. With electric starting the heater plugs are switched on prior to cranking the engine. Otherwise the power used in starting would prevent the heater plugs reaching the designed temperature. Ignition occurs when fuel is sprayed on to the surface of the heater; heater plugs do not act by raising the bulk air temperature. With electrical starting systems, low temperatures also reduce the battery performance, so adding further to any starting difficulties. A final possibility would be the use of the high-voltage surface discharge plugs used in gas turbine combustion chambers. Despite a lower energy requirement their use has not been adopted.

In multi-cylinder engines, production tolerances will give rise to variations in compression ratio between the cylinders. However, if one cylinder starts to fire, that is usually sufficient to raise the engine speed sufficiently for the remaining cylinders to fire.

The differences in the starting performance of DI and IDI Diesel engines have been reported in detail by Biddulph and Lyn (1966). They found that the increase in compression temperature once IDI Diesel engines had started was much greater than the increase in compression temperature with DI Diesel engines. Biddulph and Lyn concluded that this was a consequence of the higher levels of heat transfer that occur in IDI Diesel engines, and the greater time available for heat transfer at cranking speeds. They also pointed out that for self-ignition to occur in Diesel engines, then a combination of sufficient time and temperature is required. With IDI engines that have started firing on one cylinder, then the reduction in time available for ignition is more than compensated for by the rise in compression temperature. Thus once one cylinder fires, then all the cylinders should fire at the idling speed. Biddulph and Lyn found that this was not the case for DI Diesel engines. The smaller increase in compression temperature did not necessarily compensate for the reduced time available for self-ignition, and it was possible under marginal starting conditions, for an engine to reach idling speeds with some cylinders misfiring, and for these cylinders not to fire until the engine coolant had warmed-up. Such behaviour is obviously undesirable, and is characterised by the emission of unburned fuel as white smoke.

In turbocharged engines, the compression ratio is often reduced to limit peak pressures. This obviously has a detrimental effect on the starting performance.

5.5 Fuel injection equipment

5.5.1 Injection system overview

A typical fuel injection system is shown in figure 5.11. In general the fuel tank is below the injector pump level, and the lift pump provides a constant-pressure fuel supply (at about 0.75 bar) to the injector pump. The secondary fuel filter contains the pressure-regulating valve, and the fuel bleed also removes any air from the fuel. If air is drawn into the injection pump it cannot provide the correctly metered amount of fuel. It is essential to remove any water or other impurities from the fuel because of the fine clearances in the injection pump and injector. The injection pump contains

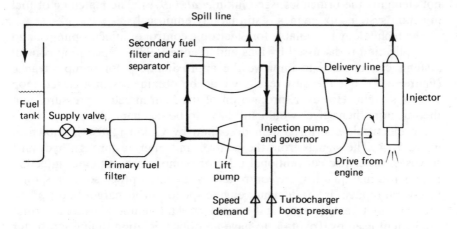

Figure 5.11   Fuel injection system for a compression ignition engine

a governor to control the engine speed. Without a governor the idle speed would vary and the engine could over-speed when the load on the engine is reduced.

The injection pump is directly coupled to the engine (half engine speed for a four-stroke engine) and the pump controls the quantity and timing of the fuel injection (figure 5.12). The quantity of fuel injected will depend on the engine load (figure 5.13). The maximum quantity of fuel that can be injected before the exhaust becomes smoky will vary with speed, and in the case of a turbocharged engine it will also vary with the boost pressure. The injection timing should vary with engine speed, and also load under some circumstances. As the engine speed increases, injection timing should be advanced to allow for the nearly constant ignition delay. As load increases, the ignition delay would reduce for a fixed injection timing, or for the same ignition delay the fuel can be injected earlier. This ensures that injection and combustion are complete earlier in the expansion stroke. In engines that have the injection advance limited by the maximum permissible cylinder pressure, the injection timing can be advanced as the load reduces. Injection timing should be accurate to 1° crank angle.

In multi-cylinder engines equal amounts of fuel should be injected to all cylinders. At maximum load the variation between cylinders should be no more than 3 per cent, otherwise the output of the engine will be limited by the first cylinder to produce a smoky exhaust. Under idling conditions the inter-cylinder variation can be larger (up to 15 per cent), but the quantities of fuel injected can be as low as 1 mm^3 per cycle. Ricardo and Hempson (1968) and Taylor (1985b) provide an introduction to fuel injection equipment, but a much more comprehensive treatment is given by Judge (1967), including detailed descriptions of different manufacturers' equipment. More recent developments include reducing the size of components, but not changing the principles, see Glikin et al. (1979). The matching of fuel injection systems to engines is still largely empirical.

The technology is available for electronic control of injector pumps, and its application is discussed by Ives and Trenne (1981). Open loop control systems can be used to improve the approximations for pump advance (figure 5.13), but the best results would be obtained with a closed loop control system. However, production of the high injection pressures and the injectors themselves would still have to be mechanical.

An electronic control system described by Glikin (1985) is shown in figure 5.14. The signals from the control unit operate through hydraulic servos in otherwise conventional injector pumps. The optimum injection timing and fuel quantity are controlled by the microprocessor in response to several inputs: driver demand, engine speed, turbocharger boost pressure, air inlet temperature and engine coolant temperature. A further refinement used by Toyota is to have an optical ignition timing sensor for providing feedback; this is described by Yamaguchi (1986). The ignition

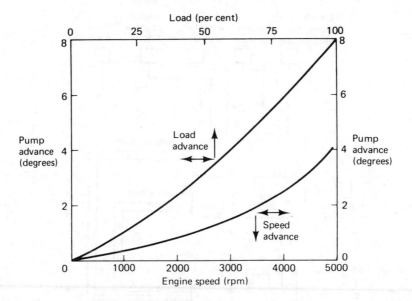

Figure 5.12   Typical fuel injection advance with speed and load

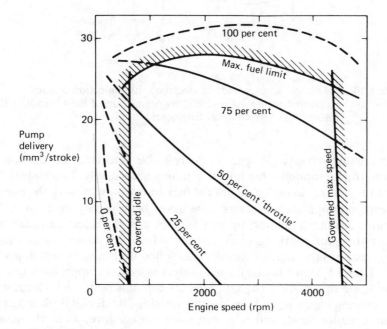

Figure 5.13   Fuel delivery map

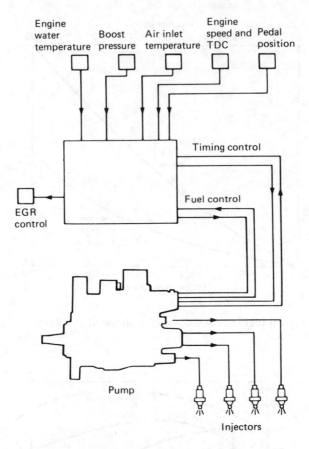

Figure 5.14   Schematic arrangement of electronic fuel injection control
(reprinted from Glikin (1985) by permission of the Council of the
Institution of Mechanical Engineers)

sensor enables changes in ignition delay to be detected, and the control
unit can then re-optimise the injection timing accordingly. This system thus
allows for changes in the quality of the fuel and the condition of the engine.

Electrically operated injectors have never been widely used, since fuel
pressures are much greater, up to 1000 bar, and the injection duration is
shorter than for spark ignition engines. However, this situation might
change as a result of an electrically controlled unit injector developed by
Lucas Diesel Systems for high-speed direct injection compression ignition
engines, see figure 5.15. The unit injector contains both the high-pressure
fuel-pumping element, and the injector nozzle. The device is placed in the
engine cylinder head, and is driven via a rocker lever from the engine

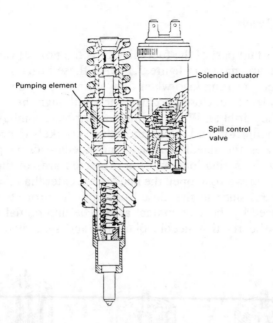

Figure 5.15   Unit injector cross-section (adapted from Frankl *et al.* (1989))

camshaft. The quantity and timing of injection are both controlled electronically through a Colenoid actuator. The Colenoid is a solenoid of
patented construction that can respond very quickly (injection periods are
typically 1 ms), to control the very high injection pressures (up to 1600
bar).

The design and performance of the Colenoid controlled unit injector is
presented in detail by Frankl *et al.* (1989). The Colenoid controls the spill
valve, and is operated at 90 V, so as to reduce the wire gauge of the
windings yet still have low resistive losses. The pumping plunger displaces
diesel fuel, and as soon as the spill valve is closed, then the fuel pressure
builds up. Injection is ended by the spill valve opening again; the injection
pressures can be as high as 1500 bar. The speed of response is such that
pilot injection can be employed, to limit the amount of fuel injected during
the ignition delay period. Frankl *et al.* point out that timing consistencies of
about 5 microseconds are required; this corresponds to the time in which
about 0.5 mm^3 of fuel is injected. They also note that all units are checked
to ensure a ± 4 per cent tolerance in fuelling level across the operating
range, and that a major source of variation is attributable to the flow
tolerance of the nozzle tip.

5.5.2   Fuel injectors

The most important part of the fuel injector is the nozzle; various types of injector nozzle are shown in figure 5.16. All these nozzles have a needle that closes under a spring load when they are not spraying. Open nozzles are used much less than closed nozzles since, although they are less prone to blockage, they dribble. When an injector dribbles, combustion deposits build up on the injector, and the engine exhaust is likely to become smoky. In closed nozzles the needle-opening and needle-closing pressures are determined by the spring load and the projected area of the needle, see figure 5.17. The pressure to open the needle is greater than that required to maintain it open, since in the closed position the projected area of the needle is reduced by the seat contact area. The differential pressures are controlled by the relative needle diameter and seat diameter. A high

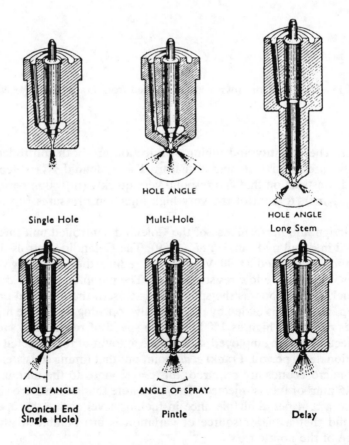

Figure 5.16   Various types of injector nozzle (courtesy of Lucas Diesel Systems)

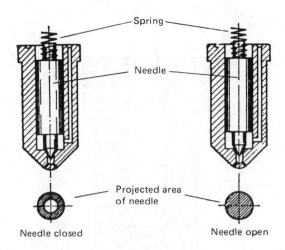

Figure 5.17   Differential action of the injector needle (from Judge (1967))

needle-closing pressure is desirable, since this keeps the nozzle holes free from blockages caused by combustion deposits. A high needle-closing pressure is also desirable since it maintains a high seat pressure, so giving a better seal. In automotive applications the nozzles are typically about 20 mm in diameter, 45 mm long, with 4 mm diameter needles.

The pintle nozzle, figure 5.18, has a needle or nozzle valve with a pin projecting through the nozzle hole. The shape of the pin controls the spray pattern and the fuel-delivery characteristics. If the pin is stepped, a small quantity of fuel is injected initially and the greater part later. Like all single-hole nozzles the pintle nozzle is less prone to blockage than a multi-hole nozzle. The Pintaux nozzle (PINTle with AUXiliary hole) injector (figure 5.19) was developed by Ricardo and CAV for improved cold starting in indirect injection engines. The spray from the auxiliary hole is directed away from the combustion chamber walls. At the very low speeds when the engine is being started, the delivery rate from the injector pump is low. The pressure rise will lift the needle from its seat, but the delivery rate is low enough to be dissipated through the auxiliary hole without increasing the pressure sufficiently to open the main hole. Once the engine starts, the increased fuel flow rate will cause the needle to lift further, and an increasing amount of fuel flows through the main hole as the engine speed increases.

In all nozzles the fuel flow helps to cool the nozzle. Leakage past the needle is minimised by the very accurate fit of the needle in the nozzle. A complete injector is shown in figure 5.20. The pre-load on the needle or nozzle valve from the compression spring is controlled by the compression screw. The cap nut locks the compression screw, and provides a connection

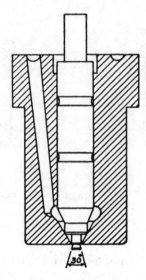

Figure 5.18   Enlarged view of a pintle nozzle (from Judge (1967))

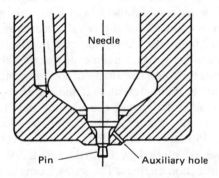

Figure 5.19   The Ricardo–CAV Pintaux nozzle (from Judge (1967))

to the spill line, in order to return any fuel that has leaked past the needle.

The spray pattern from the injector is very important, and high-speed combustion photography is very informative. The combustion can either be in an engine or in special combustion rig. Another approach is to use a water analogue model; here the larger dimensions and longer time scale make observations easier. The aim of these experiments is to develop computer models that can predict spray properties, in particular the spray penetration; such work is reported by Packer *et al*. (1983).

Increasing the injection pressure increases the spray penetration but above a certain pressure the spray becomes finely atomised and has insufficient momentum to penetrate as far. The aspect ratio of the nozzle

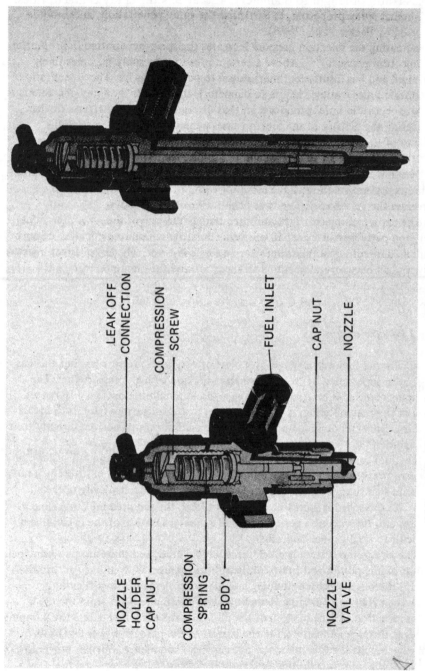

LEAK OFF
CONNECTION

COMPRESSION
SCREW

FUEL INLET

CAP NUT

NOZZLE

NOZZLE
HOLDER
CAP NUT

COMPRESSION
SPRING

BODY

NOZZLE
VALVE

Figure 5.20 Complete fuel injectors (courtesy of Lucas Diesel Systems)

holes (ratio of length to diameter), also affects the spray characteristics. The mass flow rate ($\dot{m}_f$) of fuel through the injector can be modelled by

$$\dot{m}_f = C_d A_n \sqrt{(2\rho_f \Delta p)} \tag{5.7}$$

where C_d = discharge coefficient
 A_n = nozzle flow area per hole (n holes)
 ρ_f = fuel density
 Δp = the pressure drop across the nozzle
 n = number of nozzle holes.

The nozzle hole diameters (d_n) are typically in the range of 0.2–1 mm with aspect ratios (length/diameter) from 2 to 8. Increasing the aspect ratio produces a jet that diverges less and penetrates further. The differential pressure, which is invariably greater than 300 bar, produces a high velocity jet (typically faster than 250 m/s) that becomes turbulent. The jet spreads out as a result of entraining air, and breaks up into droplets, as the jet diverges the spray velocity decreases. Two important spray parameters are: the droplet size distribution, and the spray penetration distance (S). Heywood (1988) recommends a correlation by Dent:

$$S = 3.07 \left(\frac{p}{\rho_g}\right)^{1/4} (td_n)^{1/2} \left(\frac{294}{T_g}\right)^{1/4} \tag{5.8}$$

where T_g = gas temperature (K)
 ρ_g = gas density (kg/m^3)
 t = time after the start of injection.

The fuel jet breaks-up as a result of surface waves, and it can be argued that the initial average drop diameter (D_d) is proportional to the wavelength of the most unstable surface waves:

$$D_d = C \frac{2\pi\sigma}{\rho_g v_j^2} \tag{5.9}$$

where σ = liquid-fuel surface tension (N/m)
 v_j = jet velocity (m/s)
 C is a constant in the range 3–4.5.

In practice this is a simplification, since the droplets can break-up further, and also coalesce. A convenient way of characterising a fuel spray is the Sauter mean diameter (SMD). This is the droplet size of a hypothetical spray of uniform size, such that the volume and overall volume to surface area ratio is the same

$$SMD = \Sigma\ (nD^3)/\Sigma(nD^2) \tag{5.10}$$

where n is the number of droplets in the size group of mean diameter D.

Heywood (1988) presents a useful discussion on droplet size distributions and correlations to predict droplet sizes.

An attempt to base the design process for DI engines on a more rational basis has been made by Timoney (1985). In this approach, the fuel spray trajectory during injection, within the swirling air flow, is estimated from the laws of motion and semi-empirical correlations describing the air drag forces on the spray. From the spray trajectory and the air velocity, the relative velocity between the fuel spray and the swirling air is calculated in the tangential direction. This calculation is performed at the moment that the tip of the spray impinges on the walls of the piston bowl. Timoney provided some evidence from engine tests which showed that the magnitude of the tangential relative velocity (called the crosswind velocity) correlates well with the specific fuel consumption and smoke emissions.

To summarise, increasing the density of the fuel increases penetration, but increasing the density of the air reduces jet penetration. Also, denser fuels are more viscous and this causes the jet to diverge and to atomise less. The jet penetration is also increased with increasing engine speed. The injection period occupies an approximately constant fraction of the cycle for a given load, so as engine speed increases the jet velocity (and thus penetration) also increases.

The choice of injector type and the number and size of the holes are critical for good performance under all operating conditions. Indirect injection engines and small direct injection engines have a single injector. The larger direct injection engines can have several injectors arranged around the circumference of the cylinder.

5.5.3 Injection pumps

Originally the fuel was injected by a blast of very high-pressure air, but this has long been superseded by 'solid' or airless injection of high-pressure fuel. The pumping element is invariably a piston/cylinder combination; the differences arise in the fuel metering. A possibility, discussed already (pages 202–3), is to have a unit injector — a combined pump and injector. The pump is driven directly from the camshaft so that it is more difficult to vary the timing.

The fuel can be metered at a high pressure, as in the common rail system, or at a low pressure, as in the jerk pump system. In the common rail system, a high-pressure fuel supply to the injector is controlled by a mechanically operated valve. As the speed is increased at constant load, the required injection time increases and the injection period occupies a greater fraction of the cycle; this is difficult to arrange mechanically.

The jerk pump system is much more widely used, and there are two principal types: in-line pumps and rotary or distributor pumps. With in-line (or 'camshaft') pumps there is a separate pumping and metering element for each cylinder.

A typical in-line pumping element is shown in figure 5.21, from the Lucas Minimec pump. At the bottom of the plunger stroke (a), fuel enters the pumping element through an inlet port in the barrel. As the plunger moves up (b) its leading edge blocks the fuel inlet, and pumping can commence. Further movement of the plunger pressurises the fuel, and the delivery valve opens and fuel flows towards the injector. The stroke or lift of the plunger is constant and is determined by the lift of the cam. The quantity of fuel delivered is controlled by the part of the stroke that is used for pumping. By rotating the plunger, the position at which the spill groove uncovers the spill port can be changed (c) and this varies the pumping stroke. The spill groove is connected to an axial hole in the plunger, and fuel flows back through the spill port to the fuel gallery. The rotation of the plunger is controlled by a lever, which is connected to a control rod. The control rod is actuated by the governor and throttle. An alternative arrangement is to have a rack instead of the control rod, and this engages with gears on each plunger, see figure 5.22.

The delivery valve is shown more clearly in figure 5.21. In addition to acting as a non-return valve it also partially depressurises the delivery pipe to the injector. This enables the injector needle to snap on to its seat, thus preventing the injector dribbling. The delivery valve has a cylindrical section that acts as a piston in the barrel before the valve seats on its conical face; this depressurises the fuel delivery line when the pumping element stops pumping. The effect of the delivery valve is shown in figure 5.23.

Owing to the high pumping pressures the contact stresses on the cam are very high, and a roller type cam follower is used. The fuel pumping is arranged to coincide with the early part of the piston travel while it is accelerating. The spring decelerates the piston at the end of the upstroke, and accelerates the piston at the beginning of the downstroke. The cam profile is carefully designed to avoid the cam follower bouncing.

At high speed the injection time reduces, and the injection pressures will be greater. If accurate fuel metering is to be achieved under all conditions, leakage from the pump element has to be minimal.

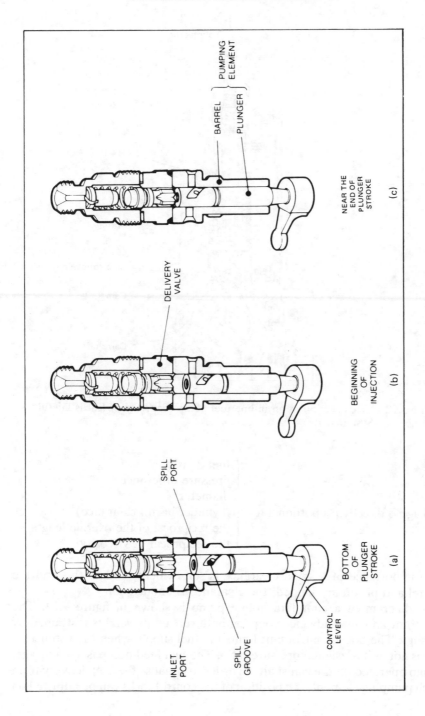

Figure 5.21 Fuel-pumping element from an in-line pump (courtesy of Lucas Diesel Systems)

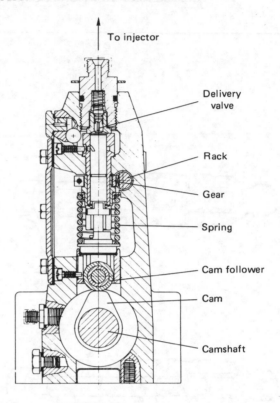

Figure 5.22   Cross-section of an in-line fuel pump (courtesy of Lucas Diesel Systems)

leakage is directly proportional to $\begin{cases} \text{fuel density} \\ \text{pressure difference} \\ \text{diameter} \\ \text{(cylinder/barrel clearance)}^3 \\ \text{the reciprocal of the overlap length} \\ \text{the reciprocal of viscosity} \end{cases}$

The importance of a small clearance is self-evident, and to this end the barrel and piston are lapped; the clearance is about 1 μm.

A diagram of a complete in-line pump is shown in figure 5.24. The governor and auto-advance coupling both rely on flyweights restrained by springs. The boost control unit limits the fuel supply when the turbocharger is not at its designed pressure ratio. The fuel feed pump is a diaphragm pump operated off the camshaft. In order to equalise the fuel delivery from each pumping element, the position of the control forks can be adjusted on

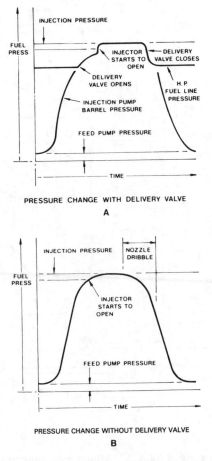

Figure 5.23   Effect of the injector delivery valve (courtesy of Lucas Diesel
              Systems)

the control rod. The control forks engage on the levers that control the
rotation of the plunger.

Rotary or distributor pumps have a single pumping element and a single
fuel-metering element. The delivery to the appropriate injector is controlled by a rotor. Such units are more compact and cheaper than an in-line
pump with several pumping and metering elements. Calibration problems
are avoided, and there are fewer moving parts. However, rotary pumps
cannot achieve the same injection pressures as in-line pumps, and only
recently have they been developed for DI engines.

The fuel system for a rotary pump is shown in figure 5.25, and figure 5.26
shows the details of the high-pressure pump and rotor for a six-cylinder

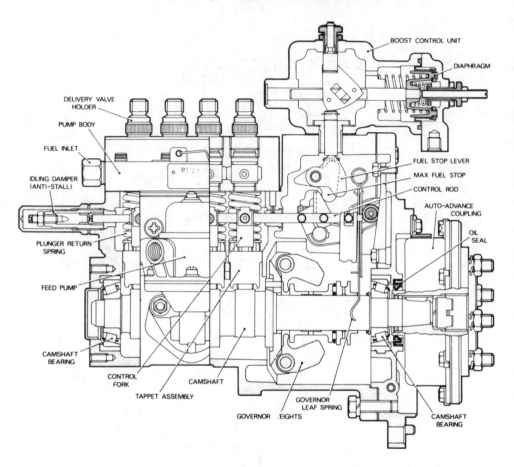

Figure 5.24 Sectional view of the CAV Minimec fuel pump (courtesy of Lucas
Diesel Systems)

engine. The transfer pump is a sliding vane pump situated at the end of the
rotor. The pump output is proportional to the rotor speed, and the transfer
pressure is maintained constant by the regulating valve. The metering
value is regulated by the governor, and controls the quantity of fuel that
flows to the rotor through the metering port, at the metering pressure.
Referring to figure 5.26, the pump plungers that produce the injection
pressures rotate in barrels in the rotor. The motion of the plungers comes
from a stationary cam with six internal lobes. The phasing is such that the
plungers move out when a charging port coincides with the fuel inlet; as the
rotor moves round, the fuel is isolated from the inlet. Further rotation
causes the distributor port to coincide with an outlet, and the fuel is
compressed by the inward movement of the pump plungers.

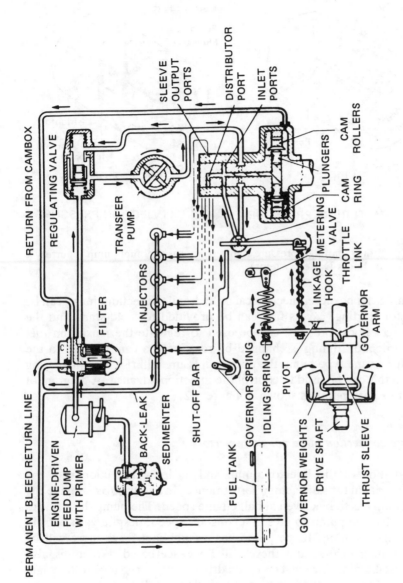

Figure 5.25 Rotary pump fuel system (with acknowledgement to Newton *et al.* (1983))

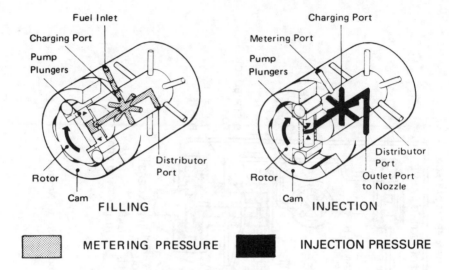

Figure 5.26   Rotor and high-pressure pump from a rotary fuel pump (courtesy of Lucas Diesel Systems)

Governing can be either mechanical or hydraulic, injection timing can be retarded for starting, excess fuel can be provided for cold start, and the turbo charger boost pressure can be used to regulate the maximum fuel delivery. Injection timing is changed by rotating the cam relative to the rotor. With a vane type pump the output pressure will rise with increasing speed, and this can be used to control the injection advance. A diagram of a complete rotary pump is shown in figure 5.27.

5.5.4   Interconnection of pumps and injectors

The installation of the injector pump and its interconnection with the injectors is critical for satisfactory performance. In a four-stroke engine the injector pump has to be driven at half engine speed. The pump drive has to be stiff so that the pump rotates at a constant speed, despite variations in torque during rotation. If the pump has a compliant drive there will be injection timing errors, and these will be exacerbated if there is also torsional oscillation. The most common drive is either a gear or roller chain system. Reinforced toothed belts can be used for smaller engines, but even so they have to be wider than gear or roller chain drives. The drives usually include some adjustment for the static injection timing.

The behaviour of the complete injection system is influenced by the compressibility effects of the fuel, and the length and size of the interconnecting pipes.

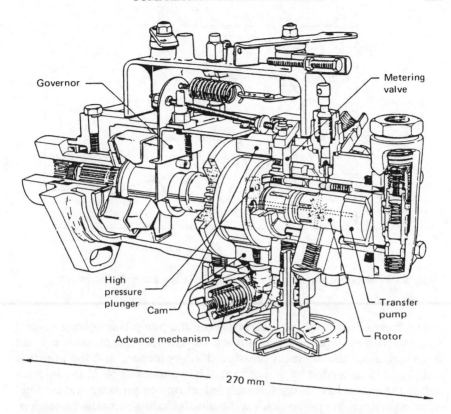

Figure 5.27 Typical rotary or distributor type fuel pump (from Ives and Trenne (1981))

The compressibility of the fuel is such that an increase in pressure of 180 bar causes a 1 per cent volume reduction. Incidentally, a 10 K rise in temperature causes about a 1 per cent increase in volume; since the fuel is metered volumetrically this will lead to a reduction in power output. Since the fuel pipes are thick walled, the change in volume is small compared with the effects due to the compressibility of fuel. Pressure (compression or rarefaction) waves are set up between the pump and the injector, and these travel at the speed of sound. The pressure waves cause pressure variations, which can influence the period of injection, the injection pressure and even cause secondary injection.

After the pump delivery valve has opened there will be a delay of about 1 ms per metre of pipe length before the fuel injection begins. To maintain the same fuel injection lag for all cylinders the fuel pipe lengths should be identical. The compression waves may be wholly or partially reflected back at the nozzle as a compression wave if the nozzle is closed, or as a rarefaction wave if the nozzle is open. During a typical injection period of a

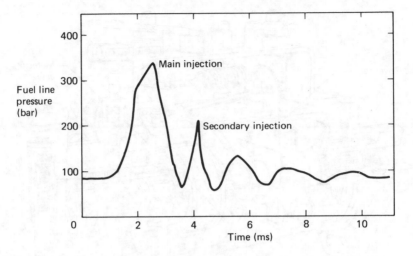

Figure 5.28   Pressure waves in the fuel line from the pump to the injector

few milliseconds, waves will travel between the pump and injector several times; viscosity damps these pressure waves. The fuel-line pressure can rise to several times the injection pressure during ejection, and the injection period can be extended by 50 per cent. The volumes of oil in the injector and at the pump have a considerable effect on the pressure waves. Fuel injection systems are prone to several faults, including secondary injection, and after-dribble.

Pressure wave effects can lead to secondary injection — fuel injected after the main injection has finished. Secondary injection can lead to poor fuel consumption, a smoky exhaust and carbon formation on the injector nozzle. Figure 5.28 shows a fuel-line pressure diagram in which there is a pressure wave after the main injection period that is sufficient to open the injector. Secondary injection can be avoided by increasing the fuel-line length, or changing the volumes of fuel at the pump or injector.

After-dribble is a similar phenomenon; in this case the pressure wave occurs as the injector should be closing. The injector does not fully close, and some fuel will enter the nozzle at too low a pressure to form a proper spray.

Problems with interconnecting pipework are of course eliminated with unit injectors, since the pump and injector are combined. Unit injectors can thus be found on large engines, since particularly long fuel pipes can be eliminated, and the bulk of the unit injector can be accommodated more easily. Techniques exist to predict the full performance of fuel injection systems in order to obtain the desired injection characteristics — see Knight (1960–61).

5.6 Diesel engine emissions

5.6.1 Emissions legislation

The worldwide variations in emissions legislation, and the introduction of stricter controls as technology improves, mean that it is difficult to give comprehensive and up-to-date information on emissions legislation. However, to provide a context in which to judge engine performance, the Western European legislation will be summarised here.

For light duty vehicles, the same test cycle is used as for gasoline powered vehicles, and table 5.1 summarises the requirements. It should be noted that the type approval value has to be met by the vehicle submitted for test, while the production conformity value has to be achieved by a randomly selected vehicle.

Table 5.1   European (ECE-R 15/04) emissions legislation for light duty vehicles (less than 3500 kg)

Emissions	Production conformity value	Type approval value
Carbon monoxide (g/test)	36	30
HC + NO_x (g/test)	10	8
Particulates (g/test)	1.4	1.1

As heavy-duty vehicles tend to be manufactured on a bespoke basis, with the customer specifying the power-train units, then it is impractical to test all possible vehicle combinations. Instead, the engines are submitted to dynamometer based tests, governed by ECE-R 49/01. This test is a series of 13 steady-state operating conditions (known as the 13-mode cycle) in which each operating point is held for 6 minutes. However, only the final minute at each operating is used for evaluating the emissions. Table 5.2 summarises the ECE-R 49/01 13 mode cycle.

Rated speed means the speed at which the maximum power is produced, and intermediate speed means the speed at which maximum torque occurs, if this is 60–75 per cent of the rated speed, otherwise the intermediate speed is set to be 60 per cent of the rated speed. The most significant shortcoming of this test is the lack of any transient, during which emissions might be expected to increase. In the USA a transient test is used that requires a computer controlled transient dynamometer.

Table 5.2   The ECE-R 49/01 13 mode cycle and emissions limits

Mode no.	Engine speed	Per cent load	Mode weighting factor
1	idle	—	0.25/3
2	intermediate	10	0.08
3	intermediate	25	0.08
4	intermediate	50	0.08
5	intermediate	75	0.08
6	intermediate	100	0.25
7	idle	—	0.25/3
8	rated	100	0.10
9	rated	75	0.02
10	rated	50	0.02
11	rated	25	0.02
12	rated	10	0.02
13	idle	—	0.25/3

Emissions	Production conformity value	Type approval value
Carbon monoxide (g/kWh)	12.3	11.2
HC (g/kWh)	2.6	2.4
NO_x (g/kWh)	15.8	14.4

5.6.2   Sources and control of emissions

Carbon monoxide emissions need not be considered here, as a correctly regulated diesel engine always has negligible emissions of carbon monoxide, since the air/fuel ratio is always lean.

The noise from Diesel engines is usually mostly attributable to the combustion noise, which originates from the high rates of pressure rise during the initial rapid combustion. A typical relationship between combustion noise and the peak rate of combustion is shown in figure 5.29. To understand the source of the rapid combustion, and how it can be controlled, it is necessary to consider the fuel injection and combustion processes.

As discussed earlier, the fuel is injected into the combustion chamber towards the end of the compression stroke. The fuel evaporates from each droplet of fuel, and mixes with air to form a flammable mixture. However, combustion does not occur immediately, and during the delay period (between the start of injection and the start of combustion) a flammable mixture is being continuously formed. Consequently, when ignition occurs it does so at many sites, and there is very rapid combustion of the mixture formed during the delay period. This rapid combustion produces the characteristic diesel knock. Evidently, the way to reduce the combustion

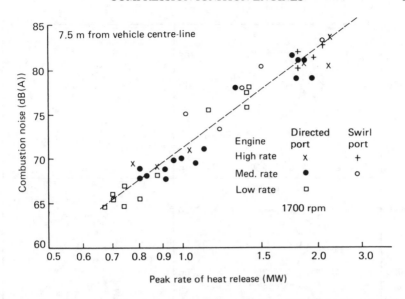

Figure 5.29   Relationship between combustion noise and peak rate of heat release (reprinted from Glikin (1985) by permission of the Council of the Institution of Mechanical Engineers)

noise is either to: reduce the quantity of mixture prepared during the delay period, and this can be achieved by: reducing the initial rate of injection (or by using pilot injection); or more commonly to reduce the duration of the delay period. The delay period is reduced by having:

(a) A higher temperature, which increases both the rate of heat transfer and the chemical reaction rates.
(b) A higher pressure, which increases the rate of heat transfer.
(c) A fuel that spontaneously ignites more readily (a higher cetane umber).

A higher cetane fuel is unlikely to be feasible, since the quality of diesel fuel is currently falling.

Higher temperatures occur in turbocharged engines and low heat-loss engines, but as will be seen later, this also leads to higher NO_x emissions. An alternative approach is to retard the injection timing, so that injection occurs closer to the end of the compression stroke. This leads to an increase in the fuel consumption and other trade-offs that are discussed later in figure 5.30, for a fixed high load.

The trade-off between specific fuel consumption and injection timing for different injection rates is shown in figure 5.30. Ideally, combustion should occur instantaneously at top dead centre. In practice, combustion commences

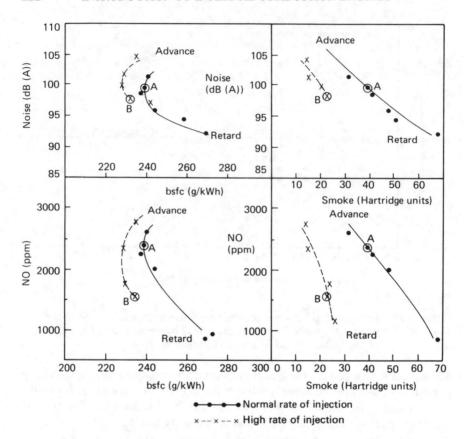

Figure 5.30   Trade-off curves between noise, smoke, NO$_x$ and specific fuel consumption for different rates of injection and injection timing (reprinted from Glikin (1985) by permission of the Council of the Institution of Mechanical Engineers)

before top dead centre and continues afterwards. Advancing the start of injection (and thus combustion) increases the compression work, but the ensuing higher pressures and temperatures at top dead centre also increase the expansion work. However, if the injection timing is advanced too much, the increase in compression work will be greater than the increase in expansion work. Clearly, faster injection leads to more rapid combustion, and this results in less advanced injection timings. There is a rate of injection above which no further gains in fuel consumption occur, and the higher the swirl, the lower this injection rate.

The black smoke from Diesel engines originates from the fuel-rich side of the reaction zone in the diffusion-controlled combustion phase. After the rapid combustion at the end of the delay period, the subsequent combustion of the fuel is controlled by the rates of diffusion of air into the fuel vapour and vice versa, and the diffusion of the combustion products away from the reaction zone. Carbon particles are formed by the thermal decomposition (cracking) of the large hydrocarbon molecules, and the soot particles form by agglomeration. The soot particles can be oxidised when they enter the lean side of the reaction zone, and further oxidation occurs during the expansion stroke, after the end of the diffusion combustion phase.

Smoke generation is increased by high temperatures in the fuel-rich zone during diffusion combustion, and by reductions in the overall air/fuel ratio. The smoke emissions can be reduced by shortening the diffusion combustion phase, since this gives less time for soot formation and more time for soot oxidation. The diffusion phase can be shortened by increased swirl, more rapid injection, and a finer fuel spray. Advancing the injection timing also reduces the smoke emissions. The earlier injection leads to higher temperatures during the expansion stroke, and more time in which oxidation of the soot particles can occur. Unfortunately advancing the injection timing leads to an increase in noise. However, if the injection rate is increased and the timing is retarded, there can be an overall reduction in both noise and smoke. One such combination of points is shown as A and B in figure 5.30, and it can also be seen that the minimum specific fuel consumption has been reduced slightly, and that there is a significant reduction in nitrogen oxide emissions.

The formation of smoke is most strongly dependent on the engine load. As the load increases, more fuel is injected, and this increases the formation of smoke for three reasons:

(a) The duration of diffusion combustion increases.
(b) The combustion temperatures increase.
(c) Less oxidation of the soot occurs during the expansion stroke since there is less time after the end of diffusion combustion, and there is also less oxygen.

On naturally aspirated engines, it is invariably the formation of smoke that limits the engine output.

As explained in chapter 3, section 3.8, the formation of NO_x is strongly dependent on temperature, the local concentration of oxygen and the duration of combustion. Thus in Diesel engines, NO_x is formed during the diffusion combustion phase, on the weak side of the reaction zone. Reducing the diffusion-controlled combustion duration by increasing the rate of injection leads to the reduction in NO_x shown in figure 5.30. Retarding the injection timing also reduces the NO_x emissions, since the later injection leads to lower temperatures, and the strong temperature dependence of the NO_x formation ensures a reduction in NO_x, despite the increase in combustion duration associated with the retarded injection. The trade-off between NO_x and smoke formation is also shown in figure 5.30.

At part load, smoke formation is negligible but NO_x is still produced; figure 5.31a shows that NO_x emissions per unit output in fact increase. In general, the emissions of NO_x are greater from an IDI engine than a DI engine, since the IDI engine has a higher compression ratio, and thus higher combustion temperatures. As with spark ignition engines, exhaust gas recirculation (EGR) at part load is an effective means of reducing NO_x emissions. The inert gases limit the combustion temperatures, thereby reducing NO_x formation. Pischinger and Cartellieri (1972) present results for a DI engine at about half load in which the NO_x emissions were halved by about 25 per cent EGR, but at the expense of a 5 per cent increase in the fuel consumption.

Unburnt hydrocarbons (HC) in a properly regulated Diesel engine come from two sources. Firstly, around the perimeter of the reaction zone there will be a mixture that is too lean to burn, and the longer the delay period, the greater the amount of HC emissions from this source. However, there is a delay period below which no further reductions in HC emissions are obtained. Under these conditions the HC emissions mostly originate from a second source, the fuel retained in the nozzle sac (the space between the nozzle seat and the spray holes) and the spray holes. Fuel from these sources can enter the combustion chamber late in the cycle, thereby producing HC emissions.

Figure 5.31b shows that HC emissions are worse for DI engines than IDI engines, especially at light load, when there is significant ignition delay in DI engines. Advancing the injection timing reduces HC emissions, but this would lead to increased NO_x (figure 5.31b) and noise. The HC emissions increase at part load, the delay period increases, and the quantity of mixture at the perimeter of the reaction zone that is too lean to burn increases. Finally, if a diesel engine is over-fuelled, there would be significant hydrocarbon emissions as well as black smoke.

The final class of emissions discussed here are particulates; these are any substance, apart from water, that can be collected by filtering diluted exhaust at a temperature of 325 K. Since the particulates are either soot or condensed hydrocarbons, any measure that reduces either the exhaust smoke or HC emissions should also reduce the level of particulates.

If additional measures are needed, the particulates can be oxidised by a catalyst incorporated into the exhaust manifold, in the manner described by Enga et al. (1982). However, for a catalyst to perform satisfactorily it has to be operating above its light-off temperature. Since Diesel engines have comparatively cool exhausts, then catalysts do not necessarily attain their light-off temperature. This has led to the development of electrically heated regenerative particulate traps, examples of which are described by Arai and Miyashita (1990), and Garrett (1990).

These particulate traps differ from spark ignition engine catalysts, as the regeneration process does not occur with the exhaust flowing through the trap. Either the exhaust flow is diverted, or the regeneration occurs when

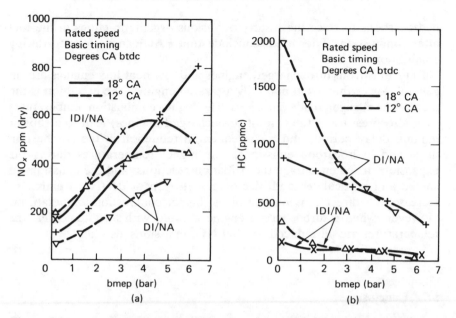

Figure 5.31   A comparison of emissions from naturally aspirated direct (DI) and indirect (IDI) injection Diesel engines. (a) Nitrogen oxide emissions; (b) hydrocarbon emissions.
From Pischinger and Cartellieri (1972). [Reprinted with permission © 1972 Society of Automotive Engineers, Inc.]

the engine is inoperative. Air is drawn into the trap, and electrical heating is used to obtain a temperature high enough for oxidation of the trapped particulate matter.

It must be remembered that the catalysts that might be fitted to Diesel engines are oxidation catalysts, which would not reduce the NO_x emissions. However, before catalyst systems can be considered, it will be necessary to reduce the levels of sulphur in the diesel fuel from the current level of 0.3 per cent by mass (table 3.5), to 0.05 per cent by mass or less. This is because an oxidation catalyst would lead to the formation of sulphur trioxide/sulphuric acid. An additional advantage of using a catalyst, is that it should lead to a reduction in the odour of diesel exhaust.

Concern has also been expressed that the exhaust from Diesel engines might contain carcinogens. Monaghan (1990) argues that the Diesel engine is probably safe in this respect, but the exhaust does contain mutagenic substances. A technique for identifying mutagens is presented by Seizinger et al. (1985), these are usually polynuclear aromatic hydrocarbons (PNAH), also known as polycylic aromatic compounds (PAC).

There has been much debate whether PAC come from unburnt fuel, unburnt lubricating oil or whether PAC can be formed during combustion. Work by Trier et al. (1990) has used radio-labelled (using carbon-14) hydrocarbons to trace the history of various organic fractions in the fuel after combustion. Trier *et al.* conclude that PAC can be formed during combustion.

Emissions from turbocharged engines and low heat loss engines are, in general, lower than from naturally aspirated engines; the exception is the increase in nitrogen oxide emissions. The higher combustion temperatures (and also pressure in the case of turbocharged engines) lead to a shorter ignition delay period, and a consequential reduction in the combustion noise and hydrocarbon emissions. The higher temperatures during the expansion stroke encourage the oxidation reactions, and the emissions of smoke and particulates both decrease. Nitrogen oxide (NO_x) emissions increase as a direct consequence of the higher combustion temperatures. However, when a turbocharged engine is fitted with an intercooler, the temperatures are all reduced, so that NO_x emissions also fall.

5.7 Conclusions

In any type of compression ignition engine it is essential to have properly matched fuel injection and air motion. These requirements are eased in the case of indirect injection engines, since the pre-chamber or swirl chamber produces good mixing of the fuel and air. Since the speed range and the air utilisation are both greater in the indirect injection engine, the output is greater than in direct injection engines. However, with divided combustion chambers there is inevitably a pressure drop and a greater heat transfer, and consequently the efficiency of indirect injection engines is less than the efficiency of direct injection engines. Thus the development of small high-speed direct injection engines is very significant.

The compression ratio of compression ignition engines is often dictated by the starting requirements, and it is likely to be higher than optimum for either maximum fuel economy or power output. This is especially true for indirect injection engines where the compression ratio is likely to be in the range 18–24:1. Even so, additional starting aids are often used with compression ignition engines, notably, excess fuel injection, heaters and special fuels.

Apart from unit injectors there are two main types of injector pump: in-line pumps and rotary or distributor pumps. Rotary pumps are cheaper, but the injection pressures are lower than those of in-line pumps. Thus rotary fuel pumps are better suited to the less-demanding requirements of indirect injection engines. The fuel injectors and nozzles are also critical components, and like the injection pumps they are usually made by specialist manufacturers.

In conclusion, the correct matching of the fuel injection to the air flow is all important. The wide variety of combustion chambers for direct injection engines shows that the actual design is less important than ensuring good mixing of the fuel and air. Since the output of any compression ignition engine is lower than that of a similar-sized spark ignition engine, turbocharging is a very important means of raising the engine output. Furthermore, the engine efficiency is also improved; this and other aspects of turbocharging are discussed in chapter 9.

5.8 Example

Using the data in figure 5.5 estimate the specification for a four-stroke, 240 kW, naturally aspirated, direct injection engine, with a maximum torque of 1200 Nm. Plot graphs of torque, power and fuel consumption against engine speed.

First calculate the total required piston area (A_t) assuming a maximum of 2.0 MW/m^2:

$$A_t = \frac{240 \times 10^3}{2.0 \times 10^6} = 0.12 \text{ m}^2$$

Assuming a maximum bmep ($\bar{p}_b$) of 8×10^5 N/m^2

$$\text{Power} = T\omega = \bar{p}_b L A_t N^*$$

where $\omega = 2N^*.2\pi$ rad/s, and

$$\text{stroke, } L = \frac{T\omega}{\bar{p}_b A_t N^*} = \frac{1200 \times 4\pi N^*}{8 \times 10^5 \times 0.12 \times N^*} = \underline{0.157\text{m}}$$

The number of pistons should be such that the piston diameter is slightly smaller than the stroke. With an initial guess of 8 cylinders:

$$A = \frac{A_t}{8} = \frac{0.12}{8} = 0.015\text{m}^2$$

$$\text{bore} = \sqrt{\left(\frac{4 \times 0.015}{\pi}\right)} = \underline{0.138\text{m}}$$

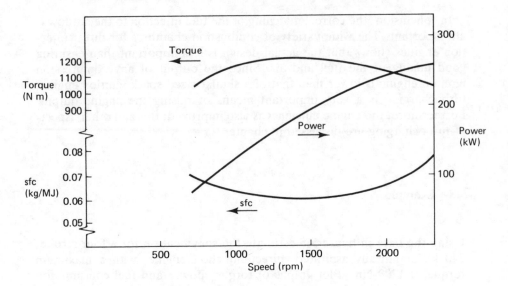

Figure 5.32   Performance curves for engine in example 5.1; 8.8 litre naturally
aspirated direct injection engine

a value that would be quite satisfactory for a stroke of 0.157 m.

$$\text{swept volume} = A_t L = 0.12 \times 0.157 = 18.84 \text{ litres}$$

For a maximum mean piston speed (V_p) of 12 m/s, the corresponding
engine speed is

$$\frac{v_p \times 60}{2L} = \frac{12 \times 60}{2 \times 0.157} = 2293 \text{ rpm}$$

The results from figure 5.5 can now be replotted as shown in figure 5.32.
The final engine specification is in broad agreement with the Rolls Royce
CV8 engine. It should be noted that the power output is controlled by the
total piston area. Increasing the stroke increases the torque, but will
reduce the maximum engine speed, thus giving no gain in power.

5.9 Problems

5.1 Contrast the advantages and disadvantages of indirect and direct
injection compression ignition engines.

5.2 Discuss the problems in starting compression ignition engines, and describe the different starting aids.

5.3 Comment on the differences between in-line and rotary fuel injection pumps.

5.4 Describe the different ways of producing controlled air motion in compression ignition engines.

5.5 An engine manufacturer has decided to change one of his engines from a spark ignition type to a compression ignition type. If the swept volume is unchanged, what effect will the change have on:

(i) maximum torque?
(ii) maximum power?
(iii) the speed at which maximum power occurs?
(iv) economy of operation?

5.6 Show that the efficiency of the Air Standard Diesel Cycle is:

$$\eta_{\text{Diesel}} = 1 - \frac{1}{r_v^{\gamma-1}} \cdot \left[\frac{\alpha^\gamma - 1}{\alpha - 1}\right] \frac{1}{\gamma}$$

(i) By considering the combustion process to be equivalent to the heat input of the Diesel cycle, derive an expression that relates the load ratio (α), to the gravimetric air/fuel ratio (F). Assume that the calorific value of the fuel is 44 MJ/kg, the compression ratio is 15:1, the air inlet temperature is 13°C, and for the temperature range involved, $c_p = 1.25$ kJ/kgK and $\gamma = 1.33$. State clearly any assumptions.

(ii) Calculate the Diesel cycle efficiencies that correspond to gravimetric air/fuel ratios of 60:1 and 20:1, and comment on the significance of the result.

5.7 Some generalised full load design data for naturally aspirated 4-stroke DI diesel engines are tabulated below:

Mean piston speed, $\bar{v}_p$ (m/s)	4	6	8	10	12
Brake specific fuel consumption, bsfc (kg/MJ)	0.068	0.063	0.061	0.064	0.076
Brake mean effective pressure, $\bar{p}_b$ (bar)	7.24	8.00	8.08	7.70	6.93
Volumetric efficiency, η_{vol} (per cent)	89.8	88.3	82.1	77.7	76.2

Ambient conditions: $p = 105$ kN/m², $T = 17$°C.

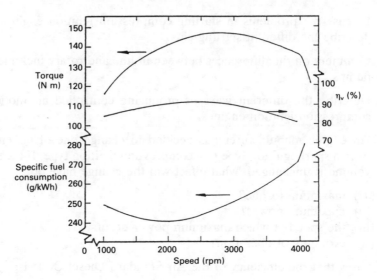

Figure 5.33

Complete the table by calculating: the brake specific power, bsp (MW/m^2), the air/fuel ratio, and the brake specific air consumption bsac (kg/MJ). State briefly what is the significance of squish and swirl in DI Diesel combustion. Why does the air/fuel ratio fall as the piston speed increases, what is the significance of the brake specific air consumption, and why is it a minimum at intermediate piston speeds? (It may be helpful to plot the data.)

5.8 A direct injection Diesel engine with a swept volume of 2.5 litres has the full load performance shown in figure 5.33. What is the maximum power output of the engine, and the corresponding air/fuel ratio?

For both the maximum power and the maximum torque, calculate the brake mean effective pressure. What is the maximum brake efficiency?

The fuel has a calorific value of 44 MJ/kg, and the ambient conditions are a temperature of 17°C and a pressure of 1.02 bar.

Identify the differences between direct (DI) and indirect injection (IDI) Diesel engines, and explain why the efficiency of the DI engine is higher. Why is IDI still used for the smallest high-speed diesel engines?

6   Induction and Exhaust Processes

6.1   Introduction

In reciprocating internal combustion engines the induction and exhaust processes are non-steady flow processes. For many purposes, such as cycle analysis, the flows can be assumed to be steady. This is quite a reasonable assumption, especially for multi-cylinder engines with some form of silencing in the induction and exhaust passages. However, there are many cases for which the flow has to be treated as non-steady, and it is necessary to understand the properties of pulsed flows and how these can interact.

Pulsed flows are very important in the charging and emptying of the combustion chambers, and in the interactions that can occur in the inlet and exhaust manifolds. This is particularly the case for two-stroke engines where there are no separate exhaust and induction strokes. An understanding of pulsed flows is also needed if the optimum performance is to be obtained from a turbocharger; this application is discussed in chapter 9. In naturally aspirated engines it is also important to design the inlet and exhaust manifolds for pulsed flows, if optimum performance and efficiency are to be attained. However, inlet and exhaust manifold designs are often determined by considerations of cost, ease of manufacture, space and ease of assembly, as opposed to optimising the flow.

There is usually some form of silencing on both the inlet and exhaust passages. Again, careful design is needed if large pressure drops are to be avoided.

In four-stroke engines the induction and exhaust processes are controlled by valves. Two-stroke engines do not need separate valves in the combustion chamber, since the flow can be controlled by the piston moving past ports in the cylinder. The different types of valve (poppet, disc, rotary and sleeve) and the different actuating mechanisms are discussed in the next section.

231

The different types of valve gear, including a historical survey, are described by Smith (1967). The gas flow in the internal combustion engine is covered in some detail by Annand and Roe (1974).

6.2   Valve gear

6.2.1   Valve types

The most commonly used valve is the mushroom-shaped poppet valve. It has the advantage of being cheap, with good flow properties, good seating, easy lubrication and good heat transfer to the cylinder head. Rotary and disc valves are still sometimes used, but are subject to heat transfer, lubrication and clearance problems.

The sleeve valve was once important, particularly for aero-engines prior to the development of the gas turbine. The sleeve valve consisted of a single sleeve or pair of sleeves between the piston and the cylinder, with inlet and exhaust ports. The sleeves were driven at half engine speed and underwent vertical and rotary oscillation. Most development was carried out on engines with a single sleeve valve. There were several advantages associated with sleeve valve engines, and these are pointed out by Ricardo and Hempson (1968), who also give a detailed account of the development work. Sleeve valves eliminated the hot spot associated with a poppet valve. This was very important when only low octane fuels were available, since it permitted the use of higher compression ratios, so leading to higher outputs and greater efficiency. The drive to the sleeve could be at crankshaft level, and this led to a more compact engine when compared with an engine that used overhead poppet valves. The piston lubrication was also improved since there was always relative motion between the piston and the sleeve. In a conventional engine the piston is instantaneously stationary, and this prevents hydrodynamic lubrication at the ends of the piston stroke. This problem is most severe at top dead centre where the pressures and temperatures are greatest.

The disadvantages of the sleeve valve were: the cost and difficulty of manufacture, lubrication and friction between the sleeve and cylinder, and heat transfer from the piston through the sleeve and oil film to the cylinder. Compression ignition engines were also developed with sleeve valves.

6.2.2   Valve-operating systems

In engines with overhead poppet valves (ohv — overhead valves), the camshaft is either mounted in the cylinder block, or in the cylinder head

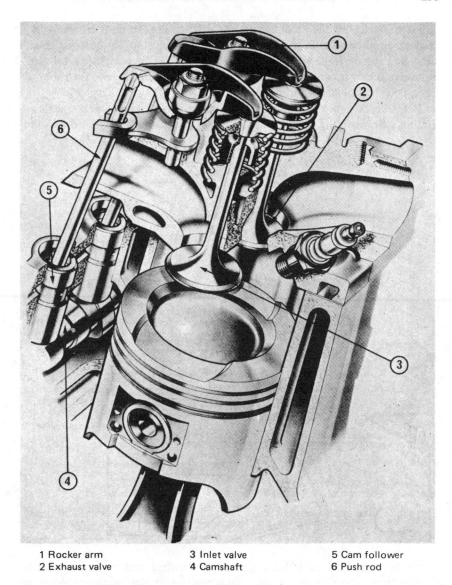

| 1 Rocker arm | 3 Inlet valve | 5 Cam follower |
| 2 Exhaust valve | 4 Camshaft | 6 Push rod |

Figure 6.1 Overhead valve engine (courtesy of Ford)

(ohc — overhead camshaft). Figure 6.1 shows an overhead valve engine in which the valves are operated from the camshaft, via cam followers, push rods and rocker arms. This is a cheap solution since the drive to the camshaft is simple (either gear or chain), and the machining is in the cylinder block. In a 'V' engine this arrangement is particularly suitable, since a single camshaft can be mounted in the valley between the two cylinder banks.

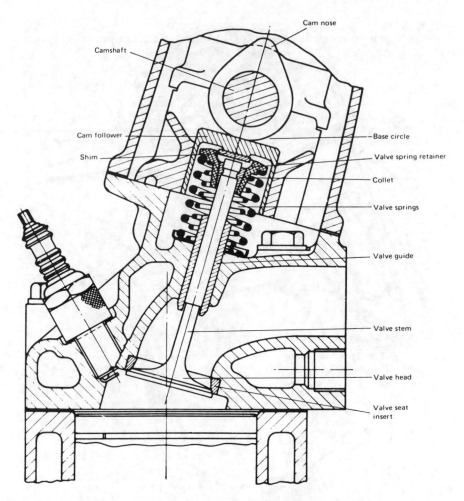

Figure 6.2   Overhead camshaft valve drive (reproduced with permission from Ricardo and Hempson (1968))

In overhead camshaft (ohc) engines the camshaft can be mounted either directly over the valve stems, as in figure 6.2, or it can be offset. When the camshaft is offset the valves are operated by rockers and the valve clearances can be adjusted by altering the pivot height. The drive to the camshaft is usually by chain or toothed belt. Gear drives are also possible, but would be expensive, noisy and cumbersome with overhead camshafts. The advantage of a toothed belt drive is that it can be mounted externally to the engine, and the rubber damps out torsional vibrations that might otherwise be troublesome.

Referring to figure 6.2, the cam operates on a follower or 'bucket'. The clearance between the follower and the valve end is adjusted by a shim. Although this adjustment is more difficult than in systems using rockers, it is much less prone to change. The spring retainer is connected to the valve spindle by a tapered split collet. The valve guide is a press-fit into the cylinder head, so that it can be replaced when worn. Valve seat inserts are used, especially in engines with aluminium alloy cylinder heads, to ensure minimal wear. Normally poppet valves rotate in order to even out any wear, and to maintain good seating. This rotation can be promoted if the centre of the cam is offset from the valve axis. Very often oil seals are placed at the top of the valve guide to restrict the flow of oil into the cylinder. This is most significant with cast iron overhead camshafts which require a copious supply of lubricant.

Not all spark ignition engines have the inlet and exhaust valves in a single line. The notable exceptions are the high-performance engines with hemispherical or pent-roof combustion chambers. Valves in such engines can be operated by various push rod mechanisms, or by twin or double overhead camshafts (dohc). One camshaft operates the inlet valves, and the second camshaft operates the exhaust valves. The disadvantages of this system are the cost of a second camshaft, the more involved machining, and the difficulty of providing an extra drive. An ingenious solution to these problems is the British Leyland 4-valve pent-roof head shown in figure 6.3. A single camshaft operates the inlet valves directly, and the exhaust valves indirectly, through a rocker. Since the same cam lobe is used for the inlet and exhaust valves, the valve phasing is dictated by the valve and rocker geometry. The use of four valves per combustion chamber is

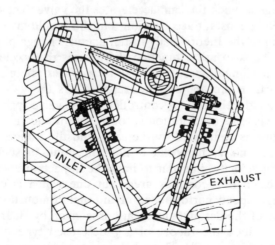

Figure 6.3   Four valve per cylinder pent-roof combustion chamber (from Campbell (1978))

quite common in high-performance spark ignition engines, and widely used in the larger compression ignition engines. The advantages of four valves per combustion chamber are: larger valve throat areas for gas flow, smaller valve forces, and larger valve seat area. Smaller valve forces occur since a lighter valve with a lighter spring can be used; this will also reduce the hammering effect on the valve seat. The larger valve seat area is important, since this is how heat is transferred (intermittently) from the valve head to the cylinder head.

6.2.3 *Dynamic behaviour of valve gear*

The theoretical valve motion is defined by the geometry of the cam and its follower. The actual valve motion is modified because of the finite mass and stiffness of the elements in the valve train. These aspects are dealt with after the theoretical valve motion.

The theoretical valve lift, velocity and acceleration are shown in figure 6.4; the lift is the integral of the velocity, and the velocity is the integral of the acceleration. Before the valve starts to move, the clearance has to be taken up. The clearance in the valve drive mechanism ensures that the valve can fully seat under all operating conditions, with sufficient margin to allow for the bedding-in of the valve. To control the impact stresses as the clearance is taken up, the cam is designed to give an initially constant valve velocity. This portion of the cam should be large enough to allow for the different clearance during engine operation. The impact velocity is limited to typically 0.5 m/s at the rated engine speed.

The next stage is when the cam accelerates the valve. The cam could be designed to give a constant acceleration, but this would give rise to shock loadings, owing to the theoretically instantaneous change of acceleration. A better practice is to use a function that causes the acceleration to rise from zero to a maximum, and then to fall back to zero; both sinusoidal and polynomial functions are appropriate examples. As the valve approaches maximum lift the deceleration is controlled by the valve spring, and as the valve starts to close its acceleration is controlled by the valve spring. The final deceleration is controlled by the cam, and the same considerations apply as before. Finally, the profile of the cam should be such as to give a constant closing velocity, in order to limit the impact stresses.

Camshaft design is a complex area, but one which is critical to the satisfactory high speed performance of internal combustion engines. A review of some of the design considerations is given by Beard (1958), but this does not include the theory of a widely used type of cam — the polydyne cam (Dudley (1948)). The polydyne cam uses a polynomial function to define the valve lift as a function of cam angle, and selects coefficients that avoid harmonics which might excite valve spring oscillations:

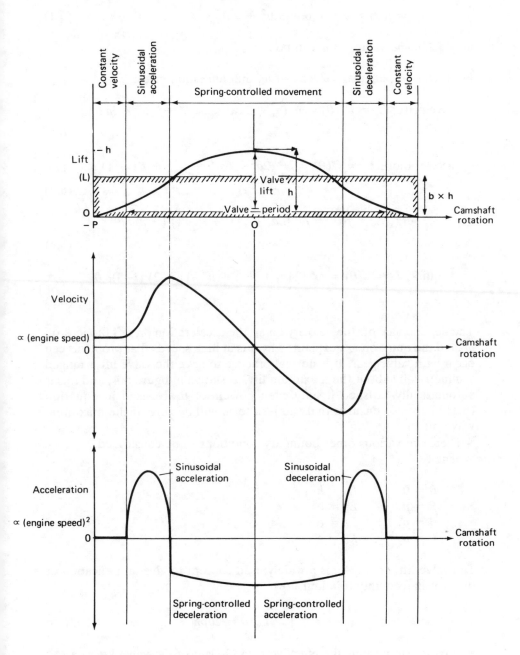

Figure 6.4   Theoretical valve motion

$$L_v = f(\theta) = a + a_1\theta + a_2\theta^2 + a_3\theta^3 + \ldots + a_i\theta^i + \ldots \quad (6.1)$$

in which some values of a_i can be zero.

For a constant angular velocity of ω, differentiation gives:

$$\text{velocity, } L_v = f'(\theta) = \omega\,(a_1 + 2a_2\theta + 3a_3\theta^2 + \ldots + ia_i\theta^{\,i-1}$$
$$+ \ldots) \quad (6.2)$$

$$\text{acceleration, } L_v = f''(\theta) = \omega^2\,(2a_2 + 6a_3\theta + \ldots + i(i-1)\,a_i\theta^{i-2}$$
$$+ \ldots) \quad (6.3)$$

$$\text{jerk, } L_v = f(\theta) = \omega^3\,(6a_3 + \ldots + i(i-1)\,(i-2)a_i\theta^{i-3}$$
$$+ \ldots) \quad (6.4)$$

$$\text{quirk, } L_v = f(\theta) = \omega^4\,(24a_4 + \ldots + i(i-1)(i-2)\,(i-3)a_i\theta^{i-4}$$
$$+ \ldots) \quad (6.5)$$

The dependence of: the velocity on ω, the acceleration on ω^2, the jerk on ω^3 and the quirk on ω^4, explains why it is at high speeds that problems can occur with valve gear. It is normal practice to have the valve lift arranged symmetrically about the maximum lift, as shown in figure 6.4, and this is automatically satisfied if only even powers of θ are used in equation (6.1). This also ensures that the jerk term will be zero at the maximum valve lift (h).

There are various other boundary conditions to be considered. When:

$$\theta = 0, \qquad L_v = h$$
$$\theta = p, \qquad L_v = 0$$
$$\theta = p, \qquad L_v = 0$$
$$\theta = p, \qquad L_v = 0$$

The valve lift 'area', A_θ, is a widely used concept to give an indication of the camshafts ability to admit flow:

$$A_\theta = \int_{-p}^{p} L_v d\theta = 2bph \quad (6.6)$$

where b represents the effective mean height of the valve lift as a
         fraction of h
        A_θ has units of $radians \times metres$.

The valve lift 'area' can be specified by b (usually >0.5) and this, along with the boundary conditions, determine that the valve lift can be defined by equation (6.1) when four even-valued power terms are used. The selection of these power terms is a complex issue discussed in some detail by Dudley.

In practice, the valve lift characteristics will also be influenced by the stiffness of the valve spring, as this has to control the deceleration prior to the maximum lift, and the acceleration that occurs after maximum lift. Ideally the spring force should be uniformly greater than the required acceleration force at the maximum design speed. The acceleration is given by equation (6.3), and the mass needs to be referred to the valve axis. For a push-rod-operated valve system:

equivalent mass, $m_e =$

$$\frac{\text{tappet} + \text{push rod mass}}{(r_r/r_v)^2} + \frac{\text{polar inertia of rocker}}{r_v^2} + \text{valve mass}$$

$$+ \text{ upper spring retainer mass} + \frac{\text{spring mass}}{3} \qquad (6.7)$$

where r_r = radius from rocker axis to cam line of action
       r_v = radius from the rocker axis to the valve axis.

In practice, when either a rocker arm or a finger-follower system is used, the values of the radii r_r and r_v will change. Due account can be taken of this to convert the valve lift to cam lift, but it must be remembered that the equivalent mass (in equation 6.7) will also become a function of the valve lift.

Additional allowances in the spring load have to be made for possible overspeeding of the engine, and friction in the valve mechanism. The force (F) at the cam/tappet interface is given by equation (6.8):

$$F = \frac{r_r}{r_v} \ (m_e \times L_v + F_0 + kL_v + F_g) \qquad (6.8)$$

where F_0 = valve spring pre-load
       F_g = gas force on valve head (normally only significant for the exhaust valve)
       k = valve spring stiffness.

Figure 6.5 shows the force at the cam/tappet interface for a range of speeds, along with the static force from the valve spring. At low speeds, the

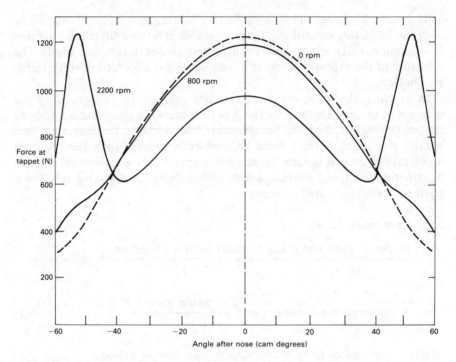

Figure 6.5   The force at the cam/tappet interface for a range of speeds

maximum force occurs at maximum valve lift, and this is because the valve spring force dominates. As the engine speed is increased, the acceleration terms dominate, and the largest force occurs just after the occurrence of the maximum acceleration.

A problem that can occur with valve springs is surge; these are intercoil vibrations. The natural frequency of the valve spring may be an order of magnitude higher than the camshaft frequency. However, as the motion of the cam is complex, there are high harmonics present and these can excite resonance of the valve spring. When this occurs the spring no longer obeys the simple force/displacement law, and the spring force will fall.

The natural frequency of a spring which has one end fixed is

$$\frac{1}{4} \sqrt{\left(\frac{k}{m}\right)} \quad \text{rad/s} \tag{6.9}$$

which can be rewritten as

$$\frac{D}{d^3} k \sqrt{\left(\frac{2}{\pi^2 G}\right)} \tag{6.10}$$

where D = coil mean diameter (m)
         d = wire diameter (m)
         G = bulk modulus (N/m^2).

Examination of the standard equations for coil springs will show that if a higher natural frequency is required, then this will lead to higher spring stresses for a prescribed spring stiffness. Thus it may not be possible to avoid surge with a single valve spring. With two concentric springs, the frequencies at which surge occurs should differ, so that surge should be less troublesome. Surge can also be controlled by: causing the coils to rub against an object to provide frictional damping, or by having a non-uniform coil pitch. If some coils in the spring close-up after lift commences, then the spring will no longer have a single natural frequency and this inhibits resonance.

The force at the cam/tappet interface (equation 6.8) also influences the contact stresses. For two elastic cylinders in contact, the Hertz stress (σ) is given by Roark and Young (1976) as

$$\sigma = \sqrt{\left(\frac{F}{\pi L} \times \frac{(1/r_1) + (1/r_2)}{[(1-v_1^2)/E_1] + [1-v_2^2)/E_2]} \right)} \tag{6.11}$$

where L = width of contact zone
         r = radius of curvature
         v = Poisson's ratio
         E = Young's modulus.

In the case of a flat follower, $1/r$ becomes zero. For surfaces with more than one curvature, appropriate formulae can be found in Roark and Young (1976). Equation (6.11) predicts line contact, and since the line is of finite width, this leads to discontinuities in the stress contours. These discontinuities have an adverse effect on the fatigue life of the contacting surfaces. The theoretical line contact is modified to a theoretical point contact, by having a tappet with a spherical surface (but very high radius of curvature) or a cam with a second radius of curvature (again a very high value, lying in the plane of the camshaft axis). The theoretical point contact will spread to an elliptical contact zone, which then has no stress discontinuities.

The valve lift has now been specified as a function of cam angle, and all that remains is to decide on the base circle diameter. Increasing the base circle diameter increases the radius of curvature at the nose of the cam, and this reduces the contact stresses. Also, when a follower with a curved surface is used, increasing the base circle diameter will also increase the sliding velocity at the cam–tappet interface.

As can be seen from figure 6.4, the theoretical valve-opening and

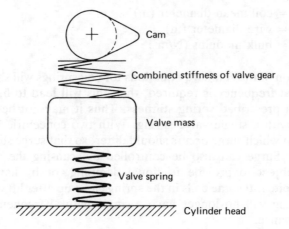

Figure 6.6   Simple valve gear model

valve-closing times will depend on the valve clearance. Consequently, the valve timing usually refers to the period between the start of the valve acceleration, and the end of the valve deceleration. The valve lift refers to the lift in the same period, and is usually limited to about $0.25D_v$, to restrict the loads in the valve mechanism.

If the force required during the spring-controlled motion is greater than that provided by the spring, then the valve motion will not follow the cam, and the valve is said to 'jump'. The accelerations will increase in proportion to the square of the engine speed, and a theoretical speed can be calculated at which the valve will jump. The actual speed at which jumping occurs will be below this, because of the elasticity of other components and the finite mass of the spring.

The actual valve motion is modified by the elasticity of the components; a simple model is shown in figure 6.6. A comparison of theoretical and actual valve motion is shown in figure 6.7. Valve bounce can occur if the seating velocity is too great, or if the pre-load from the valve spring is too small. This is likely to lead to failure, especially with the exhaust valve.

To minimise the dynamic effects, the valve should be made as light as possible, and the valve gear should be as stiff as possible. The camshaft should have a large diameter shaft with well-supported bearings, and the cams should be as wide as possible. Any intermediate components should be as light and stiff as possible.

More realistic models of valve systems can be used with a computer, which analyses the valve motion and then deduces the corrected cam profiles. The computer can then predict the response for a range of speeds.

The model should include the stiffness of the drive to the camshaft, the torsional stiffness of the camshaft, the stiffness of the camshaft mounting

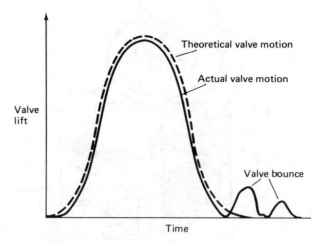

Figure 6.7   Comparison of theoretical and actual valve motion

relative to the engine structure, the stiffness of any rocker and its mount-ings, as well as the stiffness of the valve train. The valve gear is likely to be modelled by a distributed mass (connected by stiff elements). Further-more, wherever there is a stiffness there will be an associated damping factor. The damping is invariably difficult to evaluate, but small and highly significant in modifying resonances.

Finally, it should be remembered that cams can only be manufactured to finite tolerances, and that oil films and deformation at points of contact and elsewhere will modify the valve motion. Beard (1958) considers both radial and lift manufacturing tolerances, and shows that both can be significant.

Sometimes hydraulic tappets (or cam followers) are used (figure 6.8). These are designed to ensure a minimum clearance in the valve train mechanism. They offer the advantages of automatic adjustment and, owing to the compressibility of the oil, they cushion the initial valve motion. This permits the use of more rapidly opening cam profiles. The disadvantages are that they can stick (causing the valve to remain open), the valve motion is less well defined, and there are higher levels of friction unless a roller follower is also used.

6.3   Flow characteristics of poppet valves

The shape of the valve head and seat are developed empirically to produce the minimum pressure drop. Such experiments are usually carried out on steady-flow air rigs similar to the one shown in figure 6.9. The flow from a

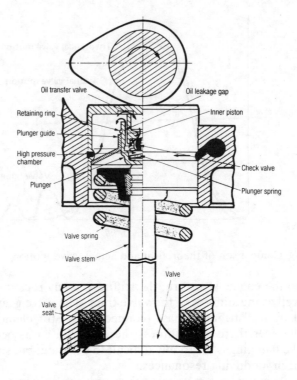

Figure 6.8   A hydraulic bucket tappet, with acknowledgement to Eureka

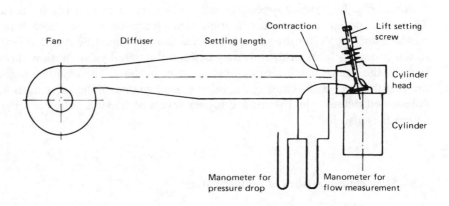

Figure 6.9   Air flow rig to determine the flow characteristics of an inlet valve
(with acknowledgement to Annand and Roe (1974))

fan is decelerated in a diffuser before entering a settling length. To help provide a uniform flow, the type of meshes used in wind tunnels may be useful. The contraction accelerates the flow, thus reducing the relative significance of any turbulence. It is essential for the contraction to match perfectly the inlet passage, otherwise turbulence and extraneous pressure drops will be introduced. The contraction also provides a means of metering the flow rate and of measuring the pressure drop across the valve. The lift setting screw enables the pressure drop to be measured for a range of valve lifts.

A similar arrangement can be used for measuring the flow characteristics of exhaust valves, the exhaust passage being connected to a suction system. Sometimes water rigs are used when the flow patterns are to be investigated. As in all experiments that measure pressure drops through orifices, slightly rounding any sharp corners can have a profound effect on the flow characteristics; care must be taken with all models.

The orifice area can be defined as a 'curtain' area, A_c:

$$A_c = \pi \, D_v \, L_v \qquad (6.12)$$

where D_v = valve diameter
and L_v = axial valve lift.

This leads to a discharge coefficient (C_D); this is defined in terms of an effective area (A_e):

$$C_D = \frac{A_e}{A_c} \qquad (6.13)$$

The effective area is a concept defined as the outlet area of an ideal frictionless nozzle which would pass the same flow, with the same pressure drop, with uniform constant-pressure flows upstream and downstream. These definitions are arbitrary, and consequently they are not universal.

For a given geometry, discharge coefficient will vary with valve lift and flow rate. These quantities can be expressed conveniently as a non-dimensional valve lift (L_v / D_v), and Reynolds number, R_e.

$$R_e = \frac{\rho v x}{\mu} \qquad (6.14)$$

where ρ = fluid density
         v = flow velocity
         x = a characteristic length
         μ = fluid viscosity.

For the inlet valve it is common practice to assume incompressible flow since the pressure drop is small, in which case the ideal velocity would be

$$v_0 = \sqrt{(2\Delta p/\rho)} \tag{6.15}$$

where v_0 = frictionless velocity
          Δ_p = pressure difference across the port and valve.

For compressible flow

$$v_0 = \left\{ \frac{2\gamma}{(\gamma-1)} \frac{p_c}{\rho_0} \left[1 - \left(\frac{p_c}{p_0} \right)^{\frac{\gamma-1}{\gamma}} \right] \right\}^{\frac{1}{2}} \tag{6.16}$$

and

$$\frac{\dot{m}}{A} = \frac{p_0}{RT_0} \left(\frac{p_c}{p_0} \right)^{\frac{1}{\gamma}} \left\{ \frac{2\gamma}{\gamma-1} \left[1 - \left(\frac{p_c}{p_0} \right)^{\frac{\gamma-1}{\gamma}} \right] \right\}^{\frac{1}{2}} \tag{6.17}$$

where p_0 = upstream pressure
          p_c = cylinder pressure
          $\dot{m}/A$ = mass flow per unit area.

The discharge coefficient (equation 6.13), can now be defined in terms of the measured volume flowrate ($\dot{V}$) compared with the ideal volume flowrate ($\dot{V}_0$):

$$A_c C_D = A_e = \frac{\dot{V}}{\dot{V}_0/A_c} = \frac{\dot{V}}{v_0} \tag{6.18}$$

It can be less ambiguous to define the valve flow performance by the effective area as a function of valve lift. This avoids the problem of how the reference flow area has been defined. For example, does the curtain area correspond to the inner or outer valve seat diameter? Sometimes the reference area is the minimum cross-sectional area for the flow. The position of this can occur in three places according to the valve lift, and this geometry is discussed in detail by Heywood (1988). Ultimately the minimum flow area will be the annulus between the valve stem diameter (D_s) and the valve inner seat diameter (D_{vi}). This value of valve lift corresponds to

$$L_v' = \frac{D_{vi}^2 - D_s^2}{4D_{vi}} \tag{6.19}$$

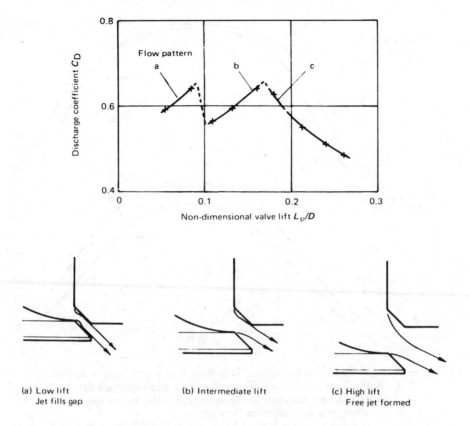

Figure 6.10   Flow characteristics of a sharp-edged inlet valve (with acknowledgement to Annand and Roe (1974))

or a non-dimensional value lift of about 0.23.

The flow characteristics of a sharp-edged inlet valve are shown in figure 6.10. At low lift the jet fills the gap and adheres to both the valve and the seat. At an intermediate lift the flow will break away from one of the surfaces, and at high lifts the jet breaks away from both surfaces to form a free jet. The transition points will depend on whether the valve is opening or closing. These points are discussed in detail by Annand and Roe (1974) along with the effects of sharp corners, radii, and valve seat width. They conclude that a 30° seat angle with a minimum width seat and 10° angle at the upstream surface gives the best results. In general, it is advantageous to round all corners on the valve and seat. For normal valve lifts the effect of Reynolds number on discharge coefficient is negligible.

When the discharge coefficient results from figure 6.10 are combined with some typical valve lift data, then it is possible to plot the effective flow area as a function of camshaft angle. This has been done in figure 6.11, and

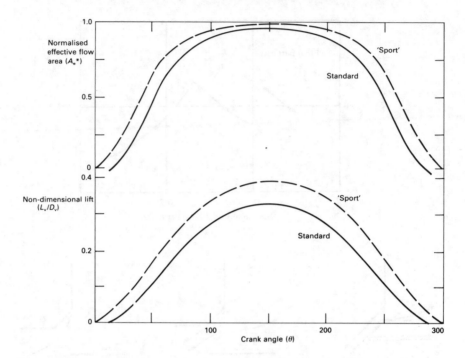

Figure 6.11   Comparison between a 'standard' cam profile and its 'sport'
counterpart with an extended period and increased lift (but with
the same maximum acceleration); also shown are the
corresponding effective flow areas

it should be noted that there is a broad maximum for the effective flow area
— this is a consequence of the effective flow area being limited by the
annular area (equation 6.19) and no longer being dependent on the valve
lift. Also shown in figure 6.11 is a 'high performance' cam profile, this has
an increased lift and valve open duration. However, the valve lift curve has
been scaled so as to give the same maximum valve train acceleration.
Increasing the valve lift has not increased the maximum effective flow area,
but a consequence of the longer duration valve event is an increase in the
width of the maximum. In other words, it is the extended duration that will
lead to an improvement in the flow performance of the valve.

Taylor (1985a) points out, that if the pressure ratio across the inlet valve
becomes too high, then there will be a rapid fall in the volumetric efficiency. Taylor characterises the flow by an inlet Mach index, Z, which is
the Mach number of a notional air velocity.

When the effective flow area in figure 6.11 is averaged, it can be divided
by the reference area (based on the valve diameter D_v), to give a mean
flow coefficient:

$$\bar{C}_d = \frac{\bar{A}_e}{\pi D_v^2 / 4} \tag{6.20}$$

The mean rate of change of the volume, is

$$\bar{v}_p \times \pi B^2/4 \tag{6.21}$$

and this leads to a notional mean velocity, which can be divided by the speed of sound (c), to give the Mach index, z:

$$Z = \frac{\bar{v}_p \times \pi B^2/4}{c \times C_d \times \pi D_v^2 / 4} = \left(\frac{B}{D_v} \right)^2 \frac{\bar{v}_p}{c \times C_d} \tag{6.22}$$

Taylor found that, for a fixed valve timing, the volumetric efficiency was only a function of the Mach index, and to maintain an acceptable volumetric efficiency the Mach index should be less than 0.6; this is illustrated by figure 6.12. Taylor (1985a) also reports on the effects of the inlet valve timing on the volumetric efficiency. Most significant is the benefit of delaying the inlet valve closure for high values (>0.6) of the Mach index.

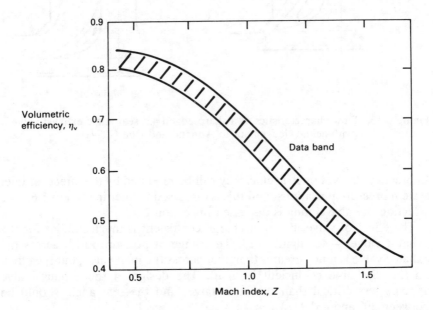

Figure 6.12 Volumetric efficiency as a function of the Mach index (adapted from Taylor (1985a))

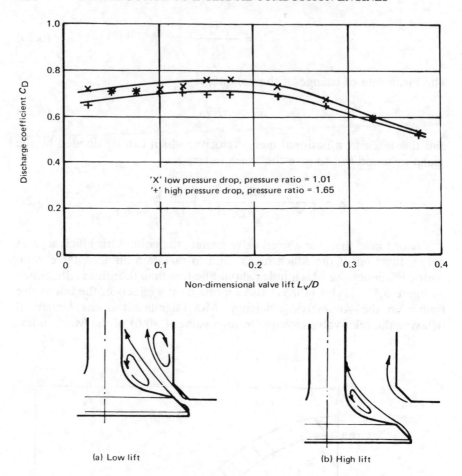

Figure 6.13   Flow characteristics of a sharp-edged 45° seat angle exhaust valve
(with acknowledgement to Annand and Roe (1974))

In practice, the volumetric efficiency will be modified by the effects of inlet
system pressure pulsations, and this is discussed further in section 6.6. The
selection of valve timing is discussed in section 6.4.

The effect of valve lift on discharge coefficient is much smaller for the
exhaust valve — see figure 6.13. The range of pressure ratios across the
exhaust valve is much greater than that across the inlet valve, but the effect
on the discharge coefficient is small. The design of the exhaust valve
appears less critical than the inlet valve, and the seat angle should be
between 30° and 45°.

When the exhaust valve opens (blowdown), there is a very high pressure
ratio across the valve. Indeed the flow can become choked if

$$\frac{p_e}{p_c} \leq \left(\frac{2}{\gamma + 1}\right)^{\frac{\gamma+1}{2(\gamma-1)}} \qquad (6.23)$$

where p_e = exhaust manifold pressure
       p_c = cylinder pressure.

For exhaust gases this critical pressure ratio is about 3 (since γ depends on the temperature and composition of the gases), and when the flow is choked, equation (6.17) becomes

$$m/A = \frac{p_c}{\sqrt{(RT_c)}} \sqrt{\gamma} \left(\frac{2}{\gamma+1}\right)^{\frac{\gamma+1}{2(\gamma-1)}} \qquad (6.24)$$

For the exhaust flow, p_c is substituted for p_o, and p_e is substituted for p_c in equation (6.17). When these flow equations are being solved as part of a cylinder filling model, it is important to remember that reverse flows can occur through both inlet and exhaust valves. Therefore, it is necessary to check the pressure ratio across the valve and then use an equation for the approriate flow direction.

   In general, the inlet valve is of larger diameter than the exhaust valve, since a pressure drop during induction has a more detrimental effect on performance than a pressure drop during the exhaust stroke. For a flat, twin valve cylinder head, the maximum inlet valve diameter is typically 44–48 per cent and the maximum exhaust valve diameter is typically 40–44 per cent of the bore diameter. With pent-roof and hemispherical combustion chambers the valve sizes can be larger.

   For a flat four-valve head, as might be used in a compression ignition engine, each inlet valve could be 39 per cent of the bore diameter, and each exhaust valve could be 35 per cent of the bore diameter. This gives about a 60 per cent increase in total valve circumference, or a 30 per cent increase in 'curtain' area (equation 6.1) for the same non-dimensional valve lift. The inlet and exhaust passages should converge slightly to avoid the risk of flow separation with its associated pressure drop. At the inlet side the division between the two valve ports should have a well-rounded nose; this will be insensitive to the angle of incidence of the flow. A knife-edge division wall would be very sensitive to flow breakaway on one side or the other. For the exhaust side the division wall can taper out to a sharp edge.

   The port arrangements in two-stroke engines are discussed in chapter 7, section 7.5.3.

6.4   Valve timing

6.4.1   Effects of valve timing

The valve timing dictated by the camshaft and follower system is modified by the dynamic effects that were discussed in section 6.2.3. Two timing diagrams are shown in figure 6.14. The first (figure 6.14a) is typical of a compression ignition engine or conventional spark ignition engine, while figure 6.14b is typical of a high-performance spark ignition engine. The greater valve overlap in the second case takes fuller advantage of the pulse effects that can be particularly beneficial at high engine speeds. Turbocharged engines also use large valve overlap periods.

In compression ignition engines the valve overlap at top dead centre is often limited by the piston to cylinder-head clearance. Also the inlet valve has to close soon after bottom dead centre, otherwise the reduction in compression ratio may make cold starting too difficult. The exhaust valve opens about 40° before bottom dead centre (bbdc) in order to ensure that all the combustion products have sufficient time to escape. This entails a slight penalty in the power stroke, but 40° bbdc represents only about 12 per cent of the engine stroke. It should also be remembered that 5° after starting to open the valve may be 1 per cent of fully open, after 10°, 5 per cent of fully open, and not fully open until 120° after starting to open.

In spark ignition engines with large valve overlap, the part throttle and idling operation suffers since the reduced induction manifold pressure causes back-flow of the exhaust. Furthermore, full load economy is poor since some urburnt mixture will pass straight through the engine when both valves are open at top dead centre. These problems are avoided in a turbocharged engine with in-cylinder fuel injection.

The level of exhaust residuals trapped in the cylinder has a significant effect on the cycle-by-cycle variations in combustion, and the emissions of NO_x. As with exhaust gas recirculation, high levels of exhaust residuals lead to lower emissions of NO_x and greater cycle-by-cycle variations in combustion.

The level of residuals increases with:

(a) decreasing absolute inlet manifold pressure
(b) reducing compression ratio
(c) increasing valve overlap
(d) decreasing speed
(e) increasing exhaust back pressure.

Comprehensive experimental results for the residual fraction have been reported by Toda *et al.* (1976). The exhaust residual (ER) levels can either

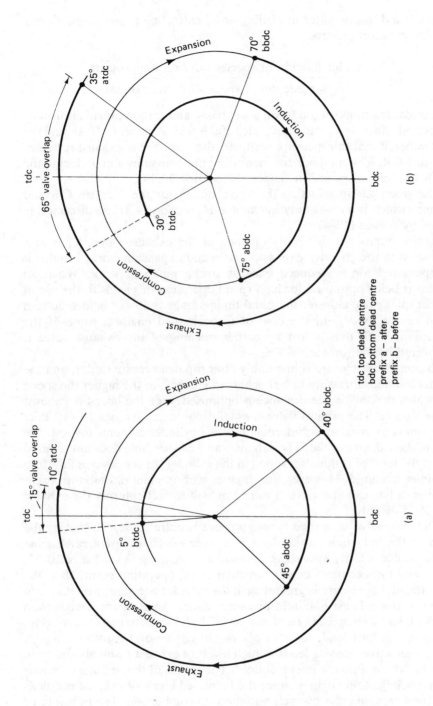

Figure 6.14 Valve-timing diagrams. (a) Small valve overlap; (b) large valve overlap

be predicted by computer modelling or by extracting a gas sample during the compression process:

$$ER = \frac{\text{molar fraction of species } i \text{ during compression}}{\text{molar fraction of species } i \text{ in the exhaust}} \qquad (6.25)$$

The molar fractions should be on a wet basis, and carbon dioxide is usually measured since it is the most plentiful species present. It should be remembered that this measurement will also include the exhaust residuals due to EGR. The methods for measuring the emissions and deducing the EGR level are discussed in chapter 13, section 13.4.6.

The trade-offs in selecting the valve timing for the 2.2 litre Chrysler engine (which is a case study in chapter 14, section 14.3) are discussed in detail by Asmus (1984).

Asmus points out that early opening of the exhaust valve, leads to a reduction in the effective expansion ratio and expansion work, but this is compensated for by reduced exhaust stroke pumping work. When an engine is being optimised for high speed operation this leads to the use of earlier inlet valve closure; the usual timing range is 40–60° before bottom dead centre (bbdc). In the case of turbocharged engines, some of the expansion work that is lost by earlier opening of the exhaust valve is recovered by the turbine.

Exhaust valve closure is invariably after top dead centre (atdc), and the higher the boost pressure in turbocharged engines, or the higher the speed for which the engine performance is optimised, then the later the exhaust valve closure. The exhaust valve is usually closed in the range 10–60° atdc. The aim is to avoid any compression of the cylinder contents towards the end of the exhaust stroke. The exhaust valve closure time does not seem to affect the level of residuals trapped in the cylinder, or the reverse flow into the inlet manifold. However, for engines with in-manifold mixture preparation, a late exhaust valve closure can lead to fuel entering the exhaust manifold directly.

The inlet valve is opened before top dead centre (btdc), so that by the start of the induction stroke there is a large effective flow area. Engine performance is fairly insensitive to inlet valve opening in the range 10–25° btdc. For turbocharged engines at their rated operating point, then the inlet manifold pressure is greater than the cylinder pressure, which in turn is above the exhaust manifold pressure: under these circumstances even earlier inlet valve opening (earlier than 30° btdc) leads to good scavenging. However, at part load, for a turbocharged engine or a throttled engine, early inlet valve opening leads to high levels of exhaust residuals and back flow of exhaust into the inlet manifold. The results of this are most obvious with spark ignition engines, since the increased levels of exhaust residuals lead to increased cycle-by-cycle variations in combustion. The influence of

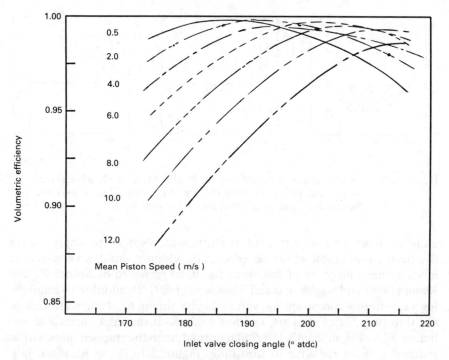

Figure 6.15   Inlet valve closing angle (°atdc)

valve overlap on the part-load performance of turbocharged Diesel engines is illustrated in chapter 10, section 10.3.3.

Inlet valve closure is invariably after bottom dead centre (abdc) and typically around 40° abdc, since at bdc the cylinder pressure is still usually below the inlet manifold pressure. This is in part a consequence of the slider crank mechanism causing the maximum piston velocity to occur after 90° bbdc. Figure 6.15 illustrates the influence of inlet valve closure angle on the volumetric efficiency. A simple model has been used here, which ignores compressibility and dynamic effects. The mean piston speed has been used as a variable, since it defines engine speed in a way that does not depend on the engine size. Figure 6.15 shows that at low speeds, a late inlet valve closure reduces the volumetric efficiency. In contrast, at high speeds an early inlet valve closure leads to a greater reduction in volumetric efficiency, and this limits the maximum power output.

6.4.2   Variable valve timing

The discussion of valve timing in the previous section has shown that there are compromises in valve timing: high speed versus low speed performance,

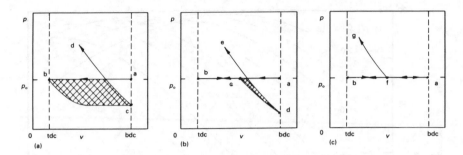

Figure 6.16   State diagrams for load control by throttling, early inlet valve closure and late inlet valve closure. In reality all processes are irreversible; 'hatching' indicates dissipated work

and full load versus part load performance. Not surprisingly, there has been considerable effort devoted to developing variable valve timing mechanisms, and some of this work has been reviewed by Stone, R. and Kwan (1989) and by Ahmad and Theobold (1989). In addition to minimising valve timing compromises, variable valve timing can be used to reduce the throttling losses in spark ignition engines. If the inlet valve is closed before bdc, or significantly later than normal, then the trapped mass will be reduced without recourse to throttling. Figure 6.16 shows idealised p–V diagrams for throttling, early inlet valve closure and late (or delayed) inlet valve closure, and it can be seen that there is potential for reducing the pumping work or throttling loss. These modified inlet valve closure angles also reduce the effective compression ratio, and it might be thought that this will lead to a lower corresponding cycle efficiency. However, study of example 2.3 will show that the expansion ratio is a more important determinant of cycle efficiency, and this is unchanged by varying the inlet valve closure angle (Stone, R. and Kwan (1989)). The lower compression ratio leads to a lower compression temperature, and this reduces the laminar burning velocity. Furthermore, Hara *et al.* (1985) also report lower levels of turbulence, so the combined effect is a reduced burn rate and a lower efficiency — this is probably why the theoretical predictions for an efficiency gain are higher than those achieved in practice. Most experimental studies of late and early inlet valve closure have used camshafts with special fixed timings, and the data from these sources has also been reviewed by Stone, R. and Kwan (1989). The conclusion is, that in the bmep range of 3–4.5 bar, there is scope for a 5–10 per cent saving in fuel consumption.

Mechanical variable valve timing systems have been devised that enable the valve events to be phased (that is, both opening and closing events are moved equally), and the valve event duration to be modified (with usually

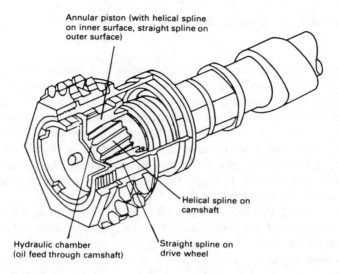

Annular piston (with helical spline
on inner surface, straight spline on
outer surface)

Helical spline on
camshaft

Hydraulic chamber
(oil feed through camshaft)

Straight spline on
drive wheel

Figure 6.17   The Alfa Romeo variable phasing system (Anon (1984b))

the ability to control the relative amount that opening and closing are
moved). Needless to say, it is the simplest devices that have been used in
production engines. The first such use was by Alfa Romeo with the
mechanism shown in figure 6.17 controlling the phasing of the inlet
camshaft. An advantage of this phasing control is that at light load and low
speeds (very significant conditions in urban driving cycles) both delaying
the inlet valve closure and delaying the inlet valve opening are beneficial.
The delayed inlet valve closure reduces the throttling losses, and the
reduced valve overlap at the light load also improves combustion. Ma
(1988) quantified the effect of valve overlap, as enabling a 200 rpm reduction in ideal speed with a 12 per cent reduction in fuel flow. He predicted
that this would lead to a 6.1 per cent reduction in the ECE-15 urban cycle
fuel consumption. The Alfa Romeo mechanism delayed the inlet valve
events by 32° ca, with a 28 per cent reduction in the urban cycle fuel
consumption (Anon (1984b)). However, it is likely that some of this
reduction in fuel consumption is attributable to a change from a carburettor to fuel injection.

Not surprisingly the timing of valve events also has an impact on exhaust
emissions. For example, control of valve overlap can be used to control the
level of exhaust residuals, thereby regulating the emissions of NO_x. Stone,
R. and Kwan (1989) discuss the potential of variable valve timing for
controlling emissions.

6.5   Unsteady compressible fluid flow

The derivation of the results for one-dimensional, unsteady compressible fluid flow can be found in many books on compressible flow, for example Daneshyar (1976); the main results will be quoted and used in this section.

Unsteady flow is treated by considering small disturbances superimposed on a steady flow. For analytical simplicity the flow is treated as adiabatic and reversible, and thus isentropic. The justification for reversibility is that, although the flow may not in fact be frictionless, the disturbances or perturbations are small. Further, the fluid properties are assumed not to change across the perturbation.

By considering the conservation of mass, momentum and energy, the propagation speed for a perturbation or small pressure wave is found to be the speed of sound, c.

For a perfect gas

$$c = \sqrt{\left(\frac{\gamma p}{\rho}\right)} = \sqrt{\gamma RT} \tag{6.26}$$

In a simple wave of finite amplitude, allowance can be made for the change in properties caused by the change in pressure. In particular, an increase in pressure causes an increase in the speed of propagation. A simple wave can be treated by considering it as a series of infinitesimal waves, each of which is isentropic. If the passage of a wave past a point increases the pressure, then it is a compression wave, while if it reduces the pressure then it is an expansion or rarefaction wave — see figure 6.18.

Recalling that an increase in pressure causes an increase in the propagation speed of a wave, then a compression wave will steepen, and an

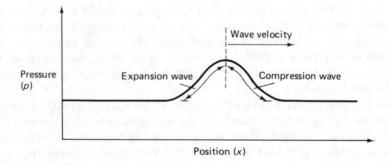

Figure 6.18   Simple pressure wave (adapted from Daneshyar (1976), © 1968 Pergamon Press Ltd)

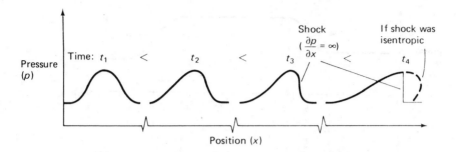

Figure 6.19   Spreading of an expansion wave, steepening of a compression
wave (adapted from Daneshyar (1976), © 1968 Pergamon Press
Ltd)

expansion wave will flatten. This is shown in figure 6.19 for a simple wave
at four successive times. When any part of the compression wave becomes
infinitely steep ($\partial p/\partial x = \infty$) then a small compression shock wave is
formed. The shock wave continues to grow and the simple theory is no
longer valid, since a shock wave is not isentropic. If the isentropic analysis
was valid then the profile shown by the dotted line would form, implying
two values of a property (for example, pressure) at a given position and
time.

The interactions between waves and boundaries can be determined from
position diagrams and state diagrams. The state diagrams are not discussed
here, but enable the thermodynamic properties, notably pressure, to be
determined. The theory and use of state diagrams in conjunction with
position diagrams is developed by books such as Daneshyar (1976).

Position diagrams are usually non-dimensional, using the duct length
(L_D) to non-dimensionalise position, and the speed of sound (c) and duct
length to non-dimensionalise time. The position diagram for two approaching compression waves is shown in figure 6.20, the slopes of the lines
correspond to the local value of the speed of sound. The values of the
pressure would be obtained from the corresponding state diagram.

For internal combustion engines it is important to know how waves
behave at boundaries. There are open ends, junctions and closed ends.
Examples of these are: an exhaust pipe entering an expansion box, a
manifold, and a closed exhaust valve in another cylinder, respectively.

At a closed end, waves are reflected in a like sense; that is, a compression wave is reflected as a compression wave. This derives from the
boundary condition of zero velocity.

At an open end there will be a complex three-dimensional motion. The
momentum of the surrounding air from the three-dimensional motion
causes pressure waves to be reflected in an unlike sense at an open (or

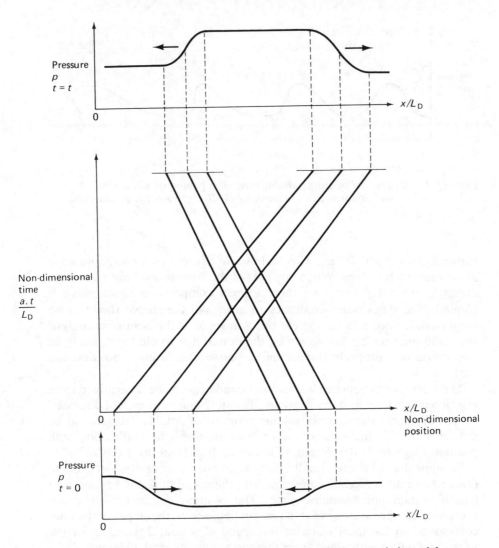

Figure 6.20   Interaction of two approaching compression waves (adapted from Daneshyar (1976), © 1968 Pergamon Press Ltd)

constant-pressure) boundary. For example, a compression wave will be reflected as an expansion wave.

A thorough simulation of the flows in the inlet or exhaust system of a reciprocating engine would need a solution to the three-dimensional unsteady compressible fluid flow equations, taking due account of turbulence effects. Typically, the three-dimensional flow equation can only be solved for steady incompressible flow, and the solution of unsteady compressible flow is restricted to one-dimensional flow.

The pressure waves in the induction and exhaust systems of an engine are of such an amplitude that they cannot be treated as sound waves. Sound waves are of sufficiently low amplitude that the local conditions are not affected, and all the waves propagate at the speed of sound. With pressure waves, the local pressure level affects the propagation velocity of that part of the wave. If any compression or expansion process associated with the passage of the wave is isentropic, and the entropy is the same anywhere in the system, then the flow is homentropic. Thus, if the flow is frictionless, adiabatic and of fixed composition, then for a constant cross-sectional area pipe, the homentropic flow equations are as presented by Benson (1982) or Heywood (1988):

continuity

$$\frac{1}{\rho} \frac{\partial \rho}{\partial t} + \frac{u}{\rho} \frac{\partial \rho}{\partial x} + \frac{\partial u}{\partial x} = 0 \tag{6.27}$$

momentum

$$\frac{1}{\rho} \frac{\partial \rho}{\partial x} + \frac{\partial u}{\partial t} + u \frac{\partial u}{\partial x} = 0 \tag{6.28}$$

energy and equation of state

$$c^2 = \left(\frac{\partial p}{\partial \rho} \right)_s \tag{6.29}$$

$$\frac{p}{p_{\text{ref}}} = \left(\frac{c}{c_{\text{ref}}} \right)^{2\gamma/(\gamma-1)} \tag{6.30}$$

where u = local flow velocity (m/s)
   x = position along price (m)
   ρ = local density (kg/m³)
   p = local pressure (N/m²).

These hyperbolic partial-differential equations can be solved by a number of techniques, and the first widely used approach was the method of characteristics.

In the method of characteristics, Riemann transformed the partial differential equations into total differential equations, which apply along so-called characteristic lines. The pressure waves propagate upstream and downstream relative to the flowing gas, at the local speed of sound. The local condition in the flow at any time and position are found from the so-called state diagram.

The method of characteristics was solved graphically by Haller (1945), and this was first applied to an engine by Jenny (1950). The first numerical solution was by Benson *et al.* (1964), who used a rectangular grid in the x and t directions. A pipe is divided into sections joined at junctions that are represented by mesh points. Initial values have to be assigned to the mesh points at $t = 0$, and the values of the Riemann variables at each mesh point can be determined for subsequent time steps.

Finite difference methods have also been applied to the direct solution of the partial differential equations (6.27)–(6.30), and the one-step Lax–Wendroff technique is often used. Heywood (1988) summarises the methodology of the Lax–Wendroff solution, and points out that some form of damping (artificial viscosity) is required to prevent instabilities developing in the solution. However, the Lax–Wendroff method is significantly faster than the method of characteristics. A comparison of the Lax–Wendroff technique with the method of characteristics, by Polini *et al.* (1987), has shown that both techniques give very similar results.

An alternative finite difference technique that has been applied by Iwamota and Deckker (1985) is the Random Choice Method: this does not require the addition of artificial viscosity terms to prevent instability in the solution. The method of characteristics can be extended to non-homentropic flow (to account for irreversibilities, such as heat transfer, fluid friction), by including a path line. This allows particle related properties (namely: temperature and entropy, but not pressure) to travel at the local fluid velocity.

Boucher and Kitsios (1986) describe how transmission line modelling techniques can be used to predict unsteady compressible flow. The right- and left-going waves in a transmission line undergo only a time delay in the line, and are otherwise unchanged in character. A fluid system is then modelled as a series of transmission lines and lumped elements (such as junctions, volumes or nozzles). The losses are most readily incorporated at the junctions, so as not to affect the transmission line solution. Boucher and Kitsios (1986) consider several applications, including the Helmholtz resonator, for which there was very close agreement in the prediction of the natural frequency.

However, it must be remembered that all solutions for unsteady flow depend on being able to define the initial and boundary conditions for the system. There are some simple boundaries, such as: open pipes, closed pipes, pipe junctions; but many junctions are highly complex: an opening or closing valve, the entry volute to a turbocharger.

In conclusion, the methods for solving unsteady one-dimensional flow may be complex, but the calculation methods are well established. It has also been shown by Polini *et al.* (1987) that when the same boundary and initial conditions are used, then both the Method of Characteristics and the Lax–Wendroff technique give consistent answers. However, a more significant problem is the definition of the initial conditions and the boundary conditions. Some of the common assumptions and a discussion of their

justification have been presented by Winterbone (1990a), who also ident-
ifies pipe boundaries (such as: turbocharger entries, pipe junctions in
manifolds) that require further study.

Owing to the complexity of solving these equations, the methods are
probably better suited to analysis rather than design. In the next section on
manifold design (6.6), acoustic methods will be described that enable
resonating systems to be designed. However, these methods do not enable
the volumetric efficiency to be quantified, so once a system has been
identified, its performance can be evaluated by an unsteady-flow analysis.

6.6 Manifold design

6.6.1 General principles

In designing the manifold system for a multi-cylinder engine, advantage
should be taken of the pulsed nature of the flow. The system should avoid
sending pulses from the separate cylinders into the same pipe at the same
time, since this will lead to increased flow losses. However, it is sensible to
have two or three cylinders that are out of phase ultimately connected to
the same manifold. When there is a junction, a compression wave will also
reflect an expansion wave back; this is shown in figure 6.21. If the
expansion wave returns to the exhaust valve at the end of the exhaust valve
opening, then it will help to scavenge the combustion products; if the inlet
valve is also open then it will help to draw in the next charge. Obviously the
cancellation of compression and expansion waves must be avoided.

A typical exhaust system for taking advantage of the pulsations from a
four-cylinder engine is shown in figure 6.22. The pipe length from each
exhaust port to its first junction is the same, and the pipes will be curved to
accommodate the specified lengths within the given distances. The length
adopted will influence the engine speed at which maximum benefit is
obtained. The manifold is such that for the given firing order (cylinders
1–3–4–2), the pressure pulses will be out of phase.

Consider the engine operating with the exhaust valve just opening on
cylinder 1. A compression wave will travel to the first junction; since the
exhaust valve on cylinder 4 is closed an expansion wave will be reflected
back to the open exhaust valve. The same process occurs 180° later in the
junction connecting cylinders 2 and 3. At the second junction the flow is
significantly steadier and ready for silencing.

Six-cylinder engines use a three-into-one connection at the first junction,
with or without a second junction. Eight-cylinder engines can be treated as
two groups of four cylinders. If a four-into-one system is used the benefits

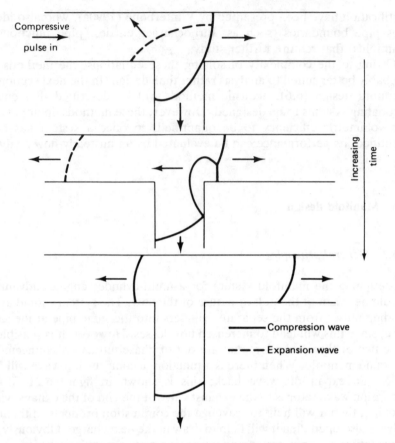

Figure 6.21   Pulsed flow at a junction (with acknowledgement to Annand and
          Roe (1974))

from pressure pulse interactions occur at much higher engine speeds. The choice of layout always depends on the firing order, and will be influenced by the layout of the engine – whether the cylinders are in-line or in 'V' formation. These points are discussed more fully by Annand and Roe (1974) and Smith (1968).

Induction systems are generally simpler than exhaust systems, especially for engines with fuel injection. The length of the induction pipe will influence the engine speed at which maximum benefit is obtained from the pulsating flow. The lengths shown in figure 6.23 are applicable to engines with fuel injection or a single carburettor per cylinder. In most cases it is impractical to accommodate the ideal length.

Inlet manifolds are usually designed for ease of production and assembly, even on turbocharged engines. When a single carburettor per

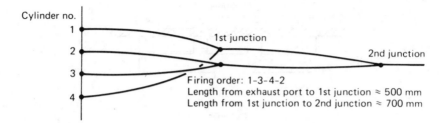

Figure 6.22   Exhaust system for a four-cylinder engine

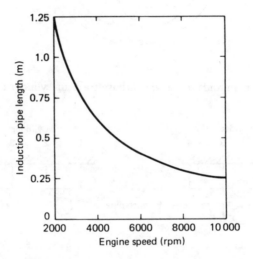

Figure 6.23   Induction pipe length for benefits from pulsating flow (based on Campbell (1978))

cylinder is used, the flow pulsations will cause a rich mixture at full throttle as the carburettor will feed fuel for flow in either direction. In engines with a carburettor supplying more than one cylinder the flow at the carburettor will be steadier because of the interaction between compression and expansion waves. The remainder of this section will deal with manifolds for carburetted spark ignition engines.

In carburetted multi-cylinder engines the carburettor is usually connected by a short inlet manifold to the cylinder head. Although a longer inlet passage would have some advantages for a pulsed flow, these would be more than offset by the added delay in response to a change in throttle, caused by fuel lag or 'hold up'. For these reasons it is difficult to devise a central carburettor arrangement for engines with horizontally opposed cylinders.

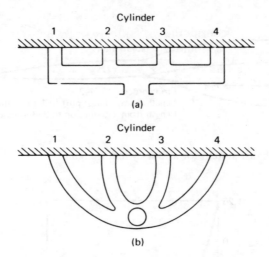

Figure 6.24   Inlet manifolds for single carburettor four-cylinder engine

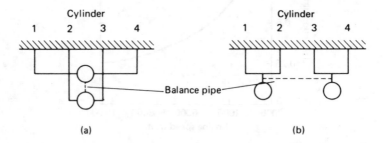

Figure 6.25   Twin carburettor arrangements for a four-cylinder engine

Even with a four-cylinder in-line engine it is difficult to design a satisfactory single carburettor installation. The manifold shown in figure 6.24a will have a poor volumetric efficiency but can be arranged to give a uniform mixture distribution. In comparison, the manifold in figure 6.24b will have a good volumetric efficiency, but is unlikely to have a uniform mixture distribution. If a twin choke carburettor or two carburettors are fitted to this engine, two of the possible manifold arrangements are shown in figure 6.25. The first arrangement has uniform inlet passages and evenly pulsed flow for the common firing order of 1–3–4–2. The second system (figure 6.25b) is more widely used as it is simpler and equally effective. The pulsed flow will be uneven in each carburettor, so that there will be a tendency for maldistribution with each pair of cylinders. However, as the inlet passages are too short to benefit from any pulse tuning, the effect is not too serious, and is further mitigated by the balance pipe. The balance pipe usually contains an orifice, and the complete geometry has to be optimised by experiment. The same considerations apply to other multi-cylinder engine arrangements. Finally, it should be remembered that the throttle plate can have an adverse effect on the mixture distribution. For example, in figure 6.25a the throttle spindle axes should be parallel to the engine axis.

6.6.2 Acoustic modelling techniques

An increasing number of spark ignition engines are using multi-point fuel injection, and this provides scope for applying induction tuning. Induction tuning is also being applied to Diesel engines, even those with turbo-chargers, since it is possible to design the induction system to improve the low speed volumetric efficiency (the operating point where turbochargers are least effective).

It has already been explained (section 6.5) how wave modelling techniques can be used to analyse the induction system performance. But it is appropriate to explain here how acoustic models can be used to design induction systems. It is well established that induction tuning or 'ramming' can lead to an improvement in volumetric efficiency. The increase in the trapped mass of air allows more fuel to be burnt, and if the air/fuel ratio is maintained constant, then there is an increase in power output. Typically an increase in output of 10–20 per cent can be obtained, with an accompanying reduction in the specific fuel consumption of 1–2 per cent. This reduction in fuel consumption occurs since the frictional losses do not increase in direct proportion to the work output. The gains are obviously dependent on: valve timing, the interaction with the exhaust process, the effectiveness of the original induction system, and the speed at which the gains in volumetric efficiency are being sought.

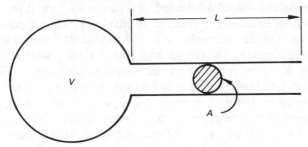

Figure 6.26   A Helmholtz resonator

The two ways of considering a tuned induction system are as an organ pipe, or as a Helmholtz resonator. A Helmholtz resonator is shown in figure 6.26, and the resonant frequency (f_H) is given by

$$f_{\mathrm{H}} = \frac{C}{2\pi} \sqrt{\left(\frac{A}{l \times V} \right)} \text{ (Hz)} \tag{6.31}$$

where C = speed of sound (m/s)
        A = pipe cross-sectional area (m²)
        l = pipe length (m)
        V = resonator volume (m³).

The way that these parameters are assigned to an engine and its induction system will be discussed later. However, even at this stage it is important to emphasise that when a system is tuned, the frequency of the oscillation is not necessarily the same as the frequency of the induction stroke ($N/120$, for a four-stroke engine, with N the engine speed (rpm)).

Alternatively the induction system can be modelled as an organ pipe, for which the fundamental resonant frequency (f_{p}) of a pipe closed at one end is given by

$$f_{\mathrm{p}} = \frac{C}{4L} \text{ (Hz)} \tag{6.32}$$

and

$$L = l + 0.3d \text{ (m)}$$

where L = effective pipe length (m)
        l = pipe length (m)
        d = pipe diameter (m).

Regardless of the model advocated, it is generally agreed that the tuned induction system benefits the volumetric efficiency, when a positive pressure pulse arrives at the inlet valve at or just before inlet valve closure. This has been demonstrated by Ohata and Ishida (1982) who modelled the effect of the inlet port pressure on the volumetric efficiency. First, Ohata and Ishida validated their cylinder filling models by using the recorded inlet port pressure as an input to their model. This gave very close agreement with the observed volumetric efficiency over the whole speed range of their four-cylinder four-stroke engine. Ohata and Ishida (1982) then considered hypothetical pressure pulse forms in the inlet system. This is illustrated here by figure 6.27, and it can be seen that when the pressure pulse occurs around top dead centre, there is a negligible effect on the volumetric efficiency. However, when the pressure pulse arrives around bottom dead centre (ivc at about 40° abdc), then there is a significant gain in volumetric efficiency. There are many examples of tuned induction system reported in

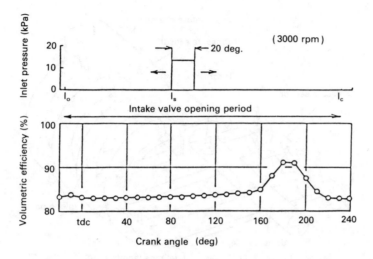

Figure 6.27   The effect of the timing of a hypothetical pressure pulse on the
computed volumetric efficiency, Ohata and Ishida (1982).
[Reprinted with permission © 1982 Society of Automotive
Engineers, Inc.]

the literature, and there are almost as many different theories that have
been contrived to explain the results. This literature (for organ pipe and
Helmholtz systems) has been reviewed by Stone, R. (1989b), so only the
main conclusions will be summarised here.

When a pipe is connected to a single cylinder, the organ pipe theory can
be used to define the resonant frequency of the pipe. Peaks in the engine
volumetric efficiency are usually found when the pipe is resonating at close
to 3, 4 or 5 times the engine cycle frequency. To maintain a resonating
system when damping (viscosity) is present requires a forcing function, and
this is provided by the part of the induction stroke in which a depression
pulse is generated in the inlet port. The induction process period distorts
the period of pressure oscillations in the inlet pipe. However, since the
induction process period is ill-defined (the inlet valve being opened and
closed at a finite rate), then this is why resonance can occur at several
engine speeds, none of which is an exact multiple of the pipe resonating
frequency. Unfortunately, between the engine speeds at which resonance
occurs there will be a minimum in the volumetric efficiency.

For multi-cylinder engines, Helmholtz resonator systems offer more
versatile and compact induction tuning systems. In particular external
resonating volumes can be used to introduce more than one degree of
freedom. Figure 6.28 shows a tuned induction system with two degrees of
freedom. Equation (6.33) defines the resonating system for the geometry
given in figure 6.28, where

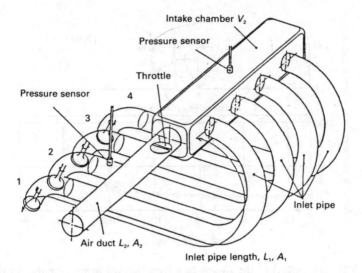

Figure 6.28   Physical arrangement of a tuned induction system with inlet pipes and a resonating volume, Ohata and Ishida (1982). [Reprinted with permission © 1982 Society of Automotive Engineers, Inc.]

$$ abL_1^2V_1^2\omega^4 - (ab + a + 1)(L_1/A_1)V_1\,\omega^2\,c^2 + c^4 = 0 \qquad (6.33) $$

for which there are two positive solutions that give the resonant frequencies:

$$ f = \frac{c}{2\pi}\sqrt{\left\{\frac{(ab + a + 1) \pm \sqrt{[(ab + a + 1)^2 - 4ab]}}{2ab(L/A)_1 V_1}\right\}} \qquad (6.34) $$

With reference to figure 6.28:

$$ a = (L/A)_2/(L/A)_1 = \frac{L_2 A_1}{L_1 A_2} $$

$$ b = V_2/V_1, \qquad V_1 = \frac{V_s}{2}\times\frac{r_v + 1}{r_v - 1} $$

$$ c = \text{speed of sound} $$

where volume V_1 corresponds to the mean cylinder volume.

The use of resonating volumes also lends itself to engines with variable geometry induction systems. Primary pipes of different lengths can be selected, resonating side limbs can be connected to the primary pipes, and resonating volumes can be connected together. Many practical examples

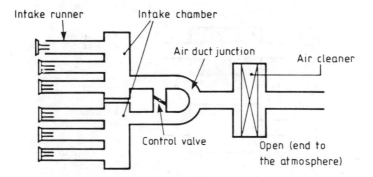

Figure 6.29   Schematic diagram of variable geometry intake manifold by Toyota (Winterbone (1990b))

are considered by Winterbone (1990b). A system that has been quite widely applied to six-cylinder engines is shown in figure 6.29.

When the control valve is closed, the system has a lower resonant frequency than when the control valve is open.

6.7   Silencing

The human ear has a logarithmic response to the magnitude of the fluctuating pressures that are sensed as sound. The ear also has a frequency response, and is most sensitive to frequencies of about 1 kHz. The most effective approach to silencing is the reduction of the peaks, especially those in the most sensitive frequency range of the ear.

The noise from the engine inlet comes from the pulsed nature of the flow, and is modified by the resonating cavities in the cylinder and inlet manifold. A high-frequency hiss is also generated by the vortices being shed from the throttle plate. The inlet noise is attenuated by the air filter and its housing. In addition to its obvious role, the air filter also acts as a flame trap if the engine back-fires.

Exhaust silencers comprise a range of types, as illustrated by Annand and Roe (1974) – see figure 6.30. In general, an exhaust system should be designed for as low a flow resistance as possible, in which case the constriction type silencer is a poor choice. Silencers work either by absorption, or by modifying the pressure waves in such a way as to lead to cancellation and a reduction in sound. Absorption silencers work by dissipating the sound energy in a porous medium. Silencers and their connecting pipes should be free of any resonances. Turbochargers tend to

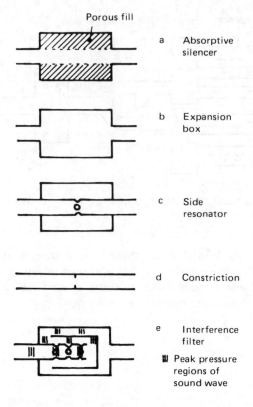

Figure 6.30　The basic silencer elements (with ackowledgement to Annand and Roe (1974))

absorb the flow pulsations from the engine exhaust, but substitute a high-frequency noise generated by the rotating blades.

6.8 Conclusions

To obtain the optimum performance from any internal combustion engine, great care is needed in the design of the induction and exhaust systems. Once the type and disposition of the valve gear have been decided, the valve timing has to be selected. The ideal valve behaviour is obviously modified by dynamic effects, owing to the finite mass and elasticity of the valve train components. The actual valve behaviour can be predicted by computer models. The valve timing will be determined by the application.

The two extremes can be generalised as: normal spark ignition engines or naturally aspirated compression ignition engines, and high output spark ignition engines or turbocharged compression ignition engines. The latter have the greater valve-opening periods. The chosen valve timing is also influenced by the design of the induction and exhaust passages.

Successful design of the induction and exhaust processes depends on a full understanding of the pulsed effects in compressible flows. The first solution methods involved a graphical approach, but these have now been superseded by computer models. Computer models can also take full account of the flow variations during the opening and closing of valves, as well as interactions between the induction and exhaust sytems. This approach is obviously very important in the context of turbochargers, the subject of the next chapter.

6.9 Problems

6.1 Two possible overhead valve combustion chambers are being considered, the first has two valves and the second design has four valves per cylinder. The diameter of the inlet valve is 23 mm for the first design and $18\frac{1}{2}$ mm for the second design. If the second design is adopted, show that the total valve perimeter is increased by 60.8 per cent. If the valve lift is restricted to the same fraction of valve diameter, calculate the increase in flow area. What are the additional benefits in using four valves per cylinder?

6.2 Describe the differences in valve timing on a naturally aspirated Diesel engine, a turbocharged Diesel engine, and a high-performance petrol engine.

6.3 Devise an induction and exhaust system for an in-line, six-cylinder, four-stroke engine with a firing order of 1–5–3–6–2–4, using: (i) twin carburettors, (ii) triple carburettors.

7 Two-stroke Engines

7.1 Introduction

The absence of the separate induction and exhaust strokes in the two-stroke engine is the fundamental difference from four-stroke engines. In two-stroke engines, the gas exchange or scavenging process can have the induction and exhaust processes occurring simultaneously. Consequently, the gas exchange processes in two-stroke engines are much more complex than in four-stroke engines, and the gas exchange process is probably the most important factor controlling the efficiency and performance of two-stroke engines.

Further complications with two-stroke engines are the way in which the gas flow performance is defined, and the terminology. The gas flow performance parameters are not always defined the same way, but the system described in section 7.2 is widely used. The terminology adopted here is shown in figure 7.1. The inlet port is sometimes called the transfer port, but this usage has been avoided here, since not all two-stroke engines use under-piston scavenging. The crankcase port can be called the inlet port, but the crankcase port is a more accurate description of its location and function.

Some background material on two-stroke engines can be found in Heywood (1988) and Taylor (1985a). The gas exchange processes are treated very comprehensively by Sher (1990). The dynamic or wave effects in the inlet system and in particular the exhaust system have a profound effect on the gas exchange processes, but they are not discussed here, as their action is no different from their action in four-stroke engines.

Two-stroke engines can be either spark ignition or compression ignition, and currently the largest and smallest internal combustion engines utilise two-stroke operation. The smallest engines are those used in models, for which compression ignition (perhaps assisted by a glow plug for starting) is more common than spark ignition. The output of these engines is frequently much less than 100 W. Small two-stroke engines with outputs in

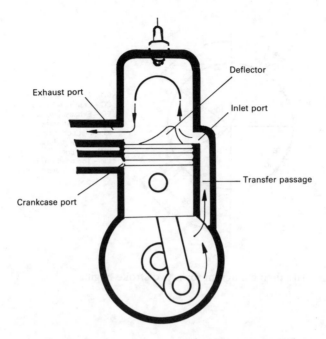

Figure 7.1   The elements of a two-stroke engine with under-piston or crankcase scavenging (adapted from Taylor (1985a))

the range of approximately 1–100 kW are usually naturally aspirated with spark ignition. The typical applications are where low weight, small bulk and high-power output are the prime considerations — for example motor cycles, outboard motors, chain saws, small generators etc. These engines require the incoming charge to be pumped in, and this is achieved most simply by using the underside of the piston and the crankcase as the pump, a method adopted by Joseph Day in 1891. This arrangement is still commonly used, and is illustrated by figure 7.1. Figure 7.2 shows the corresponding timing diagram for this under-piston or crankcase-scavenged engine. Figure 7.3 is a representation of the charging process, and some of the terms involved.

With reference to figures 7.1 and 7.2, the following processes occur in a two-stroke engine, at times determined by the piston covering and un-covering ports in the cylinder wall:

1. At about 60° before bottom dead centre, the piston uncovers the exhaust port and exhaust blowdown occurs, such that the cylinder pressure approaches the ambient pressure (EO). This is the end of the power stroke.
2. Some 5–10° later the inlet port is opened (IO), and the charge

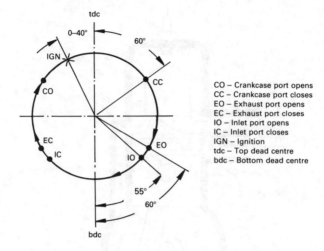

CO – Crankcase port opens
CC – Crankcase port closes
EO – Exhaust port opens
EC – Exhaust port closes
IO – Inlet port opens
IC – Inlet port closes
IGN – Ignition
tdc – Top dead centre
bdc – Bottom dead centre

Figure 7.2   The timing diagram for a two-stroke engine

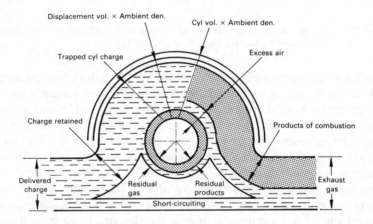

Figure 7.3   The composition of the flows in and out of a two-stroke engine and
the cylinder. [Reprinted with permission from Sher, copyright 1990,
Pergamon Press PLC]

compressed by the underside of the piston in the crankcase is able to
flow into the cylinder. With this loop scavenge arrangement (figure 7.1),
the incoming charge displaces and mixes with the exhaust gas residuals;
some incoming charge will flow directly into the exhaust system.
3. The scavenge process ends with both the crankcase and cylinder pressure close to the ambient pressure level once the inlet port is closed

(about 55° after bottom dead centre). Towards the end of the scavenge process, there can be a backflow of charge and exhaust gas residuals into the crankcase. The upward movement of the piston now reduces the pressure in the crankcase.

4. At about 60° after bottom dead centre the exhaust port is closed, and the charge is compressed by the upward motion of the piston.

5. At about 60° before top dead centre the pressure in the crankcase is significantly below the ambient pressure, and the crankcase port opens to allow the incoming charge to flow into the crankcase.

6. Ignition occurs typically within the range of 10–40° before top dead centre. Work is done by the engine on the gas until top dead centre, at which point the power stroke starts, and continues until the exhaust port opens (1).

7. The crankcase port is closed at about 60° after top dead centre, but prior to this there will be some outflow of gas from the crankcase, as the crankcase pressure will have risen above the ambient pressure level around top dead centre.

The crankcase port system is frequently replaced by a reed valve connected directly to the crankcase. This simplifies the timing diagram (figure 7.2) as the crankcase port is eliminated. Instead the reed valve will open shortly after the inlet port is closed, and the reed valve will close when the piston is near top dead centre. Furthermore, the backflow in the crankcase port as the piston moves down the cylinder is eliminated. Another alternative is to use a disc valve on the crankshaft; this is not quite the same as using a crankcase port, since the disc valve timing is not necessarily symmetric about top dead centre.

Two undesirable features of the two-stroke engine are: the mixing of the incoming charge with the exhaust residuals, and the passage of the charge direct into the exhaust system. The fuel loss due to these shortcomings is largely eliminated in Diesel engines (not considered here are model Diesel engines), as they are invariably supercharged and there is in-cylinder fuel injection.

In order to discuss the gas exchange performance of two-stroke engines, it is necessary to define several gas flow parameters.

7.2 Two-stroke gas flow performance parameters

It must be remembered that many engines are supercharged, and that for two-stroke engines it is not possible to make direct measurements for all the flow performance parameters. There are also different ways of defining the performance parameters, for example, cylinder volume can be defined as the swept volume, or as the trapped volume — the volume at the beginning of the compression process when the transfer port is closed.

The following definitions follow the SAE recommended practice (SAE Handbook, updated annually), and the term 'air' should be interpreted as 'mixture' for engines with external mixture preparation.

$$\text{delivery ratio,} \quad \lambda_d = \frac{\text{mass of delivered air}}{\text{swept volume} \times \text{ambient density}} = \frac{M_i}{M_o} \quad (7.1)$$
(or scavenge ratio)

(Ambient can refer to atmosphere conditions, or for the case of a supercharged engine inlet conditions; the delivery ratio can be calculated from direct measurements.)

$$\text{scavenging efficiency, } \eta_{sc} = \frac{\text{mass of delivered air retained}}{\text{mass of trapped cylinder charge}} = \frac{M_a}{M_a + M_b} \quad (7.2)$$

$$\text{trapping efficiency, } \eta_{tr} = \frac{\text{mass of delivered air retained}}{\text{mass of delivered air}} = \frac{M_a}{M_i} \quad (7.3)$$

The trapping efficiency is a measure of how much air flows directly into the exhaust system (short-circuiting) and how much mixing there is between the exhaust residuals and the air.

$$\text{charging efficiency, } \eta_{ch} = \frac{\text{mass of delivered air retained}}{\text{swept volume} \times \text{ambient density}} = \frac{M_a}{M_o} \quad (7.4)$$

Charging efficiency, trapping efficiency and delivery ratio are clearly related (equations 7.1, 7.3 and 7.4):

$$\eta_{ch} = \lambda_d \eta_{tr} \quad (7.5)$$

$$\text{relative charge, } \lambda_c = \frac{\text{mass of trapped cylinder charge}}{\text{swept volume} \times \text{ambient density}} = \frac{M_a + M_b}{M_o} \quad (7.6)$$
(or volumetric efficiency)

The relative charge is an indication of the degree of supercharging, and it is the ratio of the scavenging efficiency and the charging efficiency. From equations (7.2), (7.4) and (7.6):

$$\lambda_d = \eta_{ch}/\eta_{sc} \quad (7.7)$$

If a sampling system could be used to measure the composition of the trapped charge, its pressure and temperature, and the trapped charge was homogeneous, then the trapped mass of the air and the residuals could be calculated, and all the performance parameters could be evaluated. As this is not practical, a variety of experimental techniques have been developed to evaluate the gas exchange performance of two-stroke engines, and these are discussed in section 7.5.

For two-stroke spark ignition engines with under-piston or crankcase scavenging, then typical scavenge performance results are

scavenge efficiency $0.7 < \eta_{sc} < 0.9$
trapping efficiency $0.6 < \eta_{tr} < 0.8$
charging efficiency $0.5 < \eta_{ch} < 0.7$
delivery ratio $\quad\;\; 0.6 < \lambda_d < 0.95$

7.3 Scavenging systems

Scavenging is the simultaneous emptying of the burned gases and the filling with the fresh air or air/fuel mixture (operation 2 in the description of the two-stroke system in section 7.1). Ideally, the fresh charge would solely displace the burned gases, but in practice there is some mixing. There are many different scavenge arrangements, and some of the more common systems are shown in figure 7.4.

With the cross scavenge arrangement (figure 7.4a) the charge could flow directly into the exhaust system, but this tendency was reduced by the deflector on the piston. The troublesome deflector on the piston was avoided by the loop scavenging system (figure 7.4b) with the exhaust ports above the inlet ports. The incoming air or mixture is directed towards the unported cylinder wall, where it is deflected upwards by the cylinder wall and piston. A modified form of loop scavenging was devised by Schnurle in 1920 (figure 7.4c), in which two pairs of inlet ports are located symmetrically around the exhaust ports. With this system, the flow forms a 'U' shaped loop.

The final two types of scavenging system both employ a uniflow system, either by means of exhaust valve(s) (figure 7.4d), or with an opposed piston arrangement (figure 7.4e). Both these systems are particularly suited to Diesel engines, since the inlet ports can be arranged to generate swirl. Swirl is very important in promoting effective Diesel engine combustion, yet if swirl was used with either cross or loop scavenging, then there would be significant mixing of the inlet flow with the burned gases.

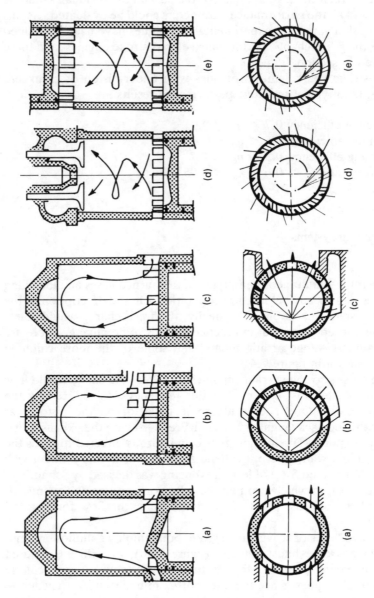

Figure 7.4   The different scavenging arrangements and the associated port geometry for two-stroke engines. (a) Cross-scavenging; (b) loop scavenging; (c) Schnurle loop scavenging; (d) uniflow scavenging with poppet exhaust valves; (e) uniflow scavenging with opposed pistons (adapted from Heywood (1988))

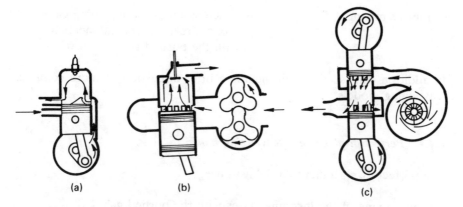

Figure 7.5   Scavenging pump arrangements. (a) Under-piston, or crankcase;
              (b) positive displacement (Roots blower); (c) radial compressor or
              fan (Taylor (1985a))

Figure 7.5 shows different scavenge pumping arrangements for two-stroke engines. The simplest arrangement for pumping the inlet gases into the cylinder is the under-piston pumping that has already been described. The swept volume of this pumping system is the same as that of the engine cylinder, yet better scavenging would be achieved with a larger displacement pump. The term 'blower' is used to describe here a low pressure ratio pump or fan that maintains close to atmospheric conditions in the cylinder. Supercharging refers to a higher pressure ratio system that generates in-cylinder pressures significantly above the atmospheric pressure. The supercharger can be a positive displacement device (such as a screw compressor) and/or a non-positive displacement device such as a radial or centrifugal compressor. These pumps can be driven from the crankshaft or by an exhaust gas turbine — that is turbocharging. Such systems are usually employed on compression ignition engines for non-automotive application, so they will not be discussed any further here.

7.4   Scavenge modelling

The two simplest models for scavenging provide an upper and lower bound for the scavenging performance of an engine. They are the:

Perfect displacement model   — which assumes no mixing of the
                                     incoming gases with the burned
                                     gases.

Complete mixing   — which assumes the incoming gases
                                     mix entirely and instantaneously
                                     with the burned gases.

In addition, this section will include a brief discussion of more complex
scavenging models.

7.4.1 Perfect displacement scavenging model

The perfect displacement model assumes:

(i)   no mixing of the incoming gases with the burned gas
(ii) the process occurs at a constant cylinder volume and pressure
(iii) there is no heat transfer from either the burned gas or the cylinder
      walls
(iv) there are no differences in ambient density.

There are two cases to consider. Firstly, with the delivery ratio less than
unity (equation 7.1), so that not all the exhaust gases are displaced. All the
air admitted will be trapped, and the trapped mass will include some
burned gases. In algebraic terms:

$$
\left.
\begin{aligned}
&\text{delivery ratio,} && \lambda_d \leq 1 && \text{(equation 7.1)}\\
&\text{then trapping efficiency,} && \eta_{tr} = 1 && \text{(equation 7.3)}\\
&\text{and scavenging efficiency,} && \eta_{sc} = \lambda_d && \text{(equation 7.2)}
\end{aligned}
\right\} \quad (7.8)
$$

Secondly, with the delivery ratio greater than unity, then all the burned
gases will be displaced, so that the scavenging efficiency is unity. However,
some of the admitted air will also be exhausted, so the trapping efficiency
will be less than unity. By inspection, for

$$
\left.
\begin{aligned}
&\text{delivery ratio,} && \lambda_d > 1 && \text{(equation 7.1)}\\
&\text{then scavenging efficiency,} && \eta_{sc} = 1 && \text{(equation 7.2)}\\
&\text{and trapping efficiency,} && \eta_{tr} = \lambda_d^{-1} && \text{(equation 7.3)}
\end{aligned}
\right\} \quad (7.9)
$$

7.4.2 Perfect mixing scavenging model

The perfect mixing model assumes:

(i)   instantaneous homogeneous mixing occurs within the cylinder
(ii) the process occurs at a constant cylinder pressure and volume
(iii) the system is isothermal, with no heat transfer from the cylinder walls

(iv) the incoming air (or mixture) and the burned gases have equal and constant ambient densities.

Thus, in the time interval dt at time t, a mass element dm_e enters the cylinder, and an equal amount of mixture dm_l leaves the cylinder. The gas leaving the cylinder comprises a mixture of previously admitted air and burned gases, of instantaneous composition x:

$$x = \frac{\text{instantaneous mass or volume of air in cylinder}}{\text{total mass or volume in cylinder}} = \frac{m_a}{M} \quad (7.10)$$

Differentiating equation (7.10) gives the change in the mass of trapped air:

$$M \, dx = dm_a \quad (7.11)$$

Also, the change in the mass of trapped air is the difference of the inflow and the outflow:

$$dm_a = dm_e - x \, dm_l \quad (7.12)$$

Combining equations (7.11) and (7.12), and recalling that $dm_e = dm_l$:

$$M \, dx = dm_e(1 - x)$$

or

$$\frac{dx}{(1 - x)} = \frac{dm_e}{M} \quad (7.13)$$

The total mass in the cylinder M is a constant and equation (7.13) can be integrated, noting the limits:

(i) At the start of scavenge there is wholly burned gas, so $x = 0$ and $M_e = 0$.
(ii) At the end of scavenge $x = \eta_{sc}$ and $M_e = \lambda_d M$.

$$\int_0^{\eta_{sc}} \frac{dx}{1 - x} = \int_0^{\lambda_d M} \frac{dm_e}{M} \quad (7.14)$$

Integrating gives

$$-\ln(1 - \eta_{sc}) = \lambda_d \quad (7.15)$$

or

$$\eta_{sc} = 1 - \exp(-\lambda_d)$$

and from equations (7.5) and (7.7):

$$\eta_{tr} = 1/\lambda_d[1 - \exp(-\lambda_d)] \tag{7.16}$$

The assumption of no density differences between the incoming gas and the cylinder contents at anytime, may appear to make this result of limited use, when there is obviously a large temperature difference between the inlet gases and the burned gas. However, Sher (1990) presents an analysis that recognises such a temperature difference, but obtains the same results for the charging efficiency, while the scavenging efficiency becomes

$$\eta_{sc} = \frac{T_s}{T_e} \eta_{ch} \tag{7.17}$$

where T_s = temperature at the end of the scavenge period. Sher assumes no variation in the specific heat capacities at constant pressure and constant volume.

The prediction of the scavenging efficiency and the trapping efficiency are plotted in figure 7.6 as a function of delivery ratio, for both the perfect displacement and the perfect isothermal mixing models. If gas dynamic effects are ignored, then for an engine with under-piston scavenging the delivery ratio has to be less than unity. This implies the potential for a high trapping efficiency but a poor scavenging efficiency. For an engine with external mixture preparation, then the high trapping efficiency is important if fuel is not to flow directly into the exhaust system. However, the low scavenging efficiency implies high levels of exhaust gas residuals which necessitate the use of rich mixtures to ensure combustion.

With externally pumped scavenge systems, the delivery ratio can be greater than unity. This leads to high scavenge efficiencies but poorer trapping efficiencies. The fuel consumption and unburnt hydrocarbon emissions will only then be acceptable if in-cylinder fuel injection is used.

Unfortunately, there are two further phenomena that compromise yet further the scavenging and trapping efficiency. Firstly, there can be pockets of burned gas that are neither diluted nor displaced, thus lowering both the scavenging efficiency and the trapping efficiency. Secondly, there can be a flow directly from the inlet port to the exhaust port; this is known as short-circuiting. The short-circuit flow does not necessarily mix with or displace the burned gas, and again, both the scavenging efficiency and the trapping efficiency are reduced.

Incorporated into Figure 7.6 is a region that envelops some experimental data obtained by Taylor (1985a) from some Diesel engines in the 1950s. In all cases the scavenging efficiency is lower than that predicted by the

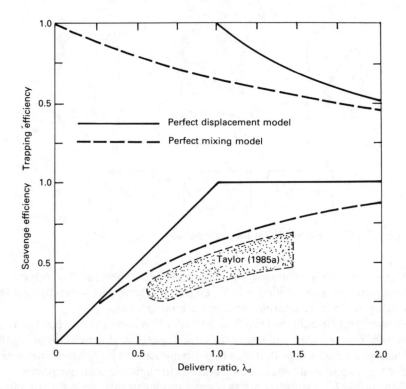

Figure 7.6   The trapping efficiency and scavenging efficiency as a function of
the delivery ratio, for the perfect displacement and perfect
isothermal mixing scavenging models. Scavenge efficiency data have
been added from Taylor (1985a) for compression ignition engines

mixing theory but as the engine speed was increased the discrepancy
always reduced. However, Sher (1990) states that for modern engines the
perfect mixing model underestimates the scavenging performance.

7.4.3   Complex scavenging models

Complex scavenging models divide the cylinder into two or more zones,
and usually allow for a fraction of the flow to be short-circuited; this
requires at least two empirical constants to be included in any such
correlation. Typical assumptions are:

 (i) the in-cylinder pressure and volume are invariant
 (ii) there is no heat or mass transfer between the zones
(iii) in each zone the temperatures may be different but are uniform
(iv) no variation in gas properties.

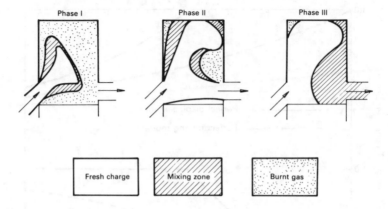

Figure 7.7   The Benson (1977) three-phase scavenging model

Maekawa (1957) proposed an isothermal two-zone model in which the entering stream was divided into three parts: a flow that short-circuits the exhaust port, a pure charge zone and a mixing zone.

Benson and Brandham (1969) also used a two zone model, but the inlet flow was divided into only two flows: a flow that short-circuited to the exhaust port, and a flow that enters a mixing zone. The second zone was a burned gas zone that would be displaced through the exhaust port.

Benson (1977) extended his two zone model to include a third zone of fresh charge. He also divided the scavenging process into the three phases shown in figure 7.7:

(i) *Phase I*   95–155° atdc, exhaust is displaced and there is mixing at the interface of the incoming charge and the burned gas.

(ii) *Phase II*   155–200° at 200° atdc, mixing at the interface continues, but some of the incoming charge is now allowed to short-circuit through the exhaust port.

(iii) *Phase III*   200–265° atdc, the short-circuit flow stops, and a homogeneous mixture of the fresh charge and the burned gas leaves the cylinder; as with the perfect mixing model, the composition is time dependent.

Sher (1990) has proposed a model that assumes the fresh charge content in the exhaust gas stream to be represented by a sigmoid ('S' type) curve. With this model, scavenging is represented by a combination of perfect displacement scavenging, followed by an isothermal mixing process. The selection of the shape factors for the sigmoid curve allows results to be produced that lie anywhere between the perfect displacement model and the perfect mixing model.

Numerical techniques have also been employed to model the scavenging process, and a short review is presented by Sher (1990). Typically, turbulence is modelled by the K–ε model; diffusion of mass heat and momentum all have to be incorporated, along with appropriate sub-models for jet mixing and propagation. In view of the complexity of the processes, it is not surprising that only limited progress seems to have been achieved.

7.5 Experimental techniques for evaluating scavenge and results for port flow coefficients

Experimental measurements have two vital roles. Firstly, to evaluate the performance of particular engines, and secondly, to enable empirical or numerical models of scavenging to be calibrated. Experiments can be conducted on either firing or non-firing engines. The techniques that can be applied to firing engines are more restricted, but they are discussed first here.

7.5.1 Firing engine tests

In section 7.2, in-cylinder sampling has already been suggested as a means of evaluating the trapped mass of the air and residuals. However, assumptions have to be made about how representative the sample of the gas is, and it is preferable to take a large sample which might affect the engine performance; it is also difficult to measure the pressure and temperature of the sample. A better technique (that can only be applied to engines with in-cylinder fuel injection) is the exhaust gas sampling system described by Taylor (1985a). The sampling valve shown in figure 7.8 is placed close to the exhaust port, and is connected to a receiver in which the pressure is controlled by the flows to a gas analyser and a bleed valve. The receiver pressure is set at a level such that the sampling valve opens during the exhaust blowdown period. This way the burned gas can be analysed, and the air/fuel ratio can be deduced. Since the quantity of fuel injected into the engine is known, then it is possible to calculate the trapped mass of air. Since the mass of delivered air can be measured directly, then it is possible to calculate: the delivery ratio, the scavenging efficiency, the trapping efficiency, and the charging efficiency (equations 7.1–7.4).

Another technique for evaluating the scavenging process is the tracer gas method. In this method, a tracer gas is introduced continuously into the incoming air so as to give a fixed percentage of the tracer gas in the air. The tracer gas has to be selected so that it will be completely burnt at the

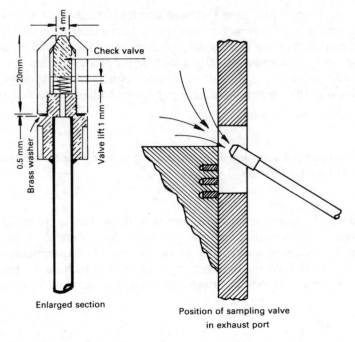

Figure 7.8   Exhaust sampling valve (Taylor (1985a))

combustion temperature, but not at the temperatures during compression or once the exhaust ports are open. Sher (1990) provides a comprehensive review of this technique, and the two most widely used gases are monomethylamine (CH_3NH_2) and carbon monoxide (CO). With monomethylamine, care is needed to ensure it is not absorbed by condensate in the exhaust. The carbon monoxide is easier to analyse (with an infra-red absorption technique) and it is also more stable in the exhaust; however carbon monoxide is also a product of combustion.

If: y is the mass fraction of tracer gas in the inlet air
     z is the mass fraction of unburned tracer gas in the exhaust and
     λ_d, the delivery ratio is found by direct measurement (equation 7.1):

$$\text{charging efficiency } \eta_{ch} = \lambda_d \eta_{tr} \text{ (equation 7.5)}$$

From the definition of trapping efficiency (equation 7.3):

$$\eta_{ch} = \lambda_d \frac{y - (M_i/M_e)z}{y} \tag{7.18}$$

7.5.2 Non-firing engine tests

There are two types of non-firing test, those that use continuous steady flow and those that motor the engine or a model — these dynamic tests are usually limited to one cycle.

The simplest steady-flow tests are those with the piston fixed at bottom dead centre, and the cylinder head removed. Such tests are widely used to investigate the flow of air into the cylinder, the aim being to have a flow that interacts with the piston and cylinder to produce the maximum axial velocity at the opposite diameter to the exhaust port. The results can be plotted as velocity contours in the manner of figure 7.9. This technique was first proposed by Jante (1968) and its subsequent development has been considered by Sher (1990).

A range of flow measurement techniques has been applied to steady-flow tests with the cylinder head in place; these techniques include laser doppler anemometry, hot wire anemometry and the flow visualisation of scavenging by water and coloured salt solutions.

Sher (1990) also presents a comprehensive review of the flow visualisation and measurement techniques applied to dynamic tests of unfired engines — mostly for a single cycle. In the tests with liquids, then only the parts of the cycle can be investigated in which the ports are open. None the less, such tests produce useful qualitative information. For example, short-circuiting flows and stagnant regions can be found. With gaseous systems, combinations of air, carbon dioxide, helium, ammonia and Freon-22 have been used to model the fresh charge and the burned gases. Some techniques give the instantaneous concentration of the species at

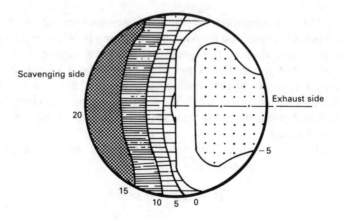

Figure 7.9   Velocity contours from a steady flow test at the open cylinder head (adapted from Sher (1990))

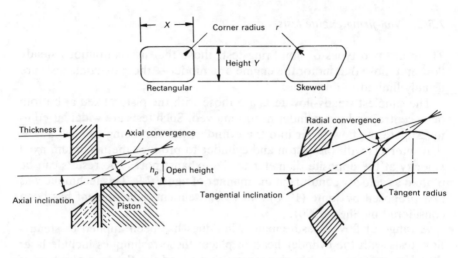

Figure 7.10   The geometric features of cylinder ports (Annand and Roe (1974))

discrete locations, although the techniques have only computed the mean concentration at the end of scavenge. Sher lists five criteria for similarity apart from geometric similarity, so it is probably impossible to obtain complete similarity between the model and the engine. Nor, of course, do these techniques provide any insight into the wave action effects that are so important in successful engine operation.

7.5.3   Port flow characteristics

A comprehensive summary of the flow phenomena through piston-controlled ports is presented by Annand and Roe (1974), who also provide a thorough description of the geometric features (figure 7.10). Beard (1984) states that up to 60–70 per cent of the bore periphery can be available as port width, but the remainder is required for carrying the axial loads and ensuring adequate lubrication for the piston and ring pack. Nor must individual ports be made too wide, otherwise the piston rings are likely to be damaged.

With reference to figure 7.10, the effective flow area when the piston completely uncovers the ports is:

$$A = XY - 0.86r^2 \qquad\qquad (7.19)$$

and when the port is partially covered by the piston (but $hp > r$), then

$$A = xhp - 0.43r^2 \qquad\qquad (7.20)$$

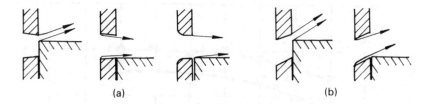

(a)   (b)

Figure 7.11   The influence of inlet port geometry on the flow patterns. (a) Port
axis perpendicular to wall; small opening and large opening with
sharp and rounded entry; (b) port axis inclined (Annand and Roe
(1974))

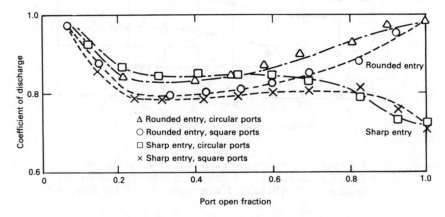

Figure 7.12   The discharge coefficient as a function of the port open fraction,
for different geometry inlet ports, Wallace, W. B. (1968).
[Reprinted with permission © 1968 Society of Automotive
Engineers, Inc.]

Figure 7.11 shows how the port geometry influences the inlet port flow
— the flow coefficients are higher than those for poppet valves; some
typical variations with the port open fraction are shown in figure 7.12.

The pressure ratio across the exhaust port varies considerably; it is high
during the exhaust blow down, but around unity during the scavenge
period. Figure 7.13 shows the effect of the pressure ratio and piston
position on the discharge coefficient for a single rectangular exhaust port.
It should be noted that even when the port is fully open, the piston position
influences the flow pattern and thus the discharge coefficient.

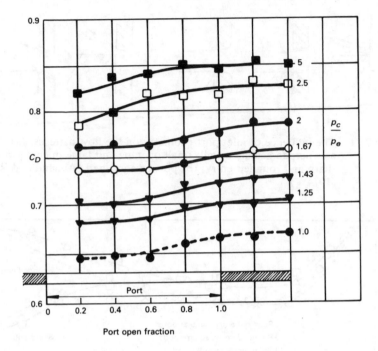

Figure 7.13   The effect of pressure ratio and port open fraction on the
discharge coefficient for a simple rectangular exhaust port,
Benson (1960). [Reprinted with permission of *The Engineer*]

7.6   Engine performance and technology

Some typical full throttle performance curves for a two-stroke spark
ignition engine are presented in figure 7.14. The maximum bmep is 6.4 bar
at about 4000 rpm. The minimum fuel consumption of 400 g/kWh occurs at
a lower speed (about 3000 rpm). For a naturally aspirated four-stroke
spark ignition engine (in fact the Rover M16 multi point injection engine)
the minimum specific fuel consumption is less than 275 g/kWh — some 31
per cent lower than that for the two-stroke engine. The maximum bmep
approaches 11 bar, so if torque or power per unit volume is considered,
then the two-stroke engine is only about 17 per cent higher. The construction of a typical two-stroke engine is shown in figure 7.15.

As mentioned in section 7.1, one of the disadvantages of the externally
carburated or fuel injected two-stroke engine is the passage of the air–fuel
mixture into the exhaust system (by short-circuiting and mixing). This
leads to the high fuel consumption and high emissions of unburned hydro-

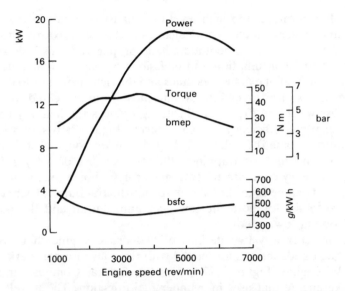

Figure 7.14   Performance characteristics of a three-cylinder, 450-cm³, two-stroke cycle, spark ignition engine. Bore = 58 mm, stroke = 56 mm, (Uchiyama *et al.* (1977))

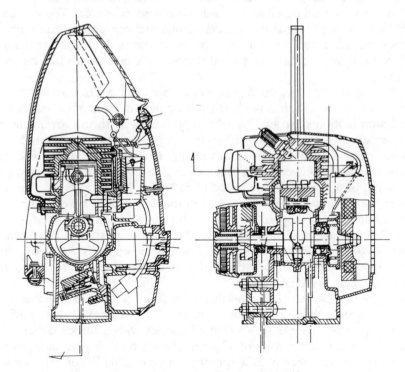

Figure 7.15   McCullock chain-saw engine (Taylor (1985b))

carbons. Furthermore, the high level of exhaust residuals in the trapped charge (low scavenging efficiency) dictates that a rich mixture is supplied, in order for combustion to be initiated and propagated: this is detrimental to both the fuel consumption and emissions of carbon monoxide, However, the high level of exhaust residuals and the rich air/fuel ratio do lead to lower emissions of nitrogen oxides than those from untreated four-stroke engines. As would be expected, throttling reduces the delivery ratio and thus the charging efficiency is also reduced. Figure 7.16 shows how the engine output (bmep) is almost solely dependent on the charging efficiency. On the plot of trapping efficiency against delivery ratio, the charging efficiency contours are rectangular hyperbolae (equation 7.5). In figure 7.16, it can be seen that at low speeds the trapping efficiency is low, and this accounts for both the low maximum output and the poor fuel consumption at low speeds.

There are currently two designs of two-stroke engine that overcome some of the disadvantages associated with the conventional spark ignition two-stroke engine. Figure 7.17 shows the Orbital Combustion Process two-stroke engine that uses in-cylinder fuel injection. The novelty lies in the fuel injection system, which uses fuel at a pressure of 5.75 bar and air blast from a supply at 5 bar to help atomise the fuel to a spray that is mostly below 10 microns. The mixture preparation allows a weak mixture to be burnt, thus controlling nitrogen oxide emissions at source. The engine is fitted with a simple oxidation catalyst. Some fuel consumption and hydrocarbon emission comparisons are shown in figure 7.18. The maximum torque from this engine occurs at 4000 rpm, and corresponds to a bmep of 7.1 bar.

The idea of using a stepped piston dates back to the 1920s and as can be seen from figure 7.19, it is most sensibly applied to pairs of cylinders with the piston motion separated by 180° crank angle. With this arrangement, the sequence of events is no different from that of a conventional two-stroke engine with under-piston scavenging. However, the area ratio of the piston affects the delivery ratio. Such an engine has been developed to give an output of 6 bar bmep at a speed of 5500 rpm (Dunn (1985)). Another advantage of this design is that conventional four-stroke wet sump type lubrication can be adopted. A subsequent development has been to employ low pressure fuel injection into the transfer port; this is shown in figure 7.20. By phasing the injection towards the end of scavenging, then the trapped charge can be stratified and losses of fuel direct into the exhaust system are eliminated.

Automotive two-stroke compression ignition engines are uncommon, but a Detroit Diesel is shown in figure 7.21. This V8 engine has a Roots-type scavenge pump operating in series with a turbocharger. The bore of 108 mm and stroke of 127 mm gives a total displacement of 9.5 litres. At its rated speed of 2100 rpm the output of 276 kW corresponds to a

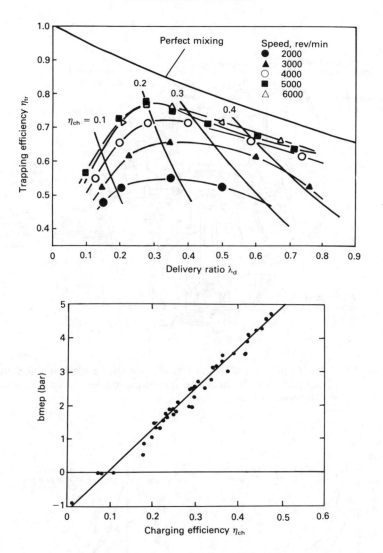

Figure 7.16   (a) The dependence of the brake mean effective pressure on the
charging efficiency (that is, the trapped mass of fuel). (b) The
trapping and charging efficiency as function of the delivery ratio
for a two-stroke, spark ignition two-cylinder engine with a 347 cm³
swept volume, Tsuchiya and Hiramo (1975). [Reprinted with
permission © 1972 Society of Automotive Engineers, Inc.]

bmep of 8.5 bar, with a specific fuel consumption of 231 g/kWh. The
specific fuel consumption is perhaps 10 per cent worse than a four-stroke
engine of comparable swept volume, but allowing for the two-stroke
operation, then the specific output is about 40 per cent higher.

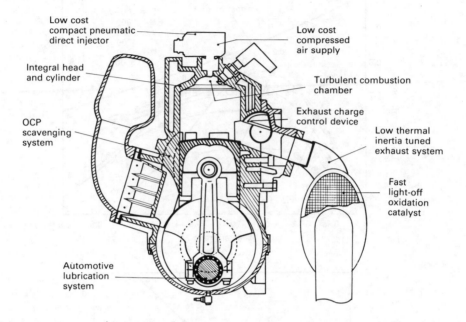

Figure 7.17   The Orbital Combustion Process two-stroke spark ignition engine
that utilises in-cylinder fuel injection, courtesy Orbital Engine
Corporation Ltd

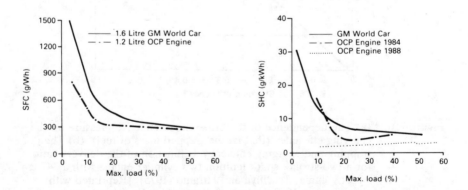

Figure 7.18   A comparison between a 1.6 litre four-stroke engine and the
Orbital Combustion Process 1.2 litre two-stroke engine. (a) The
specific fuel consumption as a function of load; (b) the specific
hydrocarbon emissions as a function of load, courtesy Orbital
Engine Corporation Ltd

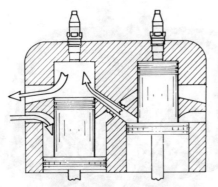

Figure 7.19   The stepped piston two-stroke engine applied to a two cylinder
              engine, Dunn (1985). [Reprinted with permission of
              The Engineer]

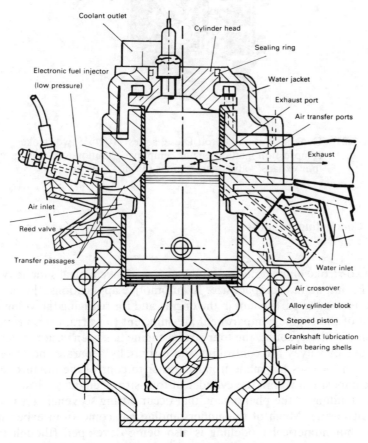

Figure 7.20   A prototype stepped piston two-stroke engine with wet-sump
              lubrication, and fuel injection into the transfer passage

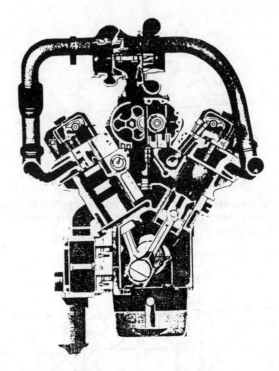

Figure 7.21   General Motors Detroit Diesel series 71 60° V two-stroke
compression ignition engine, employing uniflow scavenging, with
four exhaust valves per cylinder (Taylor (1985b))

7.7   Concluding remarks

The two-stroke engine offers the advantages of mechanical simplicity, high
specific output and compact size. In automotive applications, the potential
for reducing the parts count in the engine and for reducing the volume and
height of the engine compartment are sufficient to warrant serious investigations. Without doubt, the four-stroke engine has been easier to develop.
The use of valves for controlling the essentially separate induction and
exhaust processes has made it much easier to control the mixture motion
and composition within the cylinder of four-stroke engines. However, the
understanding of the phenomena that occur during scavenging is continually increasing. Much of this understanding has come from experimental
work, but numerical modelling is also being developed (though this of
course depends on experimental data from rig tests and firing engines for
validation).

In its most familiar form, the two-stroke spark ignition engine has crankcase or under-piston scavenging with external mixture preparation. Such engines have a poor reputation for high fuel consumption and emissions levels. The under-piston scavenging limits the delivery ratio and this can lead to high levels of exhaust residuals in the cylinder, which then dictate the induction of a rich mixture to ensure combustion. Furthermore, as the trapping efficiency is less than unity, then unburnt fuel flows directly into the exhaust, with concomitant high emissions of unburnt hydrocarbons and a high fuel consumption. At part throttle the delivery ratio is reduced, and this decreases the scavenging efficiency and the charging efficiency. Indeed, it is not unknown for some two-stroke engines to fire only on alternate revolutions, this is known as 'four-stroking'. In other words, the scavenging is so poor that in alternate revolutions the mixture in the cylinder cannot be ignited.

In section 7.6 it was discussed how these shortcomings could be overcome. The use of a stepped piston allows the designer to increase the delivery ratio. When this is coupled with in-cylinder fuel injection, or fuel injection that is synchronised towards the end of induction, then the passage of fuel directly through the exhaust port can be controlled. These developments account for the current interest of car manufacturers in two-stroke engines.

The automotive applications of two-stroke compression ignition engines are limited. However, in stationary and marine applications large two-stroke engines are widely available. The in-cylinder fuel injection ensures proper fuel utilisation, and the use of scavenge pumps and supercharging enables good scavenging and high outputs to be attained.

8 In-cylinder Motion

8.1 Introduction

The induction of air or an air–fuel mixture into the engine cylinder leads to a complex fluid motion. There can be an ordered air motion such as swirl, but always present is turbulence. The bulk air motion which approximates to a forced vortex about the cylinder axis is known as axial swirl, and it is normally associated with direct injection Diesel engines. When the air motion rotates about an axis normal to the cylinder axis (this is usually parallel to the cylinder axis), the motion is known as barrel swirl or tumble. Such a motion can be found in spark ignition engines (normally those with two inlet valves per cylinder). Barrel swirl assists the rapid and complete combustion of highly diluted mixtures, for instance those found in lean-burn engines, or engines operating with stoichiometric mixtures but high levels of exhaust gas recirculation (EGR).

Turbulence can be pictured as a random motion in three dimensions with vortices of varying size superimposed on one another, and randomly distributed within the flow. Viscous shear stresses dissipate the vortices, and in doing so also generate smaller vortices, but overall the turbulent flow energy is dissipated. Thus energy is required to generate the turbulence, and if no further energy is supplied, then the turbulence will decay. In steady flows the turbulence can be characterised by simple time averages. The local flow velocity fluctuations (u) are superimposed on the mean flow (U). For the x direction:

$$U(t) = \bar{U} + u(t) \tag{8.1}$$

For a steady flow, the mean flow is found by evaluating:

$$\bar{U} = \lim_{T \to \infty} \frac{1}{T} \int_{t_0}^{t_0 + T} U(t) \, dt \tag{8.2}$$

300

In other words the mean velocity $(\bar{U})$ is found by averaging the flow velocity $(U(t))$ over a sufficiently long time for the average to become stationary. As it will be the energy associated with the turbulence that will be of interest, the velocity fluctuations are defined by their root mean square (rms) value (u'). This is known as the turbulence or the turbulence intensity:

$$u' = \lim_{T \to \infty} \sqrt{\left(\frac{1}{T} \int_{t_0}^{t_0+T} [u(t)]^2 dt \right)} \qquad (8.3a)$$

or, as the time average $2u(t)\,\bar{U}$ is zero:

$$u' = \lim_{T \to \infty} \sqrt{\left(\frac{1}{T} \int_{t_0}^{t_0+T} \{[U(t)]^2 - U^2\} \right)} \, dt \qquad (8.3b)$$

Unfortunately, in reciprocating engines the flow is not steady, but is varying periodically. Thus measurements have to be made over a large number of engine cycles, and measurements made at a particular crank angle (or crank angle window) are averaged. This approach is known as ensemble averaging or phase averaging. Before discussing turbulence in engines further (section 8.3), it is useful to know how the flows are measured, and this is the subject of section 8.2. Section 8.4 describes how turbulence measurements are used in models that predict turbulent combustion.

The way in which the flows can be predicted computationally is beyond the scope of this chapter. Fortunately Gosman (1986) has presented a very comprehensive review of flow processes in cylinders, and this encompasses: turbulence definitions, flow measurement techniques and multi-dimensional computational fluid dynamics (CFD) techniques for flow prediction.

8.2 Flow measurement techniques

8.2.1 Background

The main flow measuring techniques for providing quantitative information are hot wire anemometry (HWA), and laser doppler anemometry or velocimetry (LDA/V); the use of these techniques has been reviewed thoroughly by Witze (1980). Other techniques that are employed are: smoke and tufts for gas motion studies, and water flow rigs with gas bubbles to identify the particle paths. In their basic forms, these techniques only provide qualitative information. However, these techniques are amenable to refinement. If the flow is illuminated by a sheet of light, then photographs can be taken as a multiple image or as a ciné film. Another technique uses Freon as the working fluid so that a model engine can be run at reduced speed, yet retain dynamic similarity for the flow. The flow can be visualised by hollow phenolic spheres (microballoons) illuminated by sheets of light. The motion can then be tracked by video cameras linked to a computer-based system to compute the flow velocities. An example of this type of work is presented by Hartmann and Mallog (1988). However, when steady-flow rigs are used for visualising or measuring flows in an engine cylinder, consideration must always be given to how the results might relate to the unsteady flow in a reciprocating engine.

8.2.2 Hot wire anemometry

Hot wire anemometry (HWA) relies on measuring the change in resistance of a fine heated wire. The usual arrangement is a constant temperature anemometer, in which a feedback circuit controls the current through the wire to maintain a constant resistance, and thus a constant temperature. The current flow then has to be related to fluid velocity. Unfortunately, the direction of the flow cannot be determined, as the wire will respond to normal flows, and to a lesser extent to an axial flow. Thus, to determine a three-dimensional flow, three independent measurements are needed, and these are most conveniently analysed if the wire orientations are orthogonal, in which case for the arrangement show in figure 8.1:

$$V^2_{\text{eft}, 1} = V^2_x + (K_2V_y)^2 + (K_3V_z)^2 \qquad (8.4)$$

$$V^2_{\text{eft}, 2} = V^2_x + (K_3V_y)^2 + (K_2V_z)^2$$

$$V^2_{\text{eft}, 3} = (K_3V_x)^2 + (K_2V_y)^2 + V^2_z$$

where $V_{\text{eft}, 1}$, $V_{\text{eft}, 2}$ and $V_{\text{eft}, 3}$ are the effective velocities calculated from the anemometer output, and K_2 and K_3 are the pitch and yaw factors that account for the reduced sensitivity in the pitch and yaw directions.

In a time-varying flow, the three measurements have to be made simultaneously, and this implies a three wire probe. Each measurement corresponds to a mean along the length of the wire, and obviously the three measurements cannot refer to exactly the same point. As a hot wire anemometer cannot deduce the direction of the flow, the computed velocity vector can lie in any of the octants. Furthermore, the presence of the wires and their supports will modify the flow structure. If the wires are

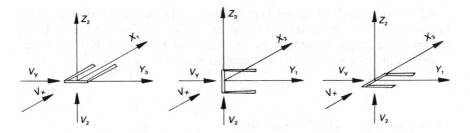

Figure 8.1   Probe positions for determining the three orthogonal velocity
components in a fixed coordinate system (X, Y, Z), with probe
coordinates: 1, 2, 3

made too short, then there will also be heat transfer to the wire supports,
as well as to the fluid stream.

Ideally, the probes have to be calibrated for the range of pressures and
temperatures that will be encountered. However, this is not convenient if
measurements are to be made in a motored engine, since it implies making
calibrations over a wide range of pressures and temperatures. Instead, a
common approach is to make corrections to a baseline calibration, using
analytical models that have been previously substantiated experimentally.

A widely used model is:

$$Nu = A + B\ Re^n \tag{8.5}$$

where $$Nu = \frac{hd}{k} = \frac{Q}{\pi kl(T_w - T_g)}$$

$$
\begin{aligned}
A, B, n &= \text{constants} \\
h &= \text{heat transfer coefficient (w/m}^2\text{K)} \\
d &= \text{wire diameter (m)} \\
k &= \text{thermal conductivity (W/mk)} \\
Q &= \text{heat flow (W)} \\
l &= \text{wire length (m)}
\end{aligned}
$$

suffixes: w — wire, g — free stream gas.

This model was first proposed by King (1914) and with $n = 0.5$, it is known
as King's Law:

$$Nu = A + B\ R_e^{0.5} \tag{8.6}$$

As with any correlation for heat transfer, there is the question of what temperature to evaluate the transport properties at. The options include: the free stream gas temperature, the wire surface temperature, or some weighted average such as the film temperature (T_f):

$$T_f = (T_w + T_g)/2 \qquad (8.7)$$

Davies and Fisher (1964) found that

$$Nu_w = 0.425\, Re_g^{\frac{1}{3}} \qquad (8.8)$$

Collis and Williams (1959) concluded that

$$Nu_f \left(\frac{T_f}{T_g}\right)^{-0.17} = 0.24 + 0.56\, Re_f^{0.45}$$
$$\text{for } 0.02 < Re < 44$$
$$= 0.48\, Re_f^{0.51}$$
$$\text{for } 44 < Re < 140 \qquad (8.9)$$

Witze (1980) recommended

$$Nu_g = A + BRe_g^n \qquad (8.10)$$

Witze points out that analytical solutions for forced convection through a laminar boundary layer invariably yield $n = 0.5$, while the fit to experimental calibration gives

$$0.24 < n < 0.47$$

the exact value being a function of the gas velocity and wire temperature.

The implication from these results is that, whenever possible, the hot wire probes should be calibrated for the range of velocities being encountered, with the tests being conducted at different pressures and temperatures.

When analytical-based corrections are being made for pressure and temperature, the transport properties that have to be predicted are: density, thermal conductivity and viscosity. Correlation for the temperature correction of dynamic viscosity (μ) and thermal conductivity (k) have been made by Collis and Williams (1959):

$$k \propto T^{0.8} \qquad (8.11)$$

$$\mu \propto T^{0.76} \qquad (8.12)$$

8.2.3 Laser Doppler anemometry

Laser Doppler anemometry (LDA) is illustrated by figure 8.2. A pair of coherent laser beams are arranged to cross at the focal point of a lens. Interference fringes are created throughout the beam intersection region or measuring volume; the fringe lines (or really planes) are parallel to the lens axis and perpendicular to the plane of the incident laser beams. When particles carried by the fluid pass through the interference fringe, they scatter light. The scattered light is detected by a photomultiplier tube, and the discontinuous output from the scattered light is called the Doppler burst. The Doppler frequency equals the component of the velocity perpendicular to the fringes (that is, parallel to the lens axis) divided by the distance between the fringes. The spacing of the fringes (Δx) is related to the wavelength of the laser light (λ) and the half-angle of the two intersecting beams (ϕ):

$$\Delta x = \frac{\lambda}{2 \sin \phi} \tag{8.13}$$

If the 514.4 nm wavelength light from an argon-ion laser is used with a half-angle of 7.37°, then the fringe spacing is 2 microns.

The measuring volume is an ellipsoid whose diameter and length depend on the half-angle and the laser beam diameter. Typically, the length (along the lens axis) is about 600 microns with a diameter of 80 microns; this will give a maximum of 40 fringes. When a particle passes through the control volume it will only pass through some of the fringes. If the velocity component of the particle is 10 m/s, then the frequency of the Doppler burst will be 5 MHz with a maximum possible duration of 8 μs.

The particle size should preferably be about a quarter of the light wavelength, and it is assumed that this is sufficiently small to move with the

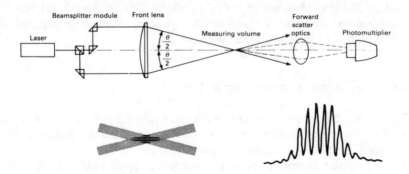

Figure 8.2 Optical arrangement for laser Doppler anemometry (LDA) showing the interference fringe pattern, collection of the light scattered by particles in the forward scatter mode, and a typical output from the photo-multiplier

flow. The flow has to be seeded with particles, for example titanium dioxide (TiO_2) is used in non-reacting flows, and zirconium fluoride (ZrF_4) is used when combustion occurs. The particles arrive in the measuring volume at 0.1 to 10 ms intervals, and the output from the photodetector can vary in amplitude and dc offset, as shown in figure 8.2. If a frequency shifter is installed in the optical path of one of the beams, then the fringe pattern will move with a constant velocity. If this velocity is greater than the negative component of any particle velocity, then the flow direction can always be resolved.

The signal processing requirements are considerable. The signal is discontinuous, varying in amplitude and dc offset; it also has a very high frequency and a short duration. A commonly used system is a frequency tracking demodulator, in which a phased locked loop latches on to the Doppler frequency. This requires a nearly continuous Doppler signal, with a limited dynamic range; as neither of these conditions occurs with internal combustion engines, then other techniques are needed. Witze mentions other signal processing techniques (namely filtering and photon correlation) but concludes that period counting seem the most appropriate technique for internal combustion engines. Period counting entails the use of a digital clock to measure the transit time of a specified number of fringe transits in a single burst. Since the counter output is digital and randomly intermittent, further digital processing is essential. For example, it can be necessary to record 2000 or more velocity/crank angle data pairs.

The optical arrangement shown in figure 8.2 is a forward scatter system, which requires a second optical access and receiving lens for focusing the scattered light on to the photomultiplier (or photodector). A second system that only requires a single optical access employs backscatter (figure 8.3). While backscatter systems are easier to implement, the signal strength is significantly weaker. Other optical arrangements can be used, for example the scattered light can be collected through a window at right angles to the main beams. In order to obtain three-dimensional velocity measurements, multiple measurements (most convenient if orthogonal) need to be made, perhaps simultaneously.

8.2.4 Comparison of anemometry techniques

Witze (1980) has provided a comprehensive comparison of hot wire and laser Doppler anemometry. The advantages and disadvantages of each techniques are almost entirely complementary.

Laser Doppler anemometry is: non-intrusive, applicable at all temperatures and pressures, linear, direction sensitive and can be used when there is no mean flow. The disadvantages of LDA are: its cost, the need for (multiple) optical access, the intermittent and poor signal-to-noise ratio of the signal and its high frequency, and the need to seed the flow. With

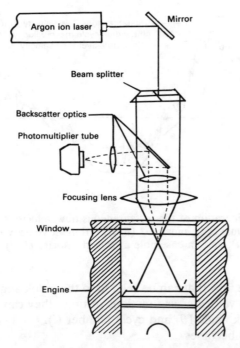

Figure 8.3   LDA system using back scatter, such that only a single optical
access is required, from Witze (1980). [Reprinted with permission
© 1980 Society of Automotive Engineers, Inc.]

LDA, it is also necessary to isolate the optical system from any engine
vibration, and to ensure that the motion of the combustion chamber (due
to vibration) is minimised — otherwise the engine motion will be superimposed on the measured flow velocity.

Laser Doppler anemometry is considered to be more accurate than hot
wire anemometry. The inaccuracy of LDA is considered to be 5–10 per
cent under favourable conditions and perhaps 10–20 per cent under adverse conditions. An over-riding advantage of LDA can be its ability to
make measurements in a firing engine. Unfortunately LDA equipment is
an order of magnitude more expensive than HWA equipment.

8.3   Turbulence

8.3.1   Turbulence definitions

In the introductory section 8.1, equations (8.1)–(8.3) were used to define
turbulence in steady flow. In a reciprocating engine there will be a periodic

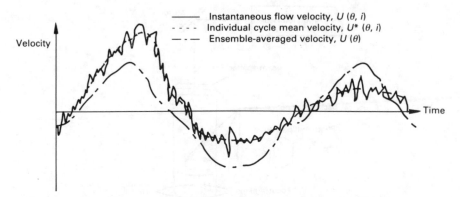

Figure 8.4   The separation of the instantaneous flow velocity signal into an
individual cycle mean U^* (θ, i); turbulent velocity fluctuations,
u' (θ, i), and an ensemble averaged velocity U (θ)

component in the flow that can be related to the crank angle. Thus instead
of expressing the velocities as functions of time, they can be expressed as
functions of crank angle (θ) and cycle number (i).

Equation (8.1) becomes

$$U(\theta, i) = \bar{U}(\theta) + u(\theta, i) \qquad (8.14a)$$

Figure 8.4 shows that there can be cycle-to-cycle variations in the mean
flow, thus equation (8.1) can alternatively be written in a way that recognises these variations separately from the turbulent fluctuations:

$$U(\theta, i) = \bar{U}(\theta, i) + u(\theta, i) \qquad (8.14b)$$

or

$$U(\theta, i) = \bar{U}(\theta) + U^*\,(\theta, i) + u(\theta, i)$$

The $U^*(\theta, i)$ term is to account for the cycle-by-cycle variations in the
mean flow. The $u(\theta, i)$ terms will be different in equations (8.14a) and
(8.14b), and a dilemma in studies of turbulence is whether or not the
low-frequency fluctuations $U^*(\theta, i)$ (that is, cycle-by-cycle variations in the
mean flow) should be considered separately from the turbulent fluctuations
— $u(\theta, i)$. Of course $U(\theta, i)$ is the only velocity that exists in the flow; the
other velocity terms are introduced to enable the flow to be characterised.

At any chosen crank angle, the ensemble-averaged velocity can be found
by averaging as many as possible velocity measurements corresponding to
that crank angle:

$$U_{ea}(\theta) = \lim_{n \to \infty} \left[\frac{1}{n} \sum_{i=1}^{n} U(\theta, i) \right] \qquad (8.15)$$

where n is the number of cycles.

The fluctuations from the mean $u(\theta, i)$ can be expressed in terms of their rms value:

$$u'_{ea}(\theta) = \lim_{n \to \infty} \left[\frac{1}{n} \sum_{i-1}^{n} u(\theta, i)^2 \right]^{\frac{1}{2}} \qquad (8.16)$$

When the velocity measurements have been taken with an LDA system, it must be remembered that the signals are not continuous. It is thus necessary to specify a crank angle window ($\pm \Delta\theta/2$) centred on a mean value of crank angle, $\bar{\theta}$, and have a system to assign the measurements to the appropriate windows — this can be called a phase resolver. The width of the crank angle window selected will influence the number of velocity measurements; in some cycles there will be no velocity measurements while in other cycles there will be several velocity measurements. Thus in equations (8.14)–(8.16)

$$\theta \text{ is replaced by } \bar{\theta} \pm \Delta\theta/2$$

and n becomes the number of velocity measurements made, not the number of cycles.

If the crank angle window is made too small, then the data rate (the rate at which velocity measurements are made) becomes too low, and a long time is required to obtain a statistically valid sample. The convergence of the averaging process within a window varies as the inverse of the square root of the number of points averaged. This implies that more measurements in total are needed as the window size is reduced. Typically, data from a few thousand engine cycles are recorded.

Conversely, if the crank angle window is too wide an error is introduced called crank angle broadening; this is due to the change in mean velocity during the crank angle window. The crank angle broadening error is discussed by Witze (1980), who suggests that a 10° window is satisfactory unless there is a dynamic event (such as combustion or fuel injection), in which case the crank angle window needs to be reduced (to say 2°).

Equations (8.15) and (8.16) can be used with equation (8.14a) to characterise the flow. However, this ignores the cycle-by-cycle variations in the mean flow that were identified by the additional term in equation (8.6). This variation is thought to be significant in the cycle-by-cycle variation in combustion observed in spark ignition engines. Gosman (1986) discusses the different techniques that can be used for separating the cycle-by-cycle variation in the mean flow (this is the U^* (θ, i) term in equation 8.14b). One approach is a hybrid time/ensemble averaging scheme, in which the averaging time (T, in equations 8.2 and 8.3) is limited to a specified crank period (say 45°). However, this period is arbitrary, and during the period the mean velocity can vary significantly. An extension of this approach is to use a low pass filter to remove the cyclic variations in the mean velocity. It is said that this makes the signal time independent (stationary), so that simple time-averaging techniques can be used. Unfortunately, this makes it impossible to resolve individual cycles.

An approach devised by Rask (1981) fits smooth curves to the velocity data from each cycle, in order to define the variation in the mean velocity, $\bar{U}(\theta, i)$. The instantaneous turbulent velocity component, $u'(\theta, i)$, is then defined as the difference between the velocity signal and the smoothed curve. This enables the following to be determined:

(a) The turbulent velocity fluctuation $u'(\theta, i)$.
(b) The cycle-by-cycle variation in the mean velocity $\bar{U}^*(\theta, i)$.
(c) The ensemble-averaged mean velocity $\bar{U}(\theta)$.

Unfortunately, the choice of the curve fitting to the original signal can slightly influence the results obtained, in an analogous manner to filtering. Some results obtained by Rask (1981) are shown in figure 8.7, which is discussed in section 8.3.2.

Gosman (1986) concludes that there is uncertainty about how to characterise the air motion in engines. This point is also made by Heywood (1988). Now that the means of obtaining turbulence data have been described, it is necessary to discuss how the measurements are used, by defining some length and time scales that are used to characterise turbulence. The largest eddies are limited by the cylinder boundaries at one extreme, and at the other by molecular diffusion. The eddies that are responsible for producing most of the turbulence during induction arise from the jet-like flow that emerges from the inlet valve. The radial and axial velocity components in the jet are an order of magnitude greater than the piston velocity, and the initial width of the jet approximates to the valve lift. Figure 8.5 shows how the shear between the jet and the cylinder contents leads to eddies. These large eddies are unstable and break down into a cascade of smaller eddies. The length and time scales in turbulence are defined and discussed comprehensively by Gosman (1986) and Heywood (1988); since the definitions are important they are also developed here.

The integral scale, l_i is a measure of the largest scale structure. If velocity measurements are made a distance apart that is significantly less than the integral scale (l_i), then the velocity measurements will correlate with each other. The spatial correlation coefficient R_x is defined in terms of the

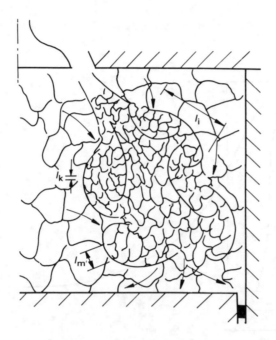

Figure 8.5   A representation of the induction process, to show the generation
of eddies by shear action, and the characteristic length scales of the
turbulence, from Gosman (1986) by permission of Oxford
University Press

velocity fluctuations at a location x_0 and the velocity fluctuations at a
distance x away:

$$R_x = \frac{u(x_0)u(x_0 + x)}{u'(x_0)u'(x_0 + x)} \qquad (8.17)$$

or $\qquad R_x = \frac{1}{n - 1} \sum_{i=1}^{n} \frac{u(x_0)u(x_0 + x)}{u'(x_0)u'(x_0 + x)}$ for n measurements

The variation of the correlation coefficient with distance is shown in
figure 8.6. The area under the curve is used to define the integral length
scale.

$$l_i = \int_0^{\infty} R_x \, dx \qquad (8.18)$$

In other words, the area under the curve is equivalent to the rectangle
bounded by the axes and the coordinates $(l_i, 1)$.

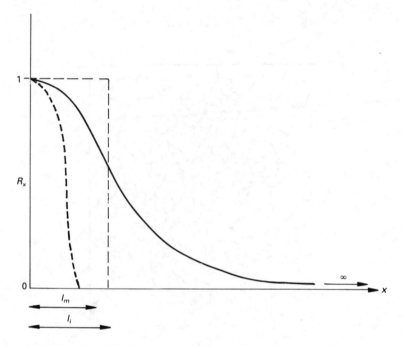

Figure 8.6   The variation of the correlation coefficient (R_x) with distance to illustrate the definitions of the integral length scale (l_i) and the Taylor microscale (l_m)

A second scale for characterising turbulence is the Taylor microscale (l_m), and this has been interpreted as the spacing between the smallest eddies. The Taylor microscale is defined by the intercept of a parabola with the x-axis, for a parabola that matches the autocorrelation function in height and curvature at $x = 0$. This is also illustrated in figure 8.6. From the equation for a parabola, it can be shown that

$$l_m^2 = -2/(\partial^2 R_x/\partial x^2)_{x=0} \tag{8.19}$$

The third and smallest length scale is the Kolmogorov scale (l_k). Since the smallest eddies respond most quickly to changes in the local flow, it is assumed that turbulence at this level is isotropic (the same in all directions). It is at this level that viscosity dissipates the kinetic energy to heat. Dimensional analysis is used to relate the Kolmogorov scale to the energy dissipation rate per unit mass, ε:

$$l_0 = \left(\frac{\nu^3}{\varepsilon}\right)^{1/4} \tag{8.20}$$

Owing to the difficulty in obtaining simultaneous flow measurements from two positions in an engine, an alternative and easier approach is to find characteristic time scales, and then see how these might be related to the length scales.

The temporal auto-correlation function (R_t) is defined in a similar way to the spatial correlation function, except that the velocity fluctuations are compared at the same point, but the measurements are taken at time t apart. With n measurements:

$$R_t = \frac{1}{n-1} \sum_{i=1}^{n} \frac{u(t_0)\, u(t_0 + t)}{u'(t_0)\, u'(t_0 + t)} \qquad (8.21)$$

This leads to a similar definition for the integral time scale, t_i:

$$t_i = \int_0^\infty R_t \, dt$$

If the turbulence is relatively weak, and the flow is convected past the measuring point without significant distortion, then it has been found that the integral time scales and the integral length scales are related by the mean flow velocity ($\bar{U}$):

$$l_i = \bar{U} t_i \qquad (8.22)$$

For flows without a mean motion, then the integral time scale (t_i) is an indication of the eddy lifetime.

For turbulence that is homogeneous and isotropic the Taylor microscale is related in a similar manner:

$$l_m = \bar{U} t_m \qquad (8.23)$$

The Kolmogorov time scale, t_k, is defined as

$$t_k = \left(\frac{v}{\varepsilon} \right)^{1/2} \qquad (8.24)$$

and it characterises the momentum diffusion within the smallest flow structures.

Heywood (1988) explains how the turbulent kinetic energy per unit mass in the large eddies is proportional to u'^2, and that as these eddies lose a substantial fraction of their energy in one 'turnover' time l_i/u' then:

$$\varepsilon \approx u'^3/l_i \qquad (8.25)$$

and from equation (8.20):

$$\frac{l_k}{l_i} \approx \left(\frac{u'l_i}{\nu}\right)^{-3/4} = Re_t^{-3/4} \tag{8.26}$$

where Re_t is the turbulent Reynolds number.

Heywood (1988) also states that

$$\frac{l_m}{l_i} = \left(\frac{15}{A}\right)^{1/2} Re_t^{-1/2} \tag{8.27}$$

where A is a constant of order unity.

8.3.2 In-cylinder turbulence

The origins of turbulence during the induction stroke were described in the introductory section 8.1, namely turbulence is generated by the shear between the jet flowing from the inlet valve and the cylinder contents. As this jet has a mean velocity that is an order of magnitude greater than the piston velocity, then the jet will impinge on the piston, as well as the cylinder walls. This leads to a toroidal vortex below the inlet valve, but offset towards the axis of the cylinder. The result of these flows is a high level of turbulence generated during induction, which then decays. However, further turbulence is generated by squish: the inward radial movement of air caused by the small clearances between the piston crown and cylinder head.

Some tabulated values of turbulence parameters from Lancaster (1976) are presented in table 8.1.

Table 8.1   Turbulence parameters in a motored CFR engine at 2000 rpm (Lancaster (1976))

	u' (m/s)	$\bar{U}$ (m/s)	l_i (mm)	l_m (mm)	l_k (mm)	t_i (ms)	t_m (ms)	t_k (ms)
Mid induction	5.0	20	4.0	1.0	0.02	0.4	0.07	0.04
Late compression	15	10	4.0	1.0	0.03	0.8	0.20	0.12

These results show that the integral, Taylor microscale and Kolmogorov lengths are almost independent of crank angle, and are all much smaller than the bore diameter or stroke (about 100 mm). The integral length scale corresponds to just under half the maximum valve lift. The timescales are in a similar ratio to one another as the length scales, and the longest time scale (the integral time scale) is significantly shorter than the duration of the induction stroke (15 ms).

Some flow measurements made at two locations by Rask (1981) are shown in figure 8.7. The engine was motored at 300 rpm, and this gave a mean piston speed of 0.76 m/s. As the engine was motored, the measurements during the expansion and exhaust strokes will not be representative of those in a fired engine. In figures 8.7(d) and 8.7(e) the differences between the smoothed ensemble and the ensemble-averaged individual-cycle variations in the mean flow, represent the cycle-by-cycle variations in the mean flow. This variation can be seen to be small in comparison with the turbulence levels during induction, but the difference becomes significant during compression. Thus the cycle-by-cycle variations in the mean flow are also significant during compression. The differences in flow for locations b and c in figure 8.7 show that the flow is not homogeneous, nor is the flow isotropic (Rask (1981)). However, these results are for an engine in which there is a strongly directed flow from the inlet valve.

Bracco and his co-workers have conducted studies with a variety of port configurations and disc-shaped combustion chambers (Liou et al. (1984); Hall and Bracco (1986)). Like Rask (1981), these studies have separated the velocity measurements into:

(i) a mean flow
(ii) a cycle-by-cycle variation in the mean flow and turbulence.

The separation was achieved through converting the velocity data into the frequency domain by a Fourier transformation, and identifying a cut-off frequency (which was a function of engine speed), and then applying inverse Fourier transformations to the separated signals. Above the cut-off frequency the velocity data represented turbulence, below the cut-off frequency, the velocity data represented the mean flow and the cycle-by-cycle variations in the mean flow.

Liou et al. (1984) report data from three different engine builds: a valved four-stroke engine (without swirl), a ported two-stroke engine with swirl, and a ported two-stroke engine without swirl. Since the combustion chambers were essentially disc-shaped, then the turbulence is mostly generated by the induction process. If the cycle-by-cycle variations in the mean flow and the turbulence intensity were not separated, then the turbulence intensity was overestimated by as much as: 300 per cent in the valved engine (no swirl), 100 per cent in the ported engine (no swirl) and 10 per cent in the ported engine with swirl. This implies that the presence of swirl reduces the cycle-by-cycle variations in the mean flow, and that this should lead to less cyclic variation in spark ignition engine combustion.

Most important, so far as combustion is concerned, is the nature of the flow around top dead centre. Liou et al. (1984) show that with open

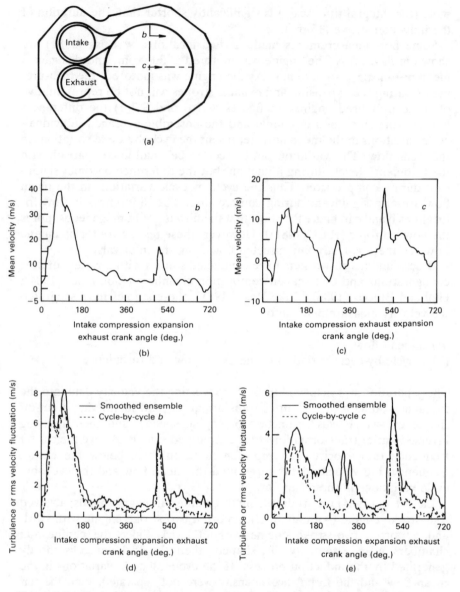

Figure 8.7   Turbulence measurements from a side valve engine motored at
300 rpm (mean piston speed of 0.76 m/s). (a) Location and
direction of the velocity measurements, note that the inlet port will
generate swirl; (b) and (c) show the ensemble-averaged mean
velocities at locations b and c; (d) and (e) show the turbulence
intensity (u'), calculated by neglecting the cycle-by-cycle variations
in the mean flow (---), and the turbulence intensity calculated when
allowance is made for the cycle-by-cycle variations in the mean flow
(---). Rask (1981), reproduced with permission of ASME

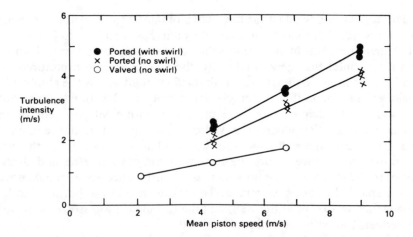

Figure 8.8   Relationship between the mean piston speed and the turbulence
            intensity at top dead centre for different induction systems (cycle-
            by-cycle variations in the mean flow are not included here in
            the turbulence intensity) (data adapted from Liou *et al.* (1984))

disc-shaped combustion chambers (for measurements at different locations
in the clearance volume around top dead centre, at the end of compression) the turbulence is relatively homogeneous (within ± 20 per cent).
In the absence of any swirl, the turbulence was also essentially isotropic
(within ± 20 per cent) near top dead centre. Data presented by Liou *et al.*
(1984) in figure 8.8, show that for individual engines, the turbulence
intensity at top dead centre increases linearly with engine speed:

$$u'_{tdc} \approx k \, \bar{v}_p \qquad\qquad (8.28)$$

with the constant of proportionality in the range $0.25 < k < 0.5$ depending
on the type of induction system. Liou *et al.* (1984) conclude, that the
maximum turbulence intensity that can be obtained at tdc in an open
chamber without swirl is equal to half the mean piston speed. Turbulence
intensity as high as 75 per cent of the mean piston speed has been reported,
but this presumably includes cycle-by-cycle variations in the mean flow.

Heywood (1988) also notes that when swirl is present, the turbulence
intensity at top dead centre is usually higher than when swirl is absent. The
variation in individual-cycle mean velocity at the end of compression also
scales with the piston speed, and this variation can be comparable in
magnitude with the turbulence intensity. With swirling flows and bowl-in
piston combustion chambers, the conservation of the moment of momentum means that the swirl speed will increase. The increase in angular
velocity (swirl speed) causes shear that can cause an increase in the
turbulence level.

The results discussed so far have been obtained from motored engines, in other words there has been no combustion. Hall and Bracco (1986) took LDA measurements in an engine with a disc-shaped combustion chamber, and a port inlet that generated swirl; the engine was both motored and fired. In the motoring tests they took readings right across the combustion chamber, and found that the tangential velocity at a location 0.5 mm from the wall (bore diameter of 82.6 mm) was within 90 per cent of the maximum tangential velocity. They also found that the turbulence intensity increased near the wall, and this was explained by shear at the wall generating turbulence. From the tests with combustion, Hall and Bracco (1986) found that there was little increase in turbulence across the flames at all the spatial locations examined. The turbulence in the burned gas was homogeneous, except at the wall, and the turbulence was isotropic (within 20 per cent) to within 1.5 mm of the wall.

The role of turbulence in spark ignition engine combustion is discussed by Kyriakides and Glover (1988) who found a strong correlation between turbulence intensity and the 10–90 per cent burn time. They investigated various means of producing air motion, and concluded that tumble (or barrel swirl) was a more effective way for generating turbulence at top dead centre than axial swirl. In this work the cycle-by-cycle variations in mean flow were not separated from the turbulence intensity.

From a knowledge of turbulence intensity and length scales it is possible to construct a turbulent entrainment model of the combustion in a spark ignition engine; this is discussed in the next section.

8.4 Turbulent combustion modelling

The combustion in a homogeneous charge spark ignition engine is commonly divided into three parts:

(a) An initial laminar burn, before the flame kernel is large enough to be influenced by turbulence; this can be considered as corresponding to the first few per cent mass fraction burned.
(b) Turbulent burning, with a comparatively wide flame front, and pockets of unburned mixture entrained behind the flame front.
(c) A final burn period, ('termination period' or 'burn-up'), in which the mixture within the thermal boundary layer is burned at a slow rate, because of a reduced fluid motion and a lower temperature.

Many different workers have published phenomenological combustion models, including: Tabaczynski (1976), Tabaczynski et al. (1977), Tabaczynski *et al.* (1980), Keck (1982) Beretta *et al.* (1983), Borgnakke (1984), Tomita and Hamamoto (1984), Keck *et al.* (1987) and James (1990).

For the majority of cases, the combustion chamber will have a complex geometric shape, in which it will be necessary to define: the enflamed volume, the flame front area, and the area wetted by the enflamed volume. Two different approaches to this problem are provided by Poulos and Heywood (1983) and Cuttler *et al.* (1987). Poulos and Heywood divide the combustion chamber surface into a series of triangular elements, and then check for interception by vectors, of random direction, radiating from the spark plug. Cuttler *et al.* fill the combustion with tetrahedra, and employ vector algebra.

The turbulent combustion models used by Keck and co-workers and Tabaczynski and co-workers both seem capable of giving good predictions of turbulent combustion. Since the information published by Tabaczynski and co-workers is more explicit (notably Tabaczynski *et al.* (1977) and (1980)), their model will be described here.

The flame is assumed to spread by a turbulent entrainment process, with burning occurring in the entrained region at a rate controlled by turbulence parameters. The flame front is assumed to entrain the mixture (that is, spread into the unburned mixture) at a rate that is governed by the turbulence intensity and the local laminar burn speed. The rate at which the mass is entrained into the flame front is given by

$$\frac{dm_e}{dt} = \rho_u A_f (u' + S_L) \tag{8.29}$$

where m_e = mass entrained into the flame front
  ρ_u = density of the unburned charge
  A_f = flame front area (excluding area in contact with surfaces)
  u' = turbulence intensity
  S_L = laminar flame speed.

The turbulence intensity is assumed to be proportional to the mean piston speed. The rate at which the mixture is burned is assumed to be proportional to the mass of unburned mixture behind the flame front:

$$\frac{dm_b}{dt} = \frac{m_e - m_b}{\tau} \tag{8.30}$$

where m_b = mass burned behind the flame front
  τ = characteristic burn time for an eddy of size l_m
  l_m = the Taylor microscale, which characterises the eddy spacing.

Since the eddies are assumed to be burned up by laminar flame propagation:

$$\tau = \frac{l_{\mathrm{m}}}{S_{\mathrm{L}}} \qquad (8.31)$$

This model assumes isotropic homogeneous turbulence throughout the combustion chamber at the time of ignition. After ignition the integral scale (l_{i}) and the turbulence intensity (u') are assumed to be governed by the conservation of the moment of momentum for coherent eddies. Thus

$$l_{\mathrm{i}} = (l_{\mathrm{i}})_{\mathrm{o}} \left(\frac{\rho_{\mathrm{ui}}}{\rho_{\mathrm{u}}} \right)^{1/3} \qquad (8.32)$$

and

$$u' = u'_{\mathrm{o}} \left(\frac{\rho_{\mathrm{u}}}{\rho_{\mathrm{ui}}} \right)^{1/3} \qquad (8.33)$$

with the suffix, 'o' referring to the values at ignition.

For isotropic turbulence:

$$l_{\mathrm{m}} = \sqrt{\frac{(15 v 1;/u;)}{u'}} \qquad (8.34)$$

where μ = kinematic viscosity.

By using equations (8.32), (8.33) and (8.34), the eddy burn-up term can be calculated at each time step (equation 8.32). However, this presupposes a knowledge of the kinematic viscosity, and the burned and unburned mixture density.

The mass within the combustion chamber (m) can be assumed constant, and this means that the only unknown is the ratio of unburned to burned gas density (D).

$$m = \rho_{\mathrm{o}} V_{\mathrm{o}} = \rho_{\mathrm{u}} V_{\mathrm{u}} + \rho_{\mathrm{b}} V_{\mathrm{b}}, \qquad \text{let } D = \frac{\rho_{\mathrm{u}}}{\rho_{\mathrm{b}}}$$

$$\rho_{\mathrm{u}} = \frac{\rho_{\mathrm{o}} V_{\mathrm{o}}}{V - V_{\mathrm{b}} + \dfrac{V_{\mathrm{b}}}{D}} \qquad (8.35)$$

where V = volume
        V_u = volume of unburned mixture
        V_b = volume of burned mixture
        V_o = volume at ignition
        ρ_o = unburned mixture density at ignition.

$$\text{mfb} = \frac{m_b}{m} = \frac{\rho_b V_b}{\rho_o V_o} \tag{8.36}$$

where   mfb = mass fraction burnt
        m_b = mass burnt.

To find V_b:

$$V_b = \frac{m_b}{m} \cdot \frac{\rho_o V_o D}{P_u} \tag{8.37}$$

The density ratio has to be calculated from the combustion thermodynamics, and the background to this has been introduced in chapter 3, section 3.3, and will be discussed further in chapter 10, section 10.2.

To calculate the density of the unburned mixture the compression process can be modelled on a stepwise basis, using a heat transfer correlation to predict the heat flow from the unburned mixture; this is discussed later in chapter 10, section 10.2.4. However, an alternative and simpler method is to use a polytropic process to describe the compression of the unburned mixture (with a user supplied value of the polytropic index). Thus

$$P = P_o \left(\frac{\rho_u}{\rho_o} \right)^k \tag{8.38}$$

and

$$T_u = \frac{P \rho_o T_o}{P_o \rho_u} \tag{8.39}$$

where T_o = temperature of (unburned) mixture at ignition.

The kinematic viscosity can be calculated from an empirical correlation proposed by Collis and Williams (1959) for the dynamic viscosity (equation 8.12).

$$v = \frac{v_o \rho_o}{\rho_u} \left(\frac{T_u}{T_o} \right)^{0.76} \tag{8.40}$$

It now remains to calculate the laminar burning velocity.

The laminar burning velocity is a function of: the equivalence ratio, pressure, initial temperature of the reactants, and exhaust residuals. The burning velocity can be represented as a parabolic function of equivalence ratio for many substances, and is given by;

$$S_L^* = B_m + B_\phi (\phi - \phi_m)^2 \qquad (8.41)$$

where S_L^* = burning velocity at datum conditions (298 K, 1 atm.)
B_m = maximum burning velocity, occurring at $\phi = \phi_m$
B_ϕ = empirical constant.

Values of B_m, B_ϕ and ϕ_m are presented in table 8.2.

Table 8.2   Laminar flame speed parameters for the range of 300–700 K and 1–8 bar

Fuel	ϕ_m	B_m (m/s)	B_ϕ (m/s)	Reference
Methanol	1.11	0.369	−1.41	Metghalchi and Keck (1980)
Propane	1.08	0.342	−1.39	Metghalchi and Keck (1980)
Iso-octanes	1.13	0.203	−0.85	Metghalchi and Keck (1980)
Gasoline	1.21	0.305	−0.55	Rhodes and Keck (1985)

The data at higher temperatures and pressures can be fitted to a power law:

$$S_L = S_{L,o} \left(\frac{T_u}{T_o} \right)^\alpha \left(\frac{P}{P_o} \right)^\beta \qquad (8.42)$$

where $\alpha = 2.18 - 0.8(\phi - 1)$
$\beta = -0.16 + 0.22 (\phi - 1)$.

For gasoline, Rhodes and Keck (1985) propose:

$$\alpha_g = 2.4 - 0.271\phi^{3.15}$$
$$\beta_g = 0.357 + 0.14\phi^{2.77} \qquad (8.43)$$

Rhodes and Keck (1985) also found that the proportional reduction in burning velocity caused by the presence of residuals, was essentially independent of: equivalence ratio, pressure, and temperature. They developed the following correlation:

$$S_L(x_b) = S_L(x_b = 0) (1 - 2.06x_b^{0.77}) \qquad (8.44)$$

where x_b = mole fraction of burned gas diluent.

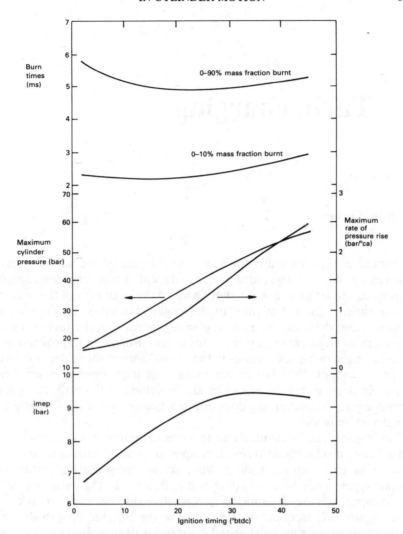

Figure 8.9   Predictions of how the burn rate, imep, maximum cylinder
              pressure, and the maximum rate of cylinder pressure rise are
              affected by ignition timing, for a Ricardo E6 engine operating at
              1500 rpm with wide open throttle and an equivalence ratio of 1.0
              using iso-octane

A model using these methods has been developed by Brown, A. G.
(1991) to investigate the combustion of different fuels under various
operating conditions. He has used this model to show how movement of
the flame nucleus around the combustion chamber can lead to cycle-
by-cycle variation in combustion. Figure 8.9 shows predictions of how the
variations in: burn rate, imep, maximum cylinder pressure, and the maximum rate of cylinder pressure rise are affected by ignition timing.

9   Turbocharging

9.1 Introduction

Turbocharging is a particular form of supercharging in which a compressor is driven by an exhaust gas turbine. The concept of supercharging, supplying pressurised air to an engine, dates back to the beginning of the century. By pressurising the air at inlet to the engine the mass flow rate of air increases, and there can be a corresponding increase in the fuel flow rate. This leads to an increase in power output and usually an improvement in efficiency since mechanical losses in the engine are not solely dependent on the power output. Whether or not there is an improvement in efficiency ultimately depends on the efficiency and matching of the turbocharger or supercharger. Turbocharging does not necessarily have a significant effect on exhaust emissions.

Compressors can be divided into two classes: positive displacement and non-positive displacement types. Examples of positive displacement compressors include: Roots, sliding vane, screw, reciprocating piston and Wankel types; some of these are shown in figure 9.1. The axial and radial flow compressors are dynamic or non-positive displacement compressors — see figure 9.2. Because of the nature of the internal flow in dynamic compressors, their rotational speed is an order of magnitude higher than internal combustion engines or positive displacement compressors. Consequently, positive displacement compressors are more readily driven from the engine crankshaft, an arrangement usually referred to as a 'supercharger'. Axial and radial compressors can most appropriately be driven by a turbine, thus forming a turbocharger. Again the turbine can be of an axial or radial flow type. The thermodynamic advantage of turbochargers over superchargers stems from their use of the exhaust gas energy during blow-down, figure 2.5.

A final type of supercharger is the Brown Boveri Comprex pressure wave supercharger shown in figure 9.3. The paddle-wheel type rotor is driven from the engine crankshaft, yet the air is compressed by the

324

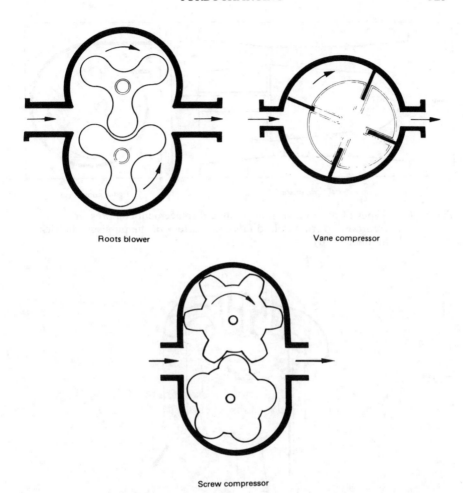

Figure 9.1   Types of positive displacement compressor (reproduced from Allard
              (1982), courtesy of the publisher Patrick Stephens Ltd)

pressure waves from the exhaust. Some mixing of the inlet and exhaust
gases will occur, but this is not significant.

The characteristics of turbochargers are fundamentally different from
those of reciprocating internal combustion engines, and this leads to
complex matching problems when they are combined. The inertia of the
rotor also causes a delay in response to changes in load — turbolag.
Superchargers have the added complication of a mechanical drive, and the
compressor efficiencies are usually such that the overall economy is reduced. However, the flow characteristics are better matched, and the
transient response is good because of the direct drive. The Comprex

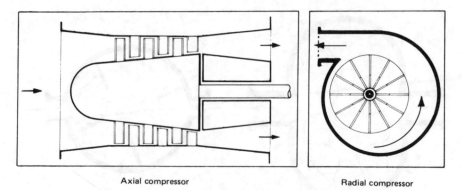

Axial compressor   Radial compressor

Figure 9.2   Types of dynamic or non-positive displacement compressor (reproduced from Allard (1982), courtesy of the publisher Patrick Stephens Ltd)

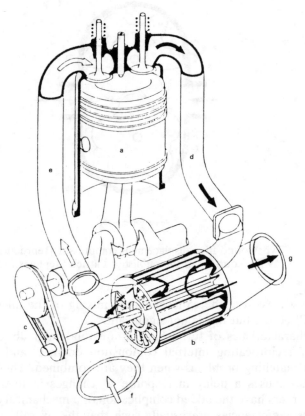

Figure 9.3   Brown Boveri Comprex pressure wave supercharger. (a) Engine; (b) cell-wheel; (c) belt drive; (d) high-pressure exhaust; (e) high-pressure air; (f) low-pressure air; (g) low-pressure exhaust

supercharger absorbs minimal power from its drive, and has a good transient response; but it is expensive to make and requires a drive. The fuel economy is worse than a turbocharger, and its thermal loading is higher. The development theory and application of the compress supercharger are covered in the Brown Boveri Review, Vol. 7 No. 8, August 1987.

Comprex superchargers have not been widely used, and superchargers are used on spark ignition engines only where the main consideration is power output. Turbochargers have been used for a long time on larger compression ignition engines, and are now being used increasingly on automotive compression ignition and spark ignition engines.

Compound engines are also likely to gain in importance. A compound engine has a turbine geared to the engine crankshaft, either the same turbine that drives the compressor or a separate power turbine. The gearing is usually a differential epicyclic arrangement, and if matching is to be optimised over a range of speeds a variable ratio drive is also needed. Such combinations are discussed by Wallace, F. J. *et al.* (1983) and by Watson and Janota (1982). Compound engines offer improvements in efficiency of a few per cent compared with conventional turbocharged diesel engines.

Another development that is most relevant to turbocharged engines is the low heat loss (so called 'adiabatic') Diesel engine.

In a naturally aspirated engine, the higher combustion chamber temperature will lead to a reduction in the volumetric efficiency, and this will offset some of the gains from the increased expansion work. The fall in volumetric efficiency is less significant in a turbocharged engine, not least since the higher exhaust temperature will lead to an increase in the work available from the turbine. This is discussed further at the end of section 9.3.

Commercial and marketing factors also influence the use of turbochargers. A turbocharged engine will fit in the existing vehicle range, and would not need the new manufacturing facilities associated with a larger engine.

Allard (1982) provides a practical guide to turbocharging and supercharging, and Watson and Janota (1982) give a rigorous treatment of turbocharging. The remainder of this chapter is devoted to turbocharging, and there is also a case study that considers a turbocharged engine at the end of chapter 10 on Computer modelling. Computer modelling is of great value in establishing the performance of turbocharged engines, since the complex performance characteristics of turbochargers make it difficult to predict their performance in conjunction with an engine.

9.2 Radial flow and axial flow machines

The turbomachinery theory applied to turbochargers is the same as for gas turbines, and is covered in books on gas turbines such as Harman (1981), and in books on turbocharging, see Watson and Janota (1982). As well as providing the theory, gas turbines also provided the materials technology for the high temperatures and stresses in turbochargers. Provided that the turbocharger is efficient enough to raise the engine inlet pressure above the exhaust pressure of the engine, the intake and exhaust processes will both benefit. This is particularly significant for engines with in-cylinder fuel injection (since unburnt fuel will not pass straight through the engine), and for two-stroke engines (since there are no separate induction and exhaust strokes).

The efficiency of turbines and compressors depends on their type (axial or radial flow) and size. Efficiency increases with size, since the losses associated with the clearances around the blades become less significant in large machines. These effects are less severe in radial flow machines, so although they are inherently less efficient than axial machines their relative efficiency is better in the smaller sizes.

Compressors are particularly difficult to design since there is always a risk of back-flow, and a tendency for the flow to separate from the blades in the divergent passages. Dynamic compressors work by accelerating the flow in a rotor, giving a rise in total or dynamic pressure, and then decelerating the flow in a diffuser to produce a static pressure rise. Radial compressors are more tolerant of different flow conditions, and they can also achieve pressure ratios of 4.5:1 in a single stage; an axial compressor would require several rotor/stator stages for the same pressure ratio.

A typical automotive turbocharger is shown in figure 9.4, with a radial flow compressor and turbine. For the large turbochargers used in marine applications, the turbine is large enough to be designed more efficiently as an axial flow turbine — see figure 9.5.

The operation of a compressor or turbine is most sensibly shown on a temperature/entropy (T–s) plot. This contrasts with the Otto and Diesel cycles which are conventionally drawn on pressure/volume diagrams. The ideal compressor is both adiabatic and reversible and is thus isentropic — a process represented by a vertical line on the T–s plot, figure 9.6. The suffix s denotes an isentropic process. Real processes are of course irreversible, and are associated with an increase in entropy; this is shown with dotted lines on figure 9.6. Expressions for work (per unit mass flow) can be found using the simplified version of the steady-flow energy equation

$$h_{in} + Q = h_{out} + W$$

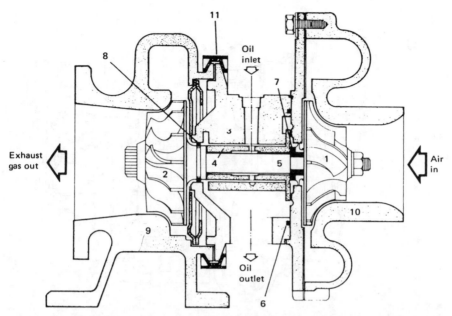

Key: 1 *Compressor wheel*. 2 *Turbine wheel*. 3 *Bearing housing*. 4 *Bearing*. 5 *Shaft*. 6 *seal ('O'*
ring). 7 *Mechanical face seal*. 8 *Piston ring seal*. 9 *Turbine housing*. 10 *Compressor housing*.
11 *'V' band clamp*.

Figure 9.4 Automotive turbocharger with radial compressor and radial turbine
(reproduced from Allard (1982), courtesy of the publisher Patrick
Stephens Ltd)

and, since the processes are treated as adiabatic

$$W = h_{in} - h_{out}$$

No assumptions about irreversibility have been made in applying the
steady-flow energy equation; thus

$$\text{turbine work, } W_t = h_3 - h_4 \qquad (9.1)$$

and defining compressor work as a negative quantity

$$W_c = h_2 - h_1$$

For real gases, enthalpy is a strong function of temperature and a weak
function of pressure. For semi-perfect gases, enthalpy is solely a function
of temperature, and this is sufficiently accurate for most purposes. Thus

a)

b)

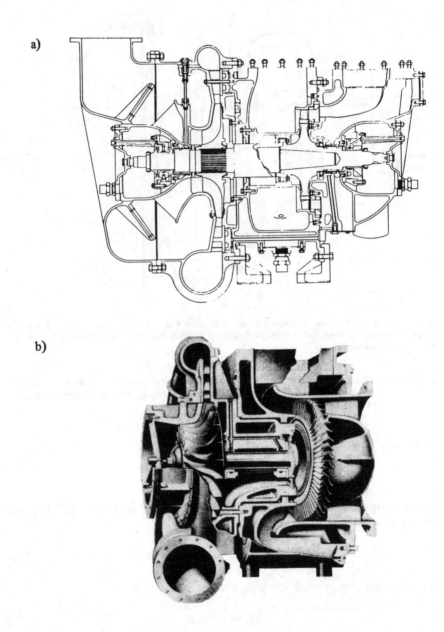

Figure 9.5   Marine turbochargers with radial compressors and axial turbines.
(a) Napier; (b) Elliot (with acknowledgement to Watson and Janota
(1982))

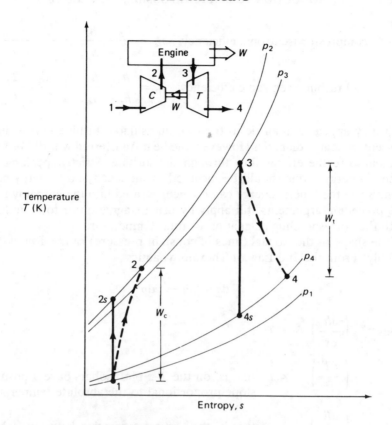

Figure 9.6   Temperature/entropy diagram for a turbocharger

$$W_c = c_p \, (T_2 - T_1) \text{ kJ/kg} \qquad (9.2)$$

and

$$W_t = c_p \, (T_3 - T_4) \text{ kJ/kg} \qquad (9.3)$$

where c_p is an appropriate mean value of the specific heat capacity. The mean specific heat capacity can be evaluated from the information in chapter 10, section 10.2.2, and such an exercise has been conducted in producing table 13.1. Consequently the T–s plot gives a direct indication of the relative compressor and turbine works.

This leads to isentropic efficiencies that compare the actual work with the ideal work.

$$\text{compressor isentropic efficiency,} \quad \eta_c = \frac{h_{2s} - h_1}{h_2 - h_1} = \frac{T_{2s} - T_1}{T_2 - T_1}$$

$$\text{and turbine isentropic efficiency,} \quad \eta_t = \frac{h_3 - h_4}{h_3 - h_{4s}} = \frac{T_3 - T_4}{T_3 - T_{4s}} .$$

It may appear unrealistic to treat an uninsulated turbine that is incandescent as being adiabatic. However, the heat transferred will still be small compared to the energy flow through the turbine. Strictly speaking, the kinetic energy terms should be included in the steady-flow energy equation. Since the kinetic energy can be incorporated into the enthalpy term, the preceding arguments still apply by using stagnation or total enthalpy with the corresponding stagnation or total temperature.

The shape of the isobars (lines of constant pressure) can be found quite readily. From the 2nd Law of Thermodynamics

$$T\, ds = dh - v dp$$

Thus $\quad \left(\dfrac{\partial h}{\partial s}\right)_P = T$

or $\quad \left(\dfrac{\partial T}{\partial s}\right)_P \propto T \quad$ that is, on the T–s plot isobars have a positive slope proportional to the absolute temperature

and $\quad \left(\dfrac{\partial h}{\partial p}\right)_s = v \quad$ that is, the vertical separation between isobars is proportional to the specific volume, and specific volume increases with temperature

Consequently the isobars diverge in the manner shown in figure 9.6.

In a turbocharger the compressor is driven solely by the turbine, and a mechanical efficiency can be defined as

$$\eta_m = \frac{W_c}{W_t} = \frac{m_{12} c_{p12} (T_2 - T_1)}{m_{34} c_{p34} (T_3 - T_4)} \tag{9.4}$$

As in gas turbines, the pressure ratios across the compressor and turbine are very important. From the pressure ratio the isentropic temperature ratio can be found:

$$\frac{T_{2s}}{T_1} = \left(\frac{p_2}{p_1}\right)^{(\gamma-1)/\gamma} \quad \text{and} \quad \frac{T_3}{T_{4s}} = \left(\frac{p_3}{p_4}\right)^{(\gamma-1)/\gamma} \tag{9.5}$$

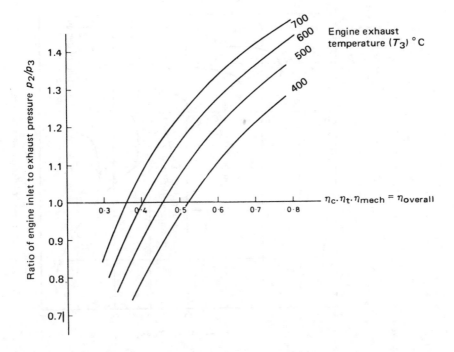

Figure 9.7   Effect of overall turbocharger on the pressure ratio between engine
inlet and exhaust manifold pressures, for a 2:1 compressor pressure
ratio ($p_2/p_1 = 2$) with different engine exhaust temperatures (with
acknowledgement to Watson and Janota (1982))

The actual temperatures, T_2 and T_4, can then be found from the respective
isentropic efficiencies.

In constant-pressure turbocharging it is desirable for the inlet pressure to
be greater than the exhaust pressure ($p_2/p_3 > 1$), in order to produce good
scavenging. This imposes limitations on the overall turbocharger efficiency
($\eta_m \cdot \eta_T \cdot \eta_c$) for different engine exhaust temperatures (T_3). This is shown in
figure 9.7. The analysis for these results originates from the above ex-
pressions, and is given by Watson and Janota (1982). Example 9.1 also
illustrates the work balance in a turbocharger.

The flow characteristics of an axial and radial compressor are compared
in figure 9.8. The isentropic efficiencies would be typical of optimum-sized
machines, with the axial compressor being much larger than the radial
compressor. Since the turbocharger compressor is very small the actual
efficiencies will be lower, especially in the case of an axial machine. The
surge line marks the region of unstable operation, with flow reversal etc.
The position of the surge line will also be influenced by the installation on

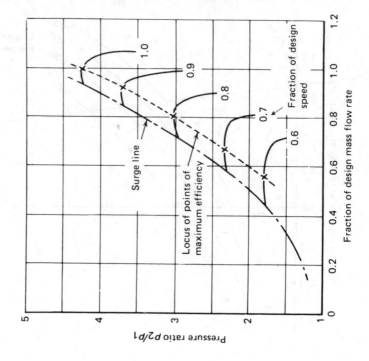

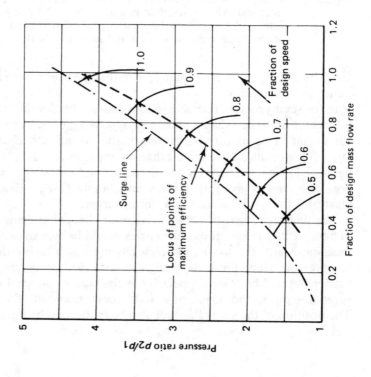

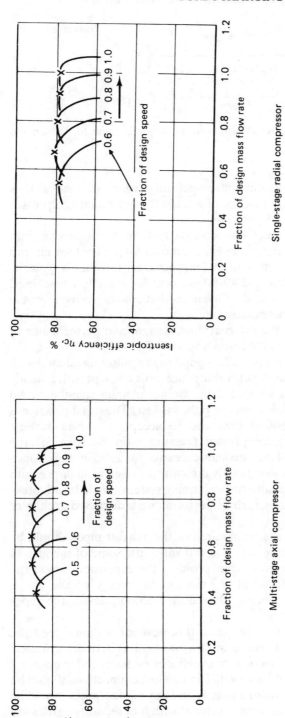

Figure 9.8 Flow characteristics of axial and radial compressors (reproduced with permission from Cohen et al. (1972))

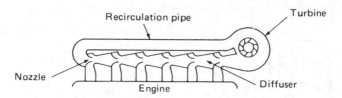

Figure 9.9   Early pulse converter system (with acknowledgement to Watson and Janota (1982))

the engine. Figure 9.9 shows the wider operating regime of a radial flow compressor. The isentropic efficiency of a turbocharger radial compressor is typically in the range 65–75 per cent.

The design of turbines is much less sensitive and the isentropic efficiency varies less in the operating range, and rises to over 90 per cent for aircraft gas turbines. The isentropic efficiency of turbocharger turbines is typically 70–85 per cent for radial flow and 80–90 per cent for axial flow machines. These are optimistic 'total to total' efficiencies that assume recovery of the kinetic energy in the turbine exhaust.

A detailed discussion of the internal flow, design, and performance of turbochargers can be found in Watson and Janota (1982).

The flow from an engine is unsteady, owing to the pulses associated with the exhaust from each cylinder, yet turbines are most efficient with a steady flow. If the exhaust flow is smoothed by using a plenum chamber, then some of the energy associated with the pulses is lost. The usual practice is to design a turbine for pulsed flow and to accept the lower turbine efficiency. However, if the compressor pressure ratio is above 3:1 the pressure drop across the turbine becomes excessive for a single stage. Since a multi-stage turbine for pulsed flow is difficult to design at high pressure ratios, a steady constant-pressure turbocharging system should be adopted. The effect of flow pulsations on turbine performance is discussed in chapter 10, section 10.3.2.

In pulse turbocharging systems the area of the exhaust pipes should be close to the curtain area of the valves at full valve lift. Some of the gain in using small exhaust pipes comes from avoiding the expansion loss at the beginning of blowdown. In addition, the kinetic energy of the gas is preserved until the turbine entry. To reduce frictional losses the pipes should be as short as possible.

For four-stroke engines no more than three cylinders should feed the same turbine inlet. Otherwise there will be interactions between cylinders exhausting at the same time. For a four-cylinder or six-cylinder engine a turbine with two inlets should be used. The exhaust connections should be such as to evenly space the exhaust pulses, and the exhaust pipes should be free of restrictions or sharp corners. Turbines with four separate entries are available, but for large engines it can be more convenient to use two separate turbochargers. For a 12-cylinder engine two turbochargers, each with a twin entry turbine, could each be connected to a group of six cylinders. This would make installation easier, and the frictional losses would be reduced by the shorter pipe lengths. For large marine diesel engines, there can be one turbocharger per pair of cylinders. While there are thermodynamic advantages in lagging the turbines and pipework, the ensuing reduction in engine room temperature may be a more important consideration.

The pressure pulses will be reflected back as compression waves and expansion waves. The exact combination of reflected waves will depend on the pipe junctions and turbine entry. The pipe lengths should be such that there are no undesirable interactions in the chosen speed range. For example, the pressure wave from an opening exhaust valve will be partially reflected as a compression wave by the small turbine entry. If the pipe length is very short the reflected wave will increase the pressure advantageously during the initial blow-down period. A slightly longer pipe, and the delayed reflected wave, will increase the pumping during the exhaust stroke — this increases the turbine output at the expense of increased piston work in the engine. An even longer pipe would cause the reflected wave to return to the exhaust valve during the period of valve overlap — this would impair the performance of a four-stroke engine and could ruin the performance of a two-stroke engine. If the pressure wave returns after the exhaust valve has closed, then it has no effect. Evidently great care is needed on engines with long exhaust pipes and large valve overlaps.

An alternative to multi-entry turbines is the use of pulse converters. An early pulse converter system is shown in figure 9.9; the idea was to use the jet from the nozzle to produce a low-pressure area around each exhaust port. The principal disadvantages are:

(1) insufficient length between the ports for efficient diffusion
(2) high frictional losses
(3) each nozzle has to be larger than the last, resulting in high manufacturing cost.

A more realistic approach is to use pulse converters to connect groups of cylinders that would otherwise be separate. For example, four cylinders could be connected to a single turbocharger entry, figure 9.10. The steadier flow can also lead to an improvement in turbine performance. The design of the pulse converter is a compromise between pressure loss and unwanted pulse propagation. Reducing the throat area increases the pressure loss, but reduces the pulse propagation from one group of cylinders to another. The optimum design will depend on the turbine, the exhaust pipe length, the valve timing, the number of cylinders, the engine speed etc.

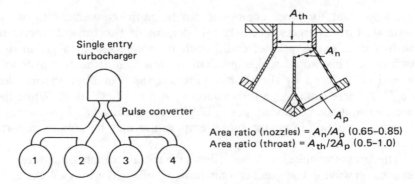

Figure 9.10   Exhaust manifold arrangement (four-cylinder engine) and pulse converter details (with acknowledgement to Watson and Janota (1982))

Constant-pressure turbocharging (that is, when all exhaust ports enter a chamber at approximately constant pressure) is best for systems with a high pressure ratio. The dissipation of the pulse energy is offset by the improved turbine efficiency. Furthermore, during blow-down the throttling loss at the exhaust valve will be reduced. However, the part load performance of a constant-pressure system is poor because of the increased piston pumping work, and the positive pressure in the exhaust system can interfere with scavenging.

9.3 Turbocharging the compression ignition engine

The purpose of turbocharging is to increase the engine output by increasing the density of the air drawn into the engine. The pressure rise across the compressor increases the density, but the temperature rise reduces the density. The lower the isentropic efficiency of the compressor, the greater the temperature rise for a given pressure ratio.

Substituting for T_{2s} from equation (9.5) into equation (9.3) and rearranging gives

$$T_2 = T_1 \left[1 + \frac{(p_2/p_1)^{(\gamma-1)/\gamma} - 1}{\eta_c} \right] \tag{9.6}$$

This result is for an ideal gas, and the density ratio can be found by applying the Gas Law, $\rho = p/RT$. Thus

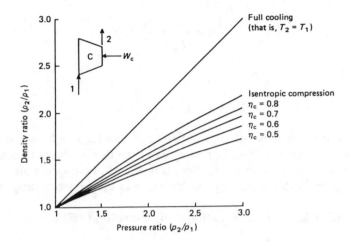

Figure 9.11 Effect of compressor efficiency on air density in the inlet manifold (with acknowledgement to Watson and Janota (1982))

$$\frac{\rho_2}{\rho_1} = \frac{p_2}{p_1}\left[1 + \frac{(p_2/p_1)^{(\gamma-1)/\gamma} - 1}{\eta_c}\right]^{-1} \qquad (9.7)$$

The effect of compressor efficiency on charge density is shown in figure 9.11; the effect of full cooling (equivalent to isothermal compression) has also been shown. It can be seen that the temperature rise in the compressor substantially decreases the density ratio, especially at high pressure ratios. Secondly, the gains in the density ratio on cooling the compressor delivery can be substantial. Finally, by ensuring that the compressor operates in an efficient part of the regime, not only is the work input minimised but the temperature rise is also minimised. Higher engine inlet temperatures raise the temperature throughout the cycle, and while this reduces ignition delay it increases the thermal loading on the engine.

The advantages of charge cooling lead to the use of inter-coolers. The effectiveness of the inter-cooler can be defined as

$$\varepsilon = \frac{\text{actual heat transfer}}{\text{maximum possible heat transfer}}$$

For the cooling medium it is obviously advantageous to use a medium (typically air or water) at ambient temperature (T_1), as opposed to the engine cooling water.

If T_3 is the temperature at exit from the inter-cooler, and the gases are perfect, then

$$\varepsilon = \frac{T_2 - T_3}{T_2 - T_1} \tag{9.8}$$

or

$$T_3 = T_2 (1 - \varepsilon) + \varepsilon T_1$$

In practice it is never possible to obtain heat transfer in a heat exchanger without some pressure drop. For many cases the two are linked linearly by Reynolds' analogy — that is, the heat transfer will be proportional to the pressure drop. In the following simple analysis the pressure drop will be ignored.

Substituting for T_2 from equation (9.6), equation (9.8) becomes

$$T_3 = T_1 \left\{ \left[1 + \frac{(p_2/p_1)^{(\gamma-1)/\gamma} - 1}{\eta_c} \right] (1 - \varepsilon) + \varepsilon \right\}$$

$$= T_1 \left[1 + (1 - \varepsilon) \frac{(p_2/p_1)^{(\gamma-1)/\gamma} - 1}{\eta_c} \right] \tag{9.9}$$

Neglecting the pressure drop in the inter-cooler, equation (9.7) becomes

$$\frac{\rho_3}{\rho_1} = \frac{p_2}{p_1} \left[1 + (1 - \varepsilon) \frac{(p_2/p_1)^{(\gamma-1)/\gamma} - 1}{\eta_c} \right]^{-1} \tag{9.10}$$

The effect of charge cooling on the density ratio is shown in figure 9.12 for a typical isentropic compressor efficiency of 70 per cent, and an ambient temperature of 20°C.

Despite the advantages of inter-cooling it is not universally used. The added cost and complexity are not justified for medium output engines, and the provision of a cooling source is troublesome. Gas to gas heat exchangers are bulky and in automotive applications would have to be placed upstream of the radiator. An additional heat exchanger could be used with an intermediate circulating liquid, but with yet more cost and complexity. In both cases energy would be needed to pump the flows. Finally, the added volume of the inter-cooler will influence the transient performance of the engine.

The effect of inter-cooling on engine performance is complex, but two cases will be considered: the same fuelling rate and the same thermal loading. Inter-cooling increases the air flow rate and weakens the air/fuel ratio for a fixed fuelling rate. The temperatures will be reduced throughout the cycle, including the exhaust stage. The turbine output will then be

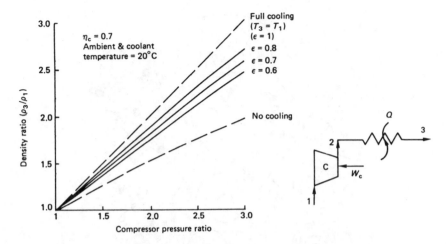

Figure 9.12 Effect of charge cooling on inlet air density (with acknowledgement to Watson and Janota (1982))

reduced, unless it is rematched, but the compressor pressure ratio will not be significantly reduced. The reduced heat transfer and changes in combustion cause an increase in bmep and a reduction in specific fuel consumption. Watson and Janota (1982) estimate both changes as 6 per cent for a pressure ratio of 2.25 and inter-cooler effectiveness of 0.7. The gains are greatest at low flow rates where the inter-cooler is most effective.

If the fuelling rate is increased to give the same thermal loading Watson and Janota (1982) estimate a gain in output of 22 per cent. The specific fuel consumption will also be improved since the mechanical losses will not have increased so rapidly as the output.

Low heat loss engines also offer scope for improving the performance of Diesel engines. Firstly, and most widely quoted, is the improvement in expansion work and the higher exhaust temperature. This leads to further gains when an engine is turbocharged. Secondly, the reduced cooling requirements allow a smaller capacity cooling system. The associated reduction in power consumed by the cooling system is, of course, most significant at part load.

Reducing the heat transfer from the combustion chamber also leads to a reduced ignition delay and hence reduced Diesel knock. However, the high combustion temperatures will lead to an increase in NO_x emissions.

Reductions in heat transfer from the combustion chamber can be obtained by redesign with existing materials, but the greatest potential here is offered by ceramics. Ceramics can be used as an insulating layer on metallic components, or more radically as a material for the complete component. These heat transfer aspects are discussed further in chapter 12, section 12.2.3.

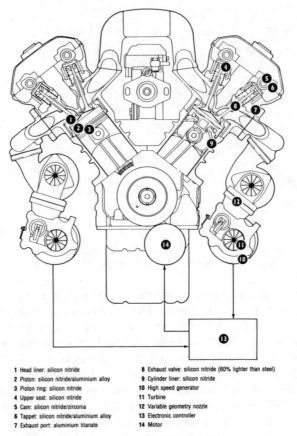

1 Head liner: silicon nitride
2 Piston: silicon nitride/aluminium alloy
3 Piston ring: silicon nitride
4 Upper seat: silicon nitride
5 Cam: silicon nitride/zirconia
6 Tappet: silicon nitride/aluminium alloy
7 Exhaust port: aluminium titanate

8 Exhaust valve: silicon nitride (60% lighter than steel)
9 Cylinder liner: silicon nitride
10 High speed generator
11 Turbine
12 Variable geometry nozzle
13 Electronic controller
14 Motor

Figure 9.13   Isuzu 2.9 litre V6 Diesel engine with many ceramic components
(Anon (1990))

Differences between the thermal coefficients of expansion of metals and ceramics mean that great care is needed in the choice of the ceramic and the metal substrate, if the insulating layer is not to separate from its substrate. None the less, ceramic insulation has been used successfully in Diesel engines — for example, the work reported by Walzer *et al.* (1985). In this turbocharged engine, 80 per cent of the combustion chamber surface was covered to an average depth of 3 mm by aluminium titanate or zirconium dioxide insulation. These measures led to a 13 per cent reduction in heat flow to the coolant, and a 5 per cent improvement in the urban cycle fuel economy (this was without re-optimisation of the cooling system).

A significant example of using ceramics in a Diesel engine has been provided by Isuzu, with a 2.9 litre V6, 24 valve engine (Anon (1990)). This engine is shown in figure 9.13, and it can be seen that there is no cooling system.

A study of a low heat loss engine by Hay *et al.* (1986) suggested that a 30 per cent reduction in heat transfer would lead to a 3.6 per cent reduction in the specific fuel consumption at the rated speed and load. Furthermore, the 2 kW reduction in the cooling fan power would lead to an additional 2.7 per cent reduction in the fuel consumption, at the rated speed and load, and correspondingly greater percentage improvements at part load. Work by Wade *et al.* (1984) in a low heat loss engine suggests that the greatest gains in efficiency are at light loads and high speeds (figure 9.14).

Alternatively, the higher exhaust temperature will allow a smaller turbine back-pressure for the same work output. Work by Hoag *et al.* (1985) indicates that a 30 per cent reduction in heat transfer would lead to a 3.4 per cent fall in volumetric efficiency, a 70 K rise in the exhaust gas temperature, and reductions in fuel consumption of 0.8 per cent for a turbocharged engine, and 2 per cent for a turbocompound engine.

So far no mention has been made of matching the turbocharger to the engine. Reciprocating engines operate over a wide speed range, and the flow range is further extended in engines with throttle control. In contrast, turbomachinery performance is very dependent on matching the gas flow

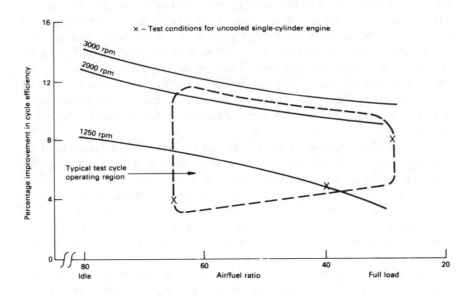

Figure 9.14   Calculated improvement in cycle efficiency for a fully insulated DI Diesel engine relative to a baseline water-cooled DI diesel engine as a function of air/fuel ratio and engine speed. Bore 80 mm, stroke 88 mm, compression ratio 21:1 (adapted from Wade *et al.* (1984))

angles to the blade angles. Consequently, a given flow rate is correct only for a specific rotor speed, and away from the 'design point' the losses increase with increasing incidence angle. Thus turbomachines are not well suited to operating over a wide flow range. However, they do have high design point efficiencies, and are small because of the high speed flows.

The first stage in matching is to estimate the air flow rate. The compressor delivery pressure will be determined by the desired bmep. The air density at entry to the engine (ρ_2) can then be calculated from equation (9.10), and this leads to the air mass flow rate:

$$\dot{m}_a \approx \rho_2 \cdot N^* \cdot V_s \cdot \eta_{vol}$$

where $\dot{m}_a$ = air mass flow rate (kg/s)
       N^* = no. of cycles per second (s^{-1})
       V_s = swept volume (m^3)
       η_{vol} = volumetric efficiency.

This will enable a preliminary choice of turbocharger to be made in terms of the 'frame size'. Within a given 'frame size', a range of compressor and turbine rotors and stators can be fitted. The compressor will be chosen in the context of the speed and load range of the engine, so that the engine will operate in an efficient flow regime of the compressor, yet still have a sufficient margin from surge. Once the compressor has been matched the turbine can be chosen. The turbine is adjusted by altering its nozzle ring, or volute if it is a radial flow machine. The turbine output is controlled by the effective flow area, hence also controlling the compressor boost pressure. Although calculations are possible, final development is invariably conducted on a test bed in the same manner as for naturally aspirated engines.

The flow characteristics (figure 9.8) can be conveniently combined by plotting contours of efficiency. The engine operating lines can then be superimposed figure 9.15. The x-axis would be dimensionless mass flow rate if multiplied by $R/A\sqrt{c_p}$, but since these are constants for a given machine they are omitted. If the engine is run at constant speed, but increasing load, then the mass flow rate will increase almost proportionally with the increasing charge density or pressure ratio. This is shown by the nearly vertical straight line in figure 9.15.

If an engine is run at constant load but increasing speed, the volumetric flow rate of air will also increase. Since the effective flow area of the turbine remains almost constant, the turbine inlet pressure rises, so increasing the turbine work. The increased turbine work increases the compressor pressure ratio. This is shown by the gently rising lines in figure 9.15.

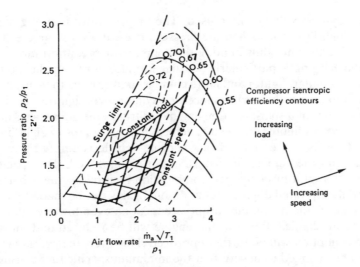

Figure 9.15   Superimposition of engine running lines on compression characteristics – constant engine load and speed lines (with acknowledgement to Watson and Janota (1982))

There must be sufficient margin between surge and the nearest operating point of the engine to allow for two factors. Firstly, the pulsating nature of the flow is likely to induce surge, and secondly the engine operating conditions may change from the datum. For example, a blocked air filter or high altitude would reduce the air flow rate, so moving the operating points closer to surge.

The turbine is tolerant of much wider flow variations than the compressor, and it is unrealistic to plot mean values for a turbine operating on a pulsed flow. Even for automotive applications with the widest flow variations, it is usually sufficient to check only the compressor operation.

The matching of two-stroke engines is simpler, since the flow is controlled by ports. These behave as orifices in series, and they have a unique pressure drop/flow characteristic. This gives an almost unique engine operating line, regardless of speed or load. However, the performance will be influenced by any scavenge pump.

In automotive applications the wide flow range is made yet more demanding by the requirement for maximum torque (or bmep) at low speeds. High torque at low speed reduces the number of gearbox ratios that are needed for starting and hill climbing. However, if the turbocharger is matched to give high torque at low speeds, then at high speeds the pressure ratio will be too great, and the turborcharger may also overspeed. This problem is particularly severe on passenger car engines and an exhaust by-pass valve is often used. The by-pass valve is spring regulated and, at high flow rates when the pressure rises, it allows some exhaust to by-pass the turbine, thus limiting the compressor pressure ratio.

Turbocharging is particularly popular for automotive applications since it enables smaller, lighter and more compact power units to be used. This is essential in cars if the performance of a compression ignition engine is to approach that of a spark ignition engine. In trucks the advantages are even greater. With a lighter engine in a vehicle that has a gross weight limit, the payload can be increased. Also, when the vehicle is empty the weight is reduced and the vehicle fuel consumption is improved. The specific fuel consumption of a turbocharged compression ignition engine is better than a naturally aspirated engine, but additional gains can be made by retuning the engine. If the maximum torque occurs at an even lower engine speed, then the mechanical losses in the engine will be reduced and the specific fuel consumption will be further improved. However, the gearing will then have to be changed to ensure that the minimum specific fuel consumption occurs at the normal operating point. Ford (1982) claim that turbocharging can reduce the weight of truck engines by 30 per cent, and improve the specific fuel consumption by from 4 to 16 per cent. Figure 9.16 shows a comparison of naturally aspirated and turbocharged truck engines of equivalent power outputs.

In passenger cars a turbocharged compression ignition engine can offer a performance approaching that of a comparably sized spark ignition engine; its torque will be greater but its maximum speed lower. Compression ignition engines can give a better fuel consumption than spark ignition engined vehicles, but this will depend on the driving pattern (Radermacher (1982)) and whether the comparison uses a volumetric or gravimetric basis (see chapter 3, section 3.7).

9.4 Turbocharging the spark ignition engine

Turbocharging the spark ignition engine is more difficult than turbocharging the compression ignition engine. The material from the previous section applies, but in addition spark ignition engines require a wider air flow range (owing to a wider speed range and throttling), a faster response, and more careful control to avoid either pre-ignition or self-ignition (knock). The fuel economy of a spark ignition engine is not necessarily improved by turbocharging. To avoid both knock and self-ignition it is common practice to lower the compression ratio, thus lowering the cycle efficiency. This may or may not be offset by the frictional losses representing a smaller fraction of the engine output.

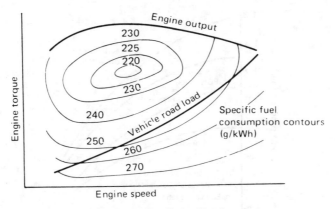

Naturally aspirated 8-cylinder Diesel engine

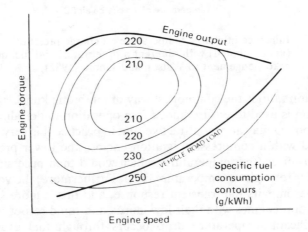

Turbocharged 6-cylinder Diesel engine

Figure 9.16   Comparison of comparably powerful naturally aspirated and
turbocharged engines (Ford (1982))

The turbocharger raises the temperature and pressure at inlet to the
spark ignition engine, and consequently pressures and temperatures are
raised throughout the ensuing processes. The effect of inlet pressure and
temperature on the knock-limited operation of an engine running at
constant speed, with a constant compression ratio, is shown in figure 9.17.
Higher octane fuels and rich mixtures both permit operation with higher
boost pressures and temperatures. Retarding the ignition timing will
reduce the peak pressures and temperatures to provide further control on
knock. Unfortunately there will be a trade-off in power and economy and
the exhaust temperature will be higher; this can cause problems with
increased heat transfer in the engine and turbocharger. Reducing the

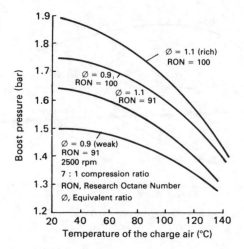

Figure 9.17   Influence of charge temperature on charge pressure (knock-limited) with different air/fuel ratios and fuel qualities (with acknowledgement to Watson and Janota (1982))

compression ratio is the commonest way of inhibiting knock and retarding the ignition is used to ensure knock-free operation under all conditions.

Inter-cooling may appear attractive, but in practice it is very rarely used. Compared with a compression ignition engine, the lower pressure ratios cause a lower charge temperature, which would then necessitate a larger inter-cooler for a given temperature drop. Furthermore, the volume of the inter-cooler impairs the transient response, and this is more significant in spark ignition engines with their low inertia and rapid response. Finally, a very significant temperature drop occurs through fuel evaporation, a process that cannot occur in compression ignition engines.

The fuel/air mixture can be prepared by either carburation or fuel injection, either before or after the turbocharger. Fuel injection systems are simplest since they deduce air mass flow rate and will be designed to be insensitive to pressure variations. In engines with carburettors it may appear more attractive to keep the carburettor and inlet manifold from the naturally aspirated engine. However, the carburettor then has to deal with a flow of varying pressure. The carburettor can be rematched by changing the jets, and the float chamber can be pressurised. Unfortunately, it is difficult to obtain the required mixture over the full range of pressures and flow rates. In general it is better to place the carburettor before the compressor for a variety of reasons. The main complication is that the compressor rotor seal needs improvement to prevent dilution of the fuel/air mixture at part load and idling conditions. The most effective solution is to replace the piston ring type seals with a carbon ring lightly loaded against a thrust face. A disadvantage of placing the carburettor before the compressor is that the volume of air and fuel between the carburettor and engine is increased. This can cause fuel hold up when the throttle is opened, and a rich mixture on over-run when the throttle is closed, as discussed in chapter 4, section 4.6.1.

The advantages of placing the carburettor before the compressor are:

(i)   the carburettor operates at ambient pressure
(ii) there is reduced charge temperature
(iii) compressor operation is further from the surge limit
(iv) there is a more homogeneous mixture at entry to the cylinders.

If the carburettor operates at ambient pressure then the fuel pump can be standard and the carburettor can be re-jetted or changed to allow for the increased volumetric flow rate.

The charge temperature will be lower if the carburettor is placed before the compressor. Assuming constant specific heat capacities, and a constant enthalpy of evaporation for the fuel, then the temperature drop across the carburettor (ΔT_{carb}) will be the same regardless of the carburettor position. The temperature rise across the compressor is given by equation (9.6).

$$T_2 = T_1 \left[1 + \frac{(p_2/p_1)^{(\gamma-1)/\gamma} - 1}{\eta_c} \right]$$

The term in square brackets is greater than unity, so that ΔT_{carb} will be magnified if the carburettor is placed before the compressor. In addition, the ratio of the specific heat capacities (γ) will be reduced by the presence of the fuel, so causing a further lowering of the charge temperature. This is illustrated by example 9.2, which also shows that the compressor work will be slightly reduced. The reduced charge temperature is very important since it allows a wider knock-free operation — see figure 9.16.

In spark ignition engines the compressor operates over a wider range of flows, and ensuring that the operation is always away from the surge line can be a greater problem than in compression ignition engines. If the carburettor, and thus the throttle, is placed before the compressor the surge margin is increased at part throttle. Consider a given compressor pressure ratio and mass flow rate and refer back to figure 9.15. The throttle does not change the temperature at inlet to the compressor (T_1), but it reduces the pressure (p_1) and will thus move the operating point to the right of the operating point when the throttle is placed after the compressor and p_1 is not reduced.

By the time a fuel/air mixture passes through the compressor it will be more homogeneous than at entry to the compressor. Furthermore, the flow from the compressor would not be immediately suitable for flow through a carburettor.

The transient response of turbocharged engines is discussed in detail by Watson and Janota (1982). The problems are most severe with spark ignition engines because of their wide speed range and low inertia; the problems are also significant with the more highly turbocharged compression ignition engines. The poor performance under changing speed or load conditions derives from the nature of the energy transfer between the engine and the turbocharger. When the engine accelerates or the load increases, only part of the energy available at the turbine appears as compressor work, the balance is used in accelerating the turbocharger rotor. Additional lags are provided by the volumes in the inlet and exhaust systems between the engine and turbocharger; these volumes should be minimised for good transient response. Furthermore, the inlet volume should be minimised in spark ignition engines to limit the effect of fuel hold-up on the fuel-wetted surfaces. Turbocharger lag cannot be eliminated without some additional energy input, but the effect can be minimised. One approach is to under-size the turbocharger, since the rotor inertia increases with $(\text{length})^5$, while the flow area increases with $(\text{length})^2$. Then to prevent undue back-pressure in the exhaust, an exhaust by-pass valve can be fitted. An alternative approach is to replace a single turbocharger by two smaller units.

The same matching procedure is used for spark ignition engines and compression ignition engines. However, the wider speed and flow range of the spark ignition engine necessitates greater compromises in the matching of turbomachinery to a reciprocating engine. If the turbocharger is matched for the maximum flow then the performance at low flows will be very poor, and the large turbocharger size will give a poor transient response. When a smaller turbocharger is fitted, the efficiency at low flow rates will be greater and the boost pressure will be higher throughout the range; the lower inertia will also reduce turbocharger lag. However, at higher flow rates the boost pressure would become excessive unless modified; two approaches are shown in figure 9.18.

The compressor pressure can be directly controlled by a relief valve, to keep the boost pressure below the knock-limited value. The flow from the relief valve does not represent a complete loss of work since the turbine work derives from energy that would otherwise be dissipated during the exhaust blow-down. The blow-off flow can be used to cool the turbine and exhaust systems. If the carburettor is placed before the compressor, then the blow-off flow has to be returned to the compressor inlet, which results in yet higher charge temperatures.

The exhaust waste gate system, figure 9.18b, is more attractive since it also permits a smaller turbine to be used, because it no longer has to be sized for the maximum flow. Turbocharger lag is reduced by the low

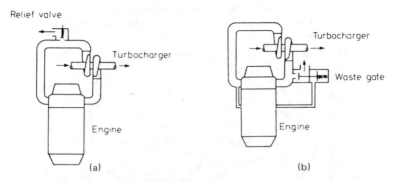

Figure 9.18   (a) Compressor pressure-relief valve control system. (b) Boost
pressure-sensitive waste control system (with acknowledgement to
Watson and Janota (1982))

inertia, and the control system ensures that the waste gate closes during
acceleration. The main difficulty is in designing a cheap reliable system that
will operate at the high temperatures. Variable area turbines, compressor
restrictors and turbine outlet restrictors can also be used to control the
boost pressure. A variable area turbine is not sufficiently better than a
waste gate, so its use is not justified; restrictors of any form are an
unsatisfactory solution.

Performance figures vary, but typically a maximum boost pressure of 1.5
bar would raise the maximum torque by 30 per cent and maximum power
by up to 60 per cent. Figure 9.19 shows the comparative specific fuel
consumption of a turbocharged and naturally aspirated spark ignition
engine. The turbocharged engine has improved fuel consumption at low
outputs, but an inferior consumption at higher outputs. The effect on
vehicle consumption would depend on the particular driving pattern.

9.5   Conclusions

Turbocharging is a very important means of increasing the output of
internal combustion engines. Significant increases in output are obtained,
yet the turbocharger system leads to only small increases in the engine
weight and volume.

The fuel economy of compression ignition engines is usually improved
by turbocharging, since the mechanical losses do not increase in direct
proportion to the gains in power output. The same is not necessarily true of
spark ignition engines, since turbocharging invariably necessitates a reduction in compression ratio to avoid knock (self-ignition of the fuel/air

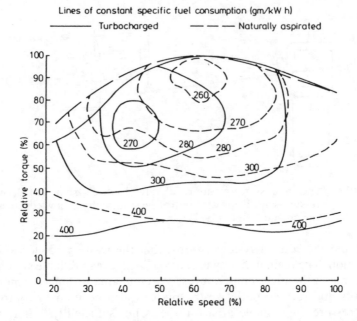

Figure 9.19   Comparative specific fuel consumption of a turbocharged and
naturally aspirated petrol engine scaled for the same maximum
torque (with acknowledgement to Watson and Janota (1982))

mixture). The reduction in compression ratio reduces the indicated efficiency and this usually negates any improvement in the mechanical
efficiency.

The relatively low flow rate in turbochargers leads to the use of radial
flow compressors and turbines. In general, axial flow machines are more
efficient, but only for high flow rates. Only in the largest turbochargers
(such as those for marine applications) are axial flow turbines used.
Turbochargers are unlike positive displacement machines, since they rely
on dynamic flow effects; this implies high velocity flows, and consequently
the rotational speeds are an order of magnitude greater than reciprocating
machines. The characteristics of reciprocating machines are fundamentally
different from those of turbochargers, and thus great care is needed in the
matching of turbochargers to internal combustion engines. The main
considerations in turbocharging matching are:

(i)   to ensure that the turbocharger is operating in an efficient regime
(ii) to ensure that the compressor is operating away from the surge line
      (surge is a flow reversal that occurs when the pressure ratio increases
      and the flow rate decreases)
(iii) to ensure a good transient response.

Turbochargers inevitably suffer from 'turbo-lag'; when either the engine load or speed increases, only part of the energy available from the turbine is available as compressor work — the balance is needed to accelerate the turbocharger rotor. The finite volumes in the inlet and exhaust system also lead to additional delays that impair the transient response.

As well as offering thermodynamic advantages, turbochargers also offer commercial advantages. In trucks, the reduced weight of a turbocharged engine gives an increase in the vehicle payload. A manufacturer can add turbocharged versions of an engine to his range more readily than producing a new engine series. Furthermore, turbocharged engines can invariably be fitted into the same vehicle range — an important marketing consideration.

9.6 Examples

Example 9.1

A Diesel engine is fitted with a turbocharger, which comprises a radial compressor driven by a radial exhaust gas turbine. The air is drawn into the compressor at a pressure of 0.95 bar and at a temperature of 15°C, and is delivered to the engine at a pressure of 2.0 bar. The engine is operating on a gravimetric air/fuel ratio of 18:1, and the exhaust leaves the engine at a temperature of 600°C and at a pressure of 1.8 bar; the turbine exhausts at 1.05 bar. The isentropic efficiencies of the compressor and turbine are 70 per cent and 80 per cent, respectively. Using the values;

$$c_{p_{air}} = 1.01 \text{ kJ/kg K}, \ \gamma_{air} = 1.4$$

and

$$c_{p_{ex}} = 1.15 \text{ kJ/kg K}, \ \gamma_{ex} = 1.33$$

calculate   (i) the temperature of the air leaving the compressor
    (ii) the temperature of the gases leaving the turbine
    (iii) the mechanical power loss in the turbocharger expressed as a percentage of the power generated in the turbine.

Solution

Referring to figure 9.20 (a new version of figure 9.6), the real and ideal temperatures can be evaluated along with the work expressions.

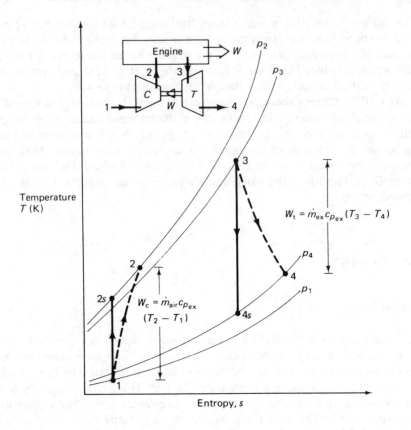

Figure 9.20 Temperature/entropy diagram for a turbocharger

(i) If the compression were isentropic, $T_{2s} = T_1 \left(\dfrac{p_2}{p_1}\right)^{(\gamma-1)/\gamma}$

$$T_{2s} = 288 \left(\frac{2.0}{0.95}\right)^{(1.4-1)/1.4} = 356 \text{ K, or } 83°C$$

From the definition of compressor isentropic efficiency, $\quad \eta_c = \dfrac{T_{2s} - T_1}{T_2 - T_1}$

$$T_2 = \frac{T_{2s} - T_1}{\eta_c} + T_1 = \frac{83 - 15}{0.7} + 15 = 113°C$$

(ii) If the turbine were isentropic, $T_{4s} = T_3 \left(\dfrac{p_4}{p_3}\right)^{(\gamma-1)/\gamma}$

$$T_{4s} = 873 \left(\frac{1.05}{1.8}\right)^{(1.33-1)/1.33} = 762.9 \text{ K or } 490°C$$

From the definition of turbine isentropic efficiency, $\eta_t = \dfrac{T_3 - T_4}{T_3 - T_{4s}}$

$$T_4 = T_3 - \eta_t (T_3 - T_{4s}) = 600 - 0.8 (600 - 490) = 512°C$$

(iii) Compressor power $\dot{W}_c = \dot{m}_{air} \, c_{p_{air}} (T_2 - T_1)$

$$= \dot{m}_{air} \, 1.01(113 - 15)\text{kW}$$

$$= \dot{m}_{air} \, 98.98 \text{ kW}$$

from the air/fuel ratio

$$\dot{m}_{ex} = \dot{m}_{air} \left(1 + \frac{1}{18}\right)$$

and turbine power

$$\dot{W}_t = \dot{m}_{ex} \, c_{p_{ex}} (T_3 - T_4)$$

$$= \dot{m}_{air} \, 1.056 \times 1.15 \, (600 - 512)$$

$$= \dot{m}_{air} \, 106.82 \text{ kW}$$

Thus, the mechanical power loss as a percentage of the power generated in the turbine is

$$\frac{106.82 - 98.98}{106.82} \times 100 = 7.34 \text{ per cent}$$

This result is in broad agreement with figure 9.7, which is for a slightly different pressure ratio and constant gas flow rates and properties.

Example 9.2

Compare the cooling effect of fuel evaporation on charge temperature in a turbocharged spark ignition engine for the following two cases:

(a) the carburettor placed before the compressor
(b) the carburettor placed after the compressor.

The specific heat capacity of the air and the latent heat of evaporation of

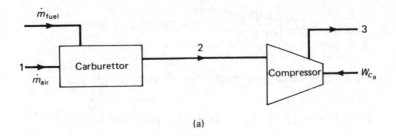

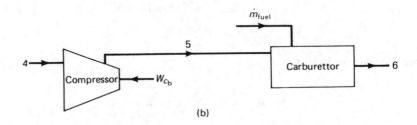

Figure 9.21   Possible arrangement for the carburettor and compressor in a
spark ignition engine. (a) Carburettor placed before the
compressor; (b) carburettor placed after the compressor

the fuel are both constant. For the air/fuel ratio of 12.5:1, the evaporation
of the fuel causes a 25 K drop in mixture temperature. The compressor
efficiency is 70 per cent for the pressure ratio of 1.5, and the ambient air is
at 15°C. Assume the following property values:
for air c_p = 1.01 kJ/kg K, γ = 1.4
for air/fuel mixture c_p = 1.05 kJ/kg K, γ = 1.34
Finally, compare the compressor work in both cases.

Solution

Both arrangements are shown in figure 9.21.

(a) T_1 = 15°C = 288 K
$T_2 = T_1 - 25$ = 263 K

If the compressor were isentropic, T_{3s} = 263 $(1.5)^{(1.34\,-\,1)/1.34}$ = 291.5 K
From the definition of compressor isentropic efficiency

$$T_3 = \frac{T_{3s} - T_2}{\eta_c} + T_2 = \frac{291.5 - 263}{0.7} + 263 = 303.7 \text{ K}$$

(b) T_4 = 288 K

For isentropic compression $T_{5s} = 288\,(1.5)^{(1.4-1)/1.4} = 323.4$ K
From the definition of compressor isentropic efficiency

$$T_5 = \frac{T_{5s} - T_4}{\eta_c} + T_4 = \frac{323.4 - 288}{0.7} + 288 = 338.5 \text{ K}$$

$$T_6 = T_5 - 25 = 338.5 - 25 = 313.5 \text{ K}$$

Since $T_6 > T_3$, it is advantageous to place the carburettor before the compressor.

Comparing the compressor power for the two cases:

$$(W_c)_a = \dot{m}_{mix}\, c_{P_{mix}}\, (T_3 - T_2)$$

$$= 1.08\, \dot{m}_{air}\, 1.05\, (303.7 - 263)$$

$$= 46.15\, \dot{m}_{air} \text{ kW}$$

$$(W_c)_b = \dot{m}_{air}\, c_{P_{air}}\, (T_5 - T_4)$$

$$= \dot{m}_{air}\, 1.01\, (338.5 - 288)$$

$$= 51.01\, \dot{m}_{air} \text{ kW}$$

Thus placing the carburettor before the compressor offers a further advantage in reduced compressor work.

The assumptions in this example are somewhat idealised. When the carburettor is placed before the compressor, the fuel will not be completely evaporated before entering the compressor, and evaporation will continue during the compression process. However, less fuel is likely to enter the cylinders in droplet form if the carburettor is placed before the compressor rather than after.

9.7 Problems

9.1 A spark ignition engine is fitted with a turbocharger that comprises a radial flow compressor driven by a radial flow exhaust gas turbine. The gravimetric air/fuel ratio is 12:1, with the fuel being injected between the compressor and the engine. The air is drawn into the compressor at a pressure of 1 bar and at a temperature of 15°C. The compressor delivery pressure is 1.4 bar. The exhaust gases from the engine enter the turbine at a pressure of 1.3 bar and a temperature of 710°C; the gases leave the turbine at a pressure of 1.1 bar. The isentropic efficiencies of the compressor and turbine are 75 per cent and 85 per cent, respectively.

Treating the exhaust gases as a perfect gas with the same properties as air, calculate:

(i) the temperature of the gases leaving the compressor and turbine
(ii) the mechanical efficiency of the turbocharger.

9.2 Why is it more difficult to turbocharge spark ignition engines than compression ignition engines? Under what circumstances might a supercharger be more appropriate?

9.3 Why do compression ignition engines have greater potential than spark ignition engines for improvements in power output and fuel economy as a result of turbocharging? When is it most appropriate to specify an inter-cooler?

9.4 Derive an expression that relates compressor delivery pressure (p_2) to turbine inlet pressure (p_3) for a turbocharger with a mechanical efficiency η_{mech}, and compressor and turbine isentropic efficiencies η_c and η_t, respectively. The compressor inlet conditions are p_1, T_1, the turbine inlet temperature is T_3 and the outlet pressure is p_4. The air/fuel ratio (AFR) and the differences between the properties of air (suffix a) and exhaust (suffix e) must all be considered.

9.5 Why do turbochargers most commonly use radial flow compressors and turbines with non-constant pressure supply to the turbine?

9.6 Why does turbocharging a compression ignition engine normally lead to an improvement in fuel economy, while turbocharging a spark ignition engine usually leads to decreased fuel economy?

9.7 Show that the density ratio across a compressor and inter-cooler is given by

$$\frac{\rho_3}{\rho_1} = \frac{P_2}{P_1} \left[1 + (1 - \varepsilon) \frac{(P_2/P_1)^{\frac{\gamma-1}{\gamma}} - 1}{\eta_c} \right]^{-1}$$

where   1   refers to compressor entry
         2   refers to compressor delivery
         3   refers to inter-cooler exit
         η_c = compressor isentropic efficiency
         ε = inter-cooler effectiveness = $(T_2 - T_3)/T_2 - T_1$.

Neglect the pressure drop in the inter-cooler, and state any assumption that you make.

Plot a graph of the density ratio against effectiveness for pressure ratios of 2 and 3, for ambient conditions of 1 bar, 300 K, if the compressor isentropic efficiency is 70 per cent.

What are the advantages and disadvantages in using an inter-cooler? Explain under what circumstances it should be used?

9.8 A turbocharged Diesel engine has an exhaust gas flow rate of 0.15 kg/s. The turbine entry conditions are 500°C at 1.5 bar, and the exit conditions are 450°C at 1.1 bar.

(a) Calculate the turbine isentropic efficiency and power output.

The engine design is changed to reduce the heat transfer from the combustion chamber, and for the same operating conditions the exhaust temperature becomes 550°C. The pressure ratio remains the same, and assume the same turbine isentropic efficiency.

(b) Calculate the increase in power output from the turbine.

How will the performance of the engine be changed by reducing the heat transfer, in terms of: economy, power output and emissions?
Assume: ratio of specific heat capacities = 1.3, and c_p = 1.15 kJ/kg K.

9.9 A compressor with the performance characteristics shown in figure 9.22 is operating with a mass flow rate (m) of 49.5 g/s at an isentropic efficiency of 60 per cent. The compressor is fitted to a turbocharged and inter-cooled Diesel engine. Assume

p (the compressor inlet pressure) is 95 kN/m²
T (the inlet temperature) is 291 K
γ (the ratio of gas specific heat capacities) is 1.4
c_p =   1.01 kJ/kg K.

Calculate the pressure ratio, the compressor speed, the compressor delivery temperature, and the compressor power.

If the inter-cooler is removed and the air/fuel ratio is kept constant, how would the compressor operating point be affected? Neglecting any change in the compressor efficiency and pressure ratio, estimate the maximum increase in output that the inter-cooler could lead to. State clearly any assumptions that you make.

9.10 A turbocharged six-cylinder four-stroke Diesel engine has a swept volume of 39 litres. The inlet manifold conditions are 2.0 bar and 53°C. The volumetric efficiency of the engine is 95 per cent, and it is operating at a load of 16.1 bar bmep, at 1200 rpm with an air/fuel

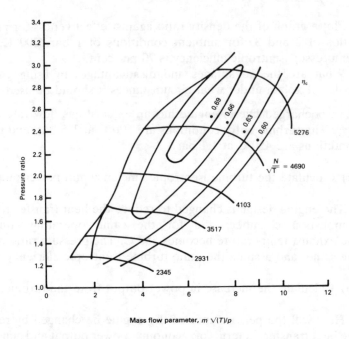

Figure 9.22   Compressor map

ratio of 21.4. The power delivered to the compressor is 100 kW, with entry conditions of 25°C and 0.95 bar. The fuel has a calorific value of 42MJ/kg.

Stating any assumptions, calculate:

(a) the power output of the engine
(b) the brake efficiency of the engine
(c) the compressor isentropic efficiency
(d) the effectiveness of the inter-cooler.

Estimate the effect of removing the inter-cooler on the power output and emissions of the engine, and the operating point of the turbocharger.

10 Engine Modelling

10.1 Introduction

The aim with modelling internal combustion engines is twofold:

(1) To predict engine performance without having to conduct tests.
(2) To deduce the performance of parameters that can be difficult to measure in tests, for example, the trapped mass of air in a two-stroke engine or a turbocharged engine.

It is obviously an advantage if engine performance can be predicted without going to the trouble of first building an engine, then instrumenting it, testing it and finally analysing the results. Modelling should lead to a saving of both time and money. Unfortunately, the processes that occur in an internal combustion engine are so complex that most of the processes cannot be modelled from first principles. Consider a turbocharged Diesel engine, for which it might be necessary to model:

(1) the compressor (and inter-cooler if fitted)
(2) unsteady-flow effects in the induction system
(3) flow through the inlet valve(s)
(4) air motion within the cylinder
(5) dynamics of the injection system
(6) fuel jet interaction with the trapped air to form a spray
(7) combustion (including the effects of the ignition delay and turbulent combustion, and possibly including the modelling of the gaseous and particulate emissions)
(8) noise generated mechanically and by combustion
(9) heat flow within the combustion chamber and to the cooling media
(10) turbine performance.

Consider the air motion within the cylinder. It is now possible to use finite difference techniques to model the gross flow details, such as the

361

distribution of swirl and axial velocities. However, predicting the turbulence levels first requires an agreement on how to define the turbulence parameters that are relevant, and also requires comprehensive experimental data for validation. Such information is vital, since the air flow interacts with the fuel during the spray formation and its subsequent combustion. The turbulence affects the heat transfer, the fuel-burning rate and thus the noise originating from combustion.

In view of these complications, it is not surprising that engine models rely heavily on experimental data and empirical correlations. For example, the turbocharger performance and the valve flow characteristics can be determined from steady-flow tests, and look-up tables can be built into the engine model, along with appropriate interpolation routines. However, it must be remembered that the flow through the engine is unsteady, and this can have a significant effect. An example of this is that the turbine operating point is non-stationary under steady-state conditions (Dale *et al.* (1988)).

Empirical correlations are used for predicting processes such as: the heat transfer, the ignition delay and the burn rate. Engine simulations that follow the approach described here are known as *zero dimensional*, *phenomenological* or *filling and emptying* models. Two simple cycle calculation models for spark ignition and compression ignition engines are listed in Fortran 4 by Benson and Whitehouse (1979). As better understanding is gained of the sub-processes (such as ignition delay), then it is possible to replace the corresponding empirical model. For example, simulations that predict turbulent combustion within a defined combustion chamber are known as *multi-dimensional* models.

A treatment of the thermodynamics and fluid mechanics of internal combustion engines, and how this knowledge is applied to engine modelling is presented very comprehensively by Benson (1982) and Horlock and Winterbone (1986).

Even if a multi-dimensional model was available that predicted performance by solving the underlying physical phenomenon, its use may not be that great for two reasons:

(1) the computational requirements might be prohibitive
(2) most engine design is derivative or evolutionary.

If a manufacturer introduces a new engine, then it is likely to have a similar combustion system to an existing engine, for which there is already experimental data, and a calibrated computer model. However, most changes to an engine can be described as development. Perhaps a different turbocharger is being considered to give a higher boost pressure and output. An existing engine model can then be adapted, to investigate how: the compression ratio, valve timing, fuel injection and other variables, might be selected to give the best performance (trade-off between fuel economy and output) within the constraints of peak cylinder pressure (typically 150 bar) and the maximum exhaust temperature (about 650°C).

Although only thermodynamic modelling is being considered here, the mechanical design of the engine also benefits from modelling. The mechanical model requires inputs of the pressure loading and thermal loading from the thermodynamics model. The mechanical model can then predict bearing loads from the engine dynamics. A finite element model is usually used for the engine structure, so that: the heat flow, the thermal and mechanical strains, the vibration modes, assembly strains, and the noise transmission can be predicted. If the motion of the valve train is to be modelled, then in addition to the information on the engine structure, the dynamic model of the valve train would need to include the stiffness of the components and the load/speed-dependent stiffness of the oil films.

When the crankshaft is being modelled then it is necessary to consider the interactions with the engine block. The engine block and crankshaft both deform under load, and this affects the bearing oil film thickness. The attitude of the piston in the bore will also be affected, and both these effects modify the damping of the crankshaft torsional oscillations. The piston and bore have to be modelled, using inputs of the temperature distributions and pressure loading, so that the piston ring/liner clearances can be predicted. This then enables the ring pack to be analysed, so that the oil film thicknesses and their contribution to friction can be found. This overview of the mechanical modelling should be sufficient to show that it is at least as complex as the thermodynamic modelling.

In subsequent sections there will be a description of how the laws of energy and mass conservation are applied to an engine, how individual processes are applied to an engine, and how individual processes can be modelled. This will be followed by an example in which SPICE (the Simulation Program for Internal Combustion Engines, written by Dr S. J. Charlton of the University of Bath) is used to investigate the influence of valve timing on engine performance.

10.2 Zero-dimensional modelling

10.2.1 Thermodynamics

As analytic functions cannot be used to describe engine processes, it is necessary to solve the governing equations on a step-wise basis; often with increments of 1° crank angle. The application of the 1st Law of Thermodynamics in modelling internal combustion engine processes is described in complementary ways by Watson and Janota (1982), Wallace, F. J. (1986a, b) and Heywood (1988).

The 1st Law of Thermodynamics can be written in differential form for an open thermodynamic system, and if changes in potential energy are neglected, then:

$$\frac{dU}{dt} = \frac{dQ}{dt} - \frac{dW}{dt} + \sum_i \frac{dH_{oi}}{dt} \tag{10.1}$$

where subscript i refers to control volume entries. Or

$$\frac{d(mu)}{dt} = \sum_s \frac{dQ_s}{dt} - p \frac{dV}{dt} + \sum_i h_{oi} \frac{dm_i}{dt} \tag{10.2}$$

where subscript s refers to surfaces at which heat transfer occurs and h_{oi} is the specific stagnation enthalpy of flows entering or leaving the system.

Consider the left-hand side term in equation (10.2). During combustion, the composition of the species present will change, but, of course, the mass is conserved. However, by expressing the internal energy of the reactants and products on an absolute basis (as in figure 3.6), then the chemical energy transferred to thermal energy during combustion does not need to be included as a separate term (*Note*: in the tables prepared by Rogers and Mayhew (1980a), the internal energy of the reactants and products is not given on an absolute basis; a datum is used of zero enthalpy at 25°C, and the enthalpy of reaction (ΔH_0) is presented for various reactions also at 25°C, so it is possible to compute the internal energy of the reactants and products on an absolute basis, see chapter 3, section 3.3 and examples 3.4 and 3.5).

Differentiating the left-hand side of equation (10.2) and dividing the combustion chamber into a series of zones gives

$$\frac{d(mu)}{dt} = \sum_j \left(m \frac{du}{dt} + u \frac{dm}{dt} \right) \tag{10.3}$$

with subscript j referring to different zones within the combustion chamber.

Within each zone, if dissociation is neglected, it can be assumed that the internal energy is only a function of the temperature and the equivalence ratio (ϕ). Thus $u = u(T, \phi)$ and equation (10.3) becomes

$$\frac{d(mu)}{dt} = \sum_j \left(m \frac{\partial u}{\partial T} \frac{dT}{dt} + m \frac{\partial u}{\partial \phi} \frac{d\phi}{dt} + u \frac{dm}{dt} \right) \tag{10.4}$$

For Diesel engines, the combustion process is frequently modelled as a single zone, and for spark ignition engines a two zone model is used, in which a flame front divides the unburned and burned zones. Thus, for a two zone model there will be mass transfer between the zones, but for a single zone the dm/dt term will be zero.

For many purposes in Diesel engine simulation, the assumption of no dissociation with a single zone model is acceptable. The assumptions reduce the computation time significantly without a serious loss of accuracy. As the combustion should always be weak of stoichiometric, this leads to temperatures at which dissociation does not have much effect on the thermodynamic performance of the engine. However, if emissions are to be considered, then a multi-zone model is required that incorporates the spatial variations in air/fuel ratio. The properties of each zone (temperature and composition) then have to be computed, as part of the modelling process for emissions.

It is less satisfactory to neglect dissociation in spark ignition engine combustion, as the mixtures are normally close to stoichiometric, and the combustion temperatures make dissociation significant. Methods for computing dissociation are presented by Benson and Whitehouse (1979), Ferguson (1986) and Baruah (1986). The prediction of emissions is also dependent on a combustion model that includes dissociation. The thermal boundary layer has to be modelled, as it quenches the flame front to generate unburnt hydrocarbon emissions. It may also be necessary to use a multi-zone model for the combustion, even if the charge is homogeneous, as the temperature of the unburnt gas will vary spatially for several reasons.

Firstly, the gas that is burned first will end up being hotter than the gas that is burned subsequently. If constant-volume combustion is considered, then the gas to be burned first contributes only a small pressure rise, while the gas burned later will contribute a greater pressure rise. Thus more work is done on the first burned gas by subsequent combustion than is done by the first burned gas on the unburned gas. This leads to temperature gradients in the burned gas that can influence emissions, especially NO_x for which the formation is strongly temperature dependent. Secondly, the combustion chamber surface temperature will vary, and this will result in different levels of heat transfer from the different regions of the combustion chamber.

However, for the current purposes a single zone combustion model without dissociation is to be developed, which can be used for simulating a Diesel engine. Substituting equation (10.4) into equation (10.2), assuming the gas behaves as a perfect gas ($pV = mRT$), gives

$$m \frac{\partial u}{\partial t} \frac{dT}{dt} + m \frac{\partial u}{\partial \phi} \frac{d\phi}{dt} + u \frac{dm}{dt} = - \frac{mRT}{V} \frac{dV}{dt} + \sum_s \frac{dQ_s}{dt} + \sum_i h_{oi} \frac{dm_i}{dt}$$

$$(10.5)$$

This equation has to be solved iteratively, and to do so, it is necessary to know:

(1) The gas properties, to calculate the internal energy as a function of temperature $u = u(T, \phi)$. Thence

$$m \frac{\partial u}{\partial T} \frac{dT}{dt} \quad \text{and} \quad m \frac{\partial u}{\partial \phi} \frac{d\phi}{dt}$$

can be found. Implicit in this, is a knowledge of how the mass fraction burned (mfb) will vary as a function of time, as the internal energies of the burned and unburned mixtures will be different functions of temperature and equivalence ratio.

(2) The mass transfer between zones (dm/dt), which will be zero here, as a single zone combustion model is being considered.

(3) The rate of change of volume, so that the displacement work can be computed:

$$- \frac{mRT}{V} \frac{dV}{dt} \qquad \text{(see equation 10.6)}$$

(4) The heat flows:

$$\sum_s \frac{dQ_s}{dt} \qquad \text{(see section 10.2.4)}$$

(5) The stagnation enthalpy of any flows in or out of the control volume, for the evaluation of

$$\sum_i h_{o,i} \frac{dm_i}{dt}$$

Equation (10.5) is a differential equation that has to be solved, so as to give temperature as a function of time or crank angle, so equation (10.5) has to be applied to each control volume. Once the temperature has been calculated, so long as the instantaneous volume and mass in the control volume are known, then the pressure can be calculated from the equation of state ($pV = mRT$).

If the control volume is of fixed volume (for example, a manifold), then the dV/dt term is zero. For the engine cylinder:

$$V = V_c + A[r(1 - \cos \theta) + \{1 - \sqrt{(l^2 - r^2 \sin^2\theta)}\}] \qquad (10.6a)$$

and by differentiation:

$$\frac{dV}{dt} = A\left[r \sin \theta \ \frac{d\theta}{dt} + (l^2 - r^2 \sin^2\theta)^{-1/2} r^2 \sin \theta \cos \theta \ \frac{d\theta}{dt}\right]$$

(10.6b)

where V_c = clearance volume at tdc

A = piston area

θ = crank angle measured from tdc

l = con-rod length

r = crank throw (stroke/2).

10.2.2 Gas properties

The gas properties are required as a function of temperature and composition.

For individual species, the internal energy can be expressed as a function of temperature by means of a polynomial expansion with either a molar or specific basis:

$$u(T) = u_0 + u_1 T + u_2 T^2 + u_3 T^3 + \ldots$$

(10.7)

The polynomial coefficients for each species can be found from several sources (such as the International Critical Tables and the JANAF tables). Benson and Whitehouse (1979) tabulate the polynomial coefficients for calculating the enthalpy, from which the internal energy can be computed.

Equation (10.7) can readily be differentiated to give du/dt as a function of temperature:

$$\frac{du(T)}{dt} = c_v(T) = u_1 + 2u_2 T + 3u_3 T^2 + \ldots$$

(10.8)

Equation 10.8 gives the specific heat capacity at constant volume as a function of temperature, and equation (10.8) could be rewritten as

$$c_v(T) = c_{v,0} + c_{v,1} T_1 + c_{v,2} T^2 + \ldots$$

(10.9)

where $c_{v,0} = u_1$

$c_{v,1} = 2u_2$

$c_{v,2} = 3u_3$ etc.

When utilising such data it is essential to check on the heat units being used, whether a molar or specific basis is being applied, and whether absolute or some other temperature scale is being used. The data provided by Reid *et al.* (1977) are for the molar heat capacity at constant pressure, and this can be converted to the molar heat capacity at constant volume, as

$$C_p - C_v = R_0$$

where the molar gas constant $R_0 = 8314.3$ J/kmol K.

The coefficients for calculating the molar heat capacity at constant volume are given in table 10.1. When equation (10.9) is integrated to give the internal energy, it is necessary to introduce an integration constant (U_0), for example, to give zero internal energy at absolute zero.

A simpler and acceptable alternative that is widely used is to consider the internal energy of the reactants and products on an absolute basis, as a function of air/fuel ratio and temperature. This is satisfactory, as most fuels have similar hydrogen/carbon ratios, and in any case the major constituent of both the reactants and products is nitrogen. Such data are presented by Gilchrist (1947) with allowance for dissociation, for a fuel containing 84 per cent carbon and 16 per cent hydrogen by mass. Gilchrist gives separate tabulations of internal energy for the reactants and products, as a function of the mixture strength and absolute temperature.

A similar approach was adopted by Krieger and Borman (1986), who provided polynomial coefficients from a curve-fit to combustion product calculations for weak mixtures ($\phi \leq 1$) of $C_n H_{2n}$ with air (85.6 per cent carbon and 14.4 per cent hydrogen by mass):

$$u = K_1(T) - K_2(T)\phi \text{ kJ/(kg of original air)} \qquad (10.10)$$

where $K_1 = 0.692T + 39.17 \times 10^{-6}T^2 + 52.9 \times 10^{-9}T^3$
$$- 228.62 \times 10^{-13}T^4 + 277.58 \times 10^{-17}T^5$$

and $K_2 = 3049.33 - 5.7 \times 10^{-2}T - 9.5 \times 10^{-5}T^2 + 21.53 \times 10^{-9}T^3$
$$- 200.26 \times 10^{-14}T^4$$

with the gas constant given by

$$R = 0.287 + 0.020\phi \text{ kJ(kg of original air)}^{-1} \text{ K}^{-1} \qquad (10.11)$$

Krieger and Borman also suggest a modification to account for dissociation if the temperature is above 1450 K:

$$u_{corr} = u + 2.32584 \exp(A + B + C) \text{ kJ/kg of original air}$$

where $A = 10.41066 + 7.85125\phi - 3.71257\phi^3$

Table 10.1 Coefficients for calculating the molar heat capacity at constant volume (kJ/kmol K) from the absolute temperature (K), derived from Reid et al. (1977)

Substance	Molar mass (kg)	$C_{v,0}$	$C_{v,1}$	$C_{v,2}$	$C_{v,3}$
N_2	28.014	22.830	-1.356×10^{-2}	26.80×10^{-6}	-11.68×10^{-9}
O_2	31.998	19.786	-3.680×10^{-6}	17.46×10^{-6}	-10.65×10^{-9}
H_2	2.016	18.824	9.272×10^{-3}	-13.81×10^{-6}	7.644×10^{-9}
CO	28.010	22.550	-1.285×10^{-2}	27.89×10^{-6}	-12.71×10^{-9}
CO_2	44.009	11.477	7.342×10^{-2}	-56.01×10^{-6}	17.15×10^{-9}
H_2O	18.015	23.928	1.924×10^{-3}	10.56×10^{-6}	3.596×10^{-9}
CH_4	16.043	10.933	5.216×10^{-2}	11.97×10^{-6}	-11.31×10^{-9}
C_3H_8	44.097	-12.579	3.062×10^{-1}	-158.6×10^{-6}	32.14×10^{-9}

$C_v(T) = C_{v,0} + C_{v,1}T^1 + C_{v,2}T^2 + C_{v,3}T^3$ (kJ/kmol K) with T(K)
or $U(T) = U_0 + C_{v,0}T + \frac{1}{2}C_{v,1}T^2 + \frac{1}{3}C_{v,2}T^3 + \frac{1}{4}C_{v,4}T^4$ (kJ/kmol).

$$B = (-15.001 - 15.838\phi + 9.613\phi^3) \times 10^3/T$$

$$C = \left[0.154226\phi^3 - 0.38656\phi - 0.10329\right.$$

$$\left. + \frac{118.27\phi - 14.763}{T}\right] \times \ln(p \times 14.503) \quad (10.12)$$

and

$$R_{corr} = R + 0.004186 \exp\left\{\left[11.98 - \frac{25442}{T}\right.\right.$$

$$\left. - 0.4354 \ln(p \times 14.503)\right]\phi$$

$$\left. + 0.2977 \ln(\phi)\right\} \text{ kJ (kg original air)}^{-1} \text{ K}^{-1} \quad (10.13)$$

For a fuel of composition $C_n H_{2n}$ the stoichiometric gravimetric fuel/air ratio is 0.0676. Thus if the internal energy of the products is wanted on a basis of per unit mass of products, it is necessary to divide equations (10.10)–(10.13) by $(1 + 0.0676\phi)$.

During combustion there will be both burned and unburned mixture present, and the next section (10.2.3) discusses how the mass fraction burned as a function of time (mfb(t)) is modelled. Prior to this, it is sensible to discuss how the internal energy of the burned and unburned mixtures are computed. Referring back to equation (10.5), it will be recalled that the internal energy of the control-volume contents are needed as a function of temperature, so that equation (10.5) can be solved in terms of temperature.

The simplest model is to have a single zone, in which the burned and unburned mixture is considered to be mixed with a uniform temperature. This was assumed in the derivation of equation (10.5). Using equation (10.10) for predicting the internal energy of the burned (suffix b) and unburned mixture (suffix u):

$$u = [(1 - \text{mfb}) K_1(T)]_u + \text{mfb}[K_1(T) - K_2(T)\phi]_b$$

$$u = K_1(T) - \text{mfb}K_2(T)\phi \quad \text{kJ/kg of original air} \quad (10.14)$$

The gas constant is used to compute the cylinder pressure (using the equation of state $pV = mRT$), from equation (10.11):

$$R = 0.287 + 0.020 \, \phi \, \text{mfb} \quad \text{kJ(kg of original air)}^{-1} \text{ K}^{-1} \quad (10.15)$$

It should be noted that in equation (10.14), the internal energy of the unburned fuel has not been considered, neither has the effect of fuel vaporisation been considered on the energy balance or the molecular balance.

10.2.3 Burn rate

The burn rate in a spark ignition engine can be modelled by a Wiebe function (chapter 3, section 3.9.2) which requires the selection of two shape parameters, and specification of the ignition timing and the combustion duration. Combustion in compression ignition engines is more complex, as it is influenced by:

(1) the delay period — the time between the start of injection and the start of combustion
(2) the amount of flammable mixture prepared during the delay period that burns rapidly once ignited
(3) the combustion of the remainder of the fuel.

Ignition delay

Ignition delay data are usually modelled by an equation that has origins in the Arrhenius equation for reaction rate:

$$\text{reaction rate} \propto \exp(-E_a/RT)$$

Ignition delay (t_{id}) is correlated by an equation of the form

$$t_{id} = A \, p^{-n} \exp(E_a/R_oT) \tag{10.16}$$

where E_a is an apparent activation energy
         A, n are constants, but functions of: fuel type, and the air/fuel mixing/motion.

Ignition delay data can be obtained from constant-volume combustion bombs, rapid compression machines, and steady flowrigs. However, Heywood (1988) shows that different workers find widely differing values for the parameters in equation (10.16):

$$0.757 < n < 2 \tag{10.17}$$
$$0.4 \times 10^{-9} < A < 0.44$$
$$4650 < E_a/R_o < 20926$$

Not surprisingly, ignition delay values predicted for engine operation also

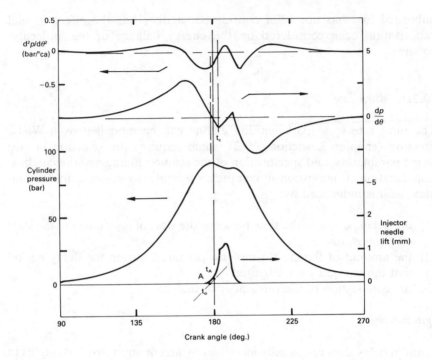

Figure 10.1   The injector needle lift, cylinder pressure, first and second
derivative of cylinder pressure from a 2.5 litre Diesel engine
operating at a speed of 2000 rpm and a bmep of 3 bar

vary widely. Belardini *et al.* (1983) reviewed many correlations for the
ignition delay in direct injection engines, and made comparisons with
ignition delay measurements made by an optical start of combustion
sensor. This leads to the question of how to measure the ignition delay
period in an engine.

The first question is how to define the start of injection. The injector can
be fitted with a needle lift transducer, and two popular approaches are to
use the change in coupling in a coil that is part of a resonating circuit, or to
use a Hall effect transducer. A typical needle lift signal is shown in figure
10.1. It should be noted that there can be noise associated with the signal
(most readily seen during the period the needle is closed). Furthermore,
the dc level of the signal can vary, especially with the resonating circuit
system. Figure 10.1 shows that initially the injector needle starts to lift very
slowly, and such movement can be difficult to distinguish from noise. This
leads to two approaches that are used to define the start of injection:

(1) as when the injector needle has lifted to (say) 10 per cent of its
maximum lift (point A and time t_a in figure 10.1)

(2) or, as when the slop at point A in figure 10.1 is extrapolated back to zero lift, to give an intercept at time t_o.

This second method is more difficult to implement, as the gradient has to be evaluated, and this can introduce errors whether undertaken digitally or with analogue electronics. Consequently, the start of injection is usually defined as when the injector needle has lifted a specified distance from its seat.

The start of combustion is more difficult to define. If combustion analysis equipment is available, it can be used to compute an apparent heat release rate that is shown in figure 10.2. (It is termed an apparent heat release rate, as no allowance is made for heat transfer. The same assumptions about gas properties have been used, as described in section 10.2.2.) The heat release rate becomes negative, owing to heat being transferred from the gas to the combustion chamber. The start of combustion can be defined as when the heat release rate becomes zero (that is, when the cumulative heat release is at a minimum), and this corresponds to 2° atdc in figure 10.2. Alternatively, the time when the cumulative heat release has returned to zero can be used to define the start of combustion; this is discussed further in chapter 13, section 13.5.3.

A simpler alternative is to differentiate the pressure with respect to time

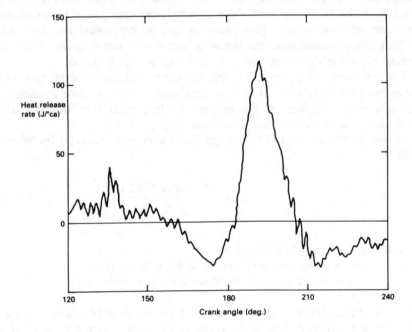

Figure 10.2   The apparent heat release rate for the pressure data shown in figure 10.1

to obtain the result already shown in figure 10.1; the second derivative of pressure has also been plotted here. The start of combustion can be defined as the minimum that occurs in the first derivative after the start of injection, time t_c in figure 10.1. The start of combustion in figure 10.1 is just after 2° atdc, and is marginally later than the start of combustion determined from the heat release rate. If the start of combustion is being evaluated by computer, it is necessary to check that a minima has indeed been found by checking that the second derivative is zero, and the third derivative of pressure is positive.

The pressure will not have been recorded on a continuous basis, but at discrete values of time (that is, crank angle). Thus the derivatives are not continuous functions, and the minimum in the first derivative has to be identified by a change in sign of the second derivative in going from negative to positive. Unless there is a correction, by means of interpolation, this implies that t_i will be detected slightly after the start of combustion.

Alternatively, the start of combustion can be sensed by an optical transducer that detects the radiation from the flame front. It has generally been found that the start of combustion detected by the optical sensor is after that detected from the pressure trace. This will no doubt be influenced by the frequency response of the optical sensors, which should be biased towards the infra-red part of the visible light spectrum. As with the trace from the needle lift transducer, there is the question of how to define the start of combustion. This problem can be by-passed if a high gain amplification is used, and the signal is allowed to saturate. In effect, the optical sensor is then being used to generate a digital signal.

The only remaining problem is that the ignition delay period is subject to cycle-by-cycle variations. It is thus necessary to record data from a sufficiently large number of consecutive cycles, such that the statistical analysis is not influenced by the sample size.

A widely used correlation for ignition delay was deduced by Watson (1979):

$$t_{id} = \frac{3.52 \exp(2100/T)}{p^{1.022}} \qquad (10.18)$$

where t_{id} = ignition delay (ms)
          p = mean pressure during ignition delay (bar)
          T = mean temperature during ignition delay (K).

This correlation has the disadvantage that mean values of temperature and pressure are required. This will require an iterative solution to be used, so as to evaluate the mean properties during the delay period.

Heywood (1988) recommends a correlation for ignition delay developed by Hardenberg and Hase (1979):

$$t_{id} = (0.36 + 0.22\, v_p)\exp\left[E_a\left(\frac{1}{R_o T} - \frac{1}{17190} \right)\left(\frac{21.2}{p-12.4} \right)^{0.63} \right] \quad (10.19)$$

where t_{id} = ignition delay (degrees crank angle)
          v_p = mean piston velocity (m/s)
          T = temperature at tdc (K)
          p = pressure at tdc (bar)

and E_a = 618 840/(CN + 25)
      where CN = fuel cetane number

This correlation has several advantages. Firstly it recognises that there is an engine speed dependence, and a dependence on the self-ignition quality of the Diesel fuel. Furthermore, the temperature and pressure are determined at a single condition. If this correlation is not being applied within a computer simulation of an engine, then the temperature and pressure can be estimated by assuming the compression can be modelled by a polytrophic process:

$$T = T_i\, r_i^{k-1}, \qquad p = p_i\, r_i^{k} \quad (10.20)$$

where suffix i refers to conditions at inlet valve closure (ivc), r_i is the volumetric compression ratio from ivc to the start of injection, and k is the polytropic index (typically 1.3).

However, the dependence on the cetane number should be treated with caution, as there is not necessarily a relation between ignition delay and cetane number for a given engine at a particular operating point. The problem arises from the way in which the cetane number is evaluated, by ignition delay measurements made in a CFR engine, with an indirect injection (IDI) combustion system. Typical values of ignition delay are within the range of 0.5–2.0 ms, though under light load low-speed operation longer ignition delays occur, especially if the engine is cold.

Diesel combustion

As explained in section 10.2.3, it is common practice to divide the combustion into two parts: the *pre-mixed* burning phase and the *diffusion* burning phase. The *pre-mixed* phase is a consequence of the mixture prepared during the ignition delay period burning rapidly; the *diffusion* burning phase accounts for the remainder of combustion. Watson *et al.* (1981) proposed:

$$\text{mfb}(t) = \beta f_1(t) + (1 - \beta)f_2(t) \qquad (10.21)$$

where   mfb(t)   = mass fraction burned
$\quad\quad\;\; \beta$ = fraction of fuel burned in the *pre-mixed* phase
$\quad\quad\;\; f_1(t)$ = *pre-mixed* burning function
$\quad\quad\;\; f_2(t)$ = (*diffusion*) burning function.

The fraction of fuel burned in the *pre-mixed* phase (β) will be a function of the ignition delay period:

$$\beta = 1 - a\phi^b/(t_{id})^c \qquad (10.22)$$

$$\phi = \text{equivalence ratio}$$

$$t_{id} = \text{ignition delay (ms)}$$

and for turbocharged DI truck engines:

$$0.8 \; < a < 0.95$$
$$0.25 < b < 0.45$$
$$0.25 < c < 0.5$$

with

$$f_1(t) = 1 - (1 - t^{K_1})^{K_2}$$
$$f_2(t) = 1 - \exp(-K_3 t^{K_4})$$

in which $K_1 = 2.00 + 1.25 \times 10^{-8}(t_{id} \times N)^{2.4}$
$\quad\quad\quad\quad K_2 = 5000$
$\quad\quad\quad\quad K_3 = 14.2/\phi^{0.644}$
$\quad\quad\quad\quad K_4 = 0.79 K_3^{0.25}$
and N is the engine speed (rpm).

A more fundamental approach has been proposed by Whitehouse, N.D. and Way (1968), and although it was derived for a single zone combustion model, it can be extended to a multi-zone combustion model. The Whitehouse–Way combustion model comprises two parts. The first part is based on the Arrhenius equation, and it predicts the reaction of fuel burn rate (FBR). The second part predicts the fuel preparation rate (FPR), based on the diffusion of oxygen into the fuel jet.

The fuel burn rate is modelled by:

$$\text{FBR} = \frac{K}{N} \times \frac{p'_{02}}{\sqrt{T}} \exp(-E_a/T) \int (\text{FPR} - \text{FBR})d\theta \qquad (10.23)$$

and the fuel preparation rate is given by

$$\text{FPR} = K' \, m_i^{(1-x)} \, m_u^x \, (p_{02}')^z \tag{10.24}$$

where p_{02}' is the partial pressure of oxygen
      m_i is the mass of fuel injected
      m_u is the mass of fuel yet to be injected.

Benson and Whitehouse (1979) also discuss the derivation of this model, and quote values of the parameters for specific two- and four-stroke engines. They lie within the following ranges:

$$0.01 < x < 1$$
$$z = 0.4$$
$$0.008 < K < 0.020 \ (\text{bar}^{-2})$$
$$1.2 \times 10^{10} < K' < 65 \times 10^{10} \ (\text{K}^{1/2}/\text{bar s})$$
$$E_a = 1.5 \times 10^4 \ (\text{K})$$

Winterbone (1986) presents cylinder pressure versus crank angle plots to show the effect of systematic variation of these parameters. The mass fraction burned (mfb(t)) is found by integration of equation (10.23).

10.2.4   Engine gas side heat transfer

Engine heat transfer is discussed later (chapter 12) in the context of heat transfer to the engine cooling media (usually aqueous ethylene glycol and lubricating oil). Typically 20–35 per cent of the fuel energy passes to the engine coolant. With rich mixtures in spark ignition engines, as little as 15 per cent of the fuel energy passes to the coolant, but with spark ignition engines at low loads a much higher proportion (over 40 per cent) of the fuel energy passes to the coolant. Of the heat flow to the coolant, about half comes from in-cylinder heat transfer, and most of the balance flows from the exhaust port. The heat flow from the exhaust port depends on its geometry, and the extent of its passage through the coolant. Exhaust ports are now sometimes insulated; this reduces the heat flow to the coolant. The higher exhaust temperature promotes further oxidation of the combustion products and increases the energy that an exhaust turbine can convert to work.

In-cylinder heat transfer

The in-cylinder heat transfer coefficient will vary with position and time. A knowledge of these spatial variations is needed if thermal stresses are to be calculated. At the opposite extreme, a simple correlation giving a position and time averaged heat transfer coefficient will be satisfactory for predicting the heat transfer to the coolant. For engine modelling, the variation in heat transfer with time is needed: the variation with position is not necessary if the emissions are not being considered.

The heat transfer from the combustion products occurs by convection and radiation. In spark ignition engines radiation may account for up to 20 per cent of the heat transfer, but it is usually subsumed into a convective heat transfer correlation, Annand (1986). However, for compression ignition engines, the radiation from soot particles during combustion can be significant, and some correlations allow for the radiation in a separate form from the contribution by convection.

Until computational fluid dynamic (CFD) techniques permit a full prediction of the in-cylinder gas motion, it is necessary to predict the in-cylinder heat transfer coefficient, by using correlations that have been derived from experimental measurements. Fortunately, the predictions of the engine output and efficiency are not very sensitive to the predictions of heat transfer. Typically, a 10 per cent error in the prediction of in-cylinder heat transfer leads to a 1 per cent error in the engine performance prediction.

One of the earliest correlations for in-cylinder heat transfer was developed by Eichelberg (1939):

$$Q_s/A_s = 2.43v_p(pT)^{\frac{1}{2}}\{T - T_s\} \text{ W/m}^2 \qquad (10.25)$$

where v_p = mean piston speed m/s
p = instantaneous cylinder pressure (bar)
T = instantaneous bulk gas temperature (K)
T_s = mean surface temperature (K)
A_s = instantaneous surface area (m²)
Q_s = instantaneous heat flow rate (W)

thus the heat transfer coefficient (h) is given by

$$h = \frac{Q_s}{A_s(T - T_s)} \qquad (10.26)$$

This correlation has the advantage of simplicity, as it only requires the user to specify a surface temperature (a value of around 350 K is often assumed).

However, the Eichelberg correlation is not dimensionally consistent, and it has been argued that its generality is therefore suspect. This has led to correlations of the form

$$Nu = aRe^b \qquad (10.27)$$

where Nu = Nusselt number = hx/k
$\quad\quad\;\; Re$ = Reynolds number = $\rho u x/\mu$
$\quad\quad\;\; h$ = gas thermal conductivity (w/mK)
$\quad\quad\;\; \rho$ = gas density (kg/m^3)
$\quad\quad\;\; \mu$ = gas dynamic viscosity (kg/m s)
$\quad\quad\;\; x$ = characteristic dimension (m)
$\quad\quad\;\; u$ = characteristic velocity (m/s).

As the details of the fluid motion are not known, then arbitrary but convenient definitions are used, with the characteristic length (x) assumed to be the bore diameter (B), and the characteristic velocity (u) assumed to be the mean piston speed ($\bar{v}_p$).

When a separate radiation term is included, an equation of the following form results. This was first proposed by Annand (1963) and then refined further by Annand and Ma (1971):

$$\frac{Q_s}{A_s} = c\,\frac{k}{B}\,Re^b(T - T_s) + d(T^4 - T_s^4) \qquad (10.28)$$

Watson and Janota (1982) suggested that for a compression ignition engine:

$$
\begin{aligned}
b &= 0.7 \\
0.25 &< c < 0.8 \\
d &= 0.576\sigma \\
\sigma &= \text{Stefan–Bolzmann constant}
\end{aligned}
$$

During the intake compression and exhaust processes the radiation term should be zero ($d = 0$) and for a spark ignition engine $d = 0.075\sigma$. Watson and Janota comment that the range of values for c suggests that some important underlying variables are being ignored.

Woschni (1967) pointed out that equation (10.27) could be expanded:

$$h = (ak/x)(\rho u x/\mu)^b$$

$$= (ak)\,\frac{p^b}{RT}\,(k/\mu)^b\,u^b x^{b-1}$$

Woschni also assumed that

$$k \propto T^{0.75} \quad \text{and} \quad \mu \propto T^{0.62}$$

thus

$$h \propto p^b u^b x^{b-1} T^{(0.75\,-\,1.62b)} \qquad (10.29)$$

Woschni assumed and subsequently justified the value of b as 0.8, and took the characteristic length (x) to be the piston bore (B). He argued that the characteristic velocity would comprise two parts:

 (i) a contribution due to piston motion
 (ii) a contribution due to combustion.

Woschni expressed the contribution due to combustion as a function of the pressure rise due to combustion, that is

$$p - p_m$$

where p_m is the pressure (motoring pressure) that would occur without combustion.

Equation (10.29) becomes:

$$h = 129.8 \, p^{0.8} \, u^{0.8} \, B^{-0.2} \, T^{-0.55} \text{ W/m}^2\text{K} \qquad (10.30)$$

where p = instantaneous cylinder pressure (bar)
       B = bore diameter (m)
       T = instantaneous gas temperature (K)

and $u = C_1 \bar{v}_p + C_2 \dfrac{V_s T_r}{p_r V_r} (p - p_m)$

with V_s = swept volume
       V_r, T_r, p_r evaluated at any reference condition, such as inlet
            valve closure.

   Watson and Janota (1982) suggest that the motoring pressure is evaluated by assuming the compression and expansion to be modelled by a polytropic process:

$$p_m = p_r \left(\frac{V_r}{V} \right)^k \qquad (10.31)$$

a typical value of k being around 1.3.

   Values suggested by Woschni for C_1 and C_2 are

For gas exchange	$C_1 = 6.18$	$C_2 = 0$
For compression	$C_1 = 2.28$	$C_2 = 0$
For combustion and expansion	$C_1 = 2.28$	$C_2 = 3.24 \times 10^{-3}$
For IDI engines		$C_2 = 6.22 \times 10^{-3}$

Sihling and Woschni (1979) suggested modified values for the coefficients

C_1 for higher speed direct injection engines with swirl:

For gas exchange $\quad C_1 = 6.18 + 0.417 \dfrac{B\omega_p/2}{\bar{v}_p}$

elsewhere $\qquad C_1 = 2.28 + 0.308 \dfrac{B\omega_p/2}{\bar{v}_p}$

where $\quad \omega_p = \quad$ paddle wheel angular velocity (rad/s) in steady-flow swirl tests.

Hohenberg (1979) proposed a simplified form of equation (10.30):

$$h = \frac{129.8p^{0.8}}{V_s^{0.06}T^{0.4}}(\bar{v}_p + 1.4) \tag{10.32}$$

for which the piston area A_{pis} is given by

$$A_{pis} = A(\text{piston crown}) + A^{0.3}(\text{piston top land})$$

Not surprisingly, the different correlations for heat transfer coefficient can lead to widely varying predictions of the instantaneous and the mean heat transfer coefficient; this is illustrated by figure 10.3. However, the variation is not too important, since (as pointed out in the introduction to this section) errors in predicting the heat transfer coefficient only have a small effect on the engine performance prediction.

Heat transfer during the gas exchange processes

The heat transfer from the combustion chamber walls to the incoming charge is small compared with the total heat transfer. However, as with heat transfer from the inlet valve and inlet port, it can have a significant effect on the volumetric efficiency. The significance of the heat flow to the exhaust valve and port (up to half the heat flow to the coolant) has already been discussed in the introduction to section 10.2.4.

The Woschni correlation (equation 10.30) includes terms for the heat flow within the cylinder. There appear to be few correlations and data for predicting the heat transfer in other parts of the gas exchange process, but Annand (1986) provides a comprehensive review.

A summary of Annand's recommendations follows:

(1) Heat flow to the incoming flow from an inlet valve:

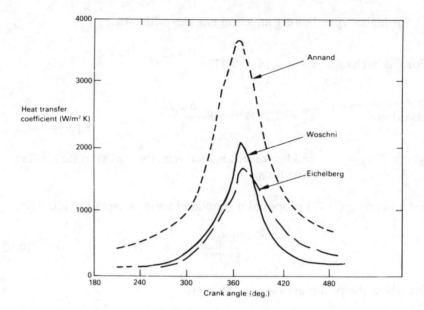

Figure 10.3   Comparison between the heat transfer correlations predicted by different correlations (adapted from Woschni (1967))

$$Nu = a\,Re^b \tag{10.33}$$

$$\text{where} \quad Nu = \frac{\dot{Q}}{0.25\pi d_v^2 \Delta T}$$

$$\text{and} \quad Re = \frac{4\dot{m}}{\pi d_v \mu}$$

The value of a is high at low value lifts, but decreases to 0.9 in the range $0.1 < L_v/d_v < 0.4$. The Reynolds number index is in the range of 0.6–0.9, and this might be a function of the port geometry.

(2) Heat flow to the exhaust valve from an outflowing charge. Annand suggests using equation (10.33) but with

$$a = 0.4\, d_v/L_v \tag{10.34}$$

(3) Heat flow to the exhaust port. Annand recommends

$$Nu = a\,Re^{0.8} \tag{10.35}$$

with the value of a depending on the valve seat angle ($a = 0.258$ for a
45° seat angle), but with no apparent dependence on the port curvature.

10.2.5 Induction and exhaust processes

Flow through the valves

The flow processes through the inlet and exhaust valves have already been
discussed in chapter 6, section 6.3. It will be recalled that the flow
coefficients are a function of:

> flow direction
> valve lift
> pressure ratio
> valve seat angle
> port geometry
> valve face angle
> whether or not the flow is treated as compressible

For inlet valves the pressure ratio is usually close to unity, but for
exhaust valves the pressure ratio can vary widely, and during exhaust
blowdown the flow can be supersonic. Care also needs to be taken in
deciding how the reference area for the discharge coefficient has been
defined. For this reason, it is less ambiguous to talk about effective flow
areas, even though this does not make comparisons between different
engines very easy. Values for flow coefficients have been presented in
chapter 6, section 6.3.

Scavenging

In the case of naturally aspirated four-stroke engines scavenging is not
usually significant. However, in the cases of:

(i) two-stroke engines, naturally aspirated or turbocharged
(ii) four-stroke engines, turbocharged with large valve overlap periods

then scavenging is significant. The different ways of modelling scavenging
have been reviewed in chapter 7, section 7.4 and comprehensive treat-
ments of scavenging in the context of engine modelling are provided by
Horlock and Winterbone (1986) and Wallace, F. J. (1986a, b). Watson and
Janota (1982) suggest that using a perfect mixing model is the best compromise. This gives a pessimistic estimate of the scavenging performance,

but it seems that the energy available at the turbine is largely independent of the model used. The displacement scavenging model predicts a smaller exhaust flow with a higher temperature, while the perfect mixing model predicts a larger mass flow but with a lower temperature.

Scavenging will also be influenced by the flow pulsations that occur in the inlet and exhaust systems. These pulsations have the greatest effect on the scavenging of two-stroke engines, but they also affect the gas exchange processes in four-stroke engines. However, their treatment is outside the present scope; a definitive exposition of flow pulsation effects, and the method of their simulation, is presented by Benson (1982).

Turbochargers

Turbochargers are most easily modelled by using turbine and compressor maps, which show the speed and isentropic efficiency as functions of the pressure ratio and mass flow.

The compressor maps are usually produced by conducting steady-flow experiments, while the turbine maps are often produced by modelling techniques. The turbine and compressor data are stored in arrays, typically as a series of discrete turbocharger speeds for which the: pressure ratio, mass flow parameter, and isentropic efficiency are tabulated.

An immediate question is: how representative is the assumption of steady flow in the turbocharger? The conditions in the compressor are steady, certainly in comparison with the pulsations that occur in the exhaust system of an engine using pulse turbocharging. The usual compromise is to calculate the instantaneous pressure difference across the turbine at each crank angle increment, and to compute the turbine performance. In other words, the flow is treated as quasi-steady, such that the turbine performs under non-steady flow conditions as it would if the instantaneous flow was a steady flow. This leads to turbocharger speed fluctuations (thus it is necessary to know the moment of inertia of the turbocharger) and a locus of operating points for a single engine operating condition. There are thus flow variations in the compressor within an engine cycle. Watson and Janota (1982) suggest that these errors will not exceed 5 per cent.

Another source of errors occurs with multi-entry turbochargers for which there will be a different pattern of flow pulsations at each entry — this can lead to so-called partial admission losses. Dale *et al.* (1988) describe a test facility for measuring the instantaneous performance in a turbine subject to pulsating flow; they could also investigate the effect of partial admission losses. Dale *et al.* (1988) found that the operating locus with pulsating flow gave a smaller mean mass flow for a given pressure ratio (figure 10.4), and that the efficiency was also lower (figure 10.5). They found that for a twin entry turbine with flow pulsations arriving out-of-

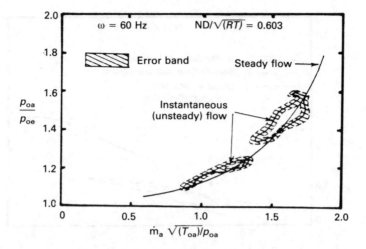

Figure 10.4   Locus of instantaneous flow under pulsed inlet conditions,
             superimposed on steady-flow data (for two mean inlet pressures,
             single-entry turbine) (Dale *et al.* (1988), reproduced by permission
             of the Council of the Institution of Mechanical Engineers)

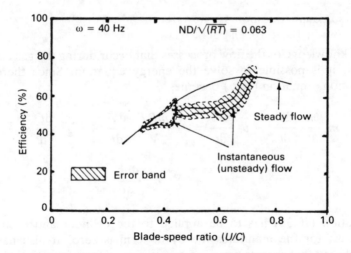

Figure 10.5   Locus of instantaneous efficiency under pulsed inlet conditions,
             superimposed on steady-flow data (for two mean inlet pressures,
             single-entry turbine) (Dale *et al.* (1988), reproduced by permission
             of the Council of the Institution of Mechanical Engineers)

phase at each inlet, the instantaneous operating locus was closer to the
operating line for partial admission steady flow, than the operating line for
steady full admission (figure 10.6).

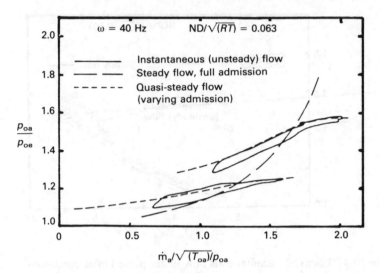

Figure 10.6   Instantaneous flow locus, at one entry, with out-of-phase flows at each entry of a twin-entry turbine (for two mean inlet pressures) (Dale *et al.* (1988), reproduced by permission of the Council of the Institution of Mechanical Engineers)

Energy balance

With a knowledge of the flow processes that occur during the gas exchange process, it is possible to solve the energy equation. Since there is no combustion, equation (10.5) becomes

$$m \frac{\partial u}{\partial t} \frac{dT}{dt} + u \frac{dm}{dt} = - \frac{mRT}{V} \frac{dV}{dt} + \sum_s \frac{dQ_s}{dt}$$

$$+ \sum_i h_{oi} \frac{dm_i}{dt} \qquad (10.36)$$

Equation (10.36) has to be applied to the cylinder and each of the manifolds; for the manifolds the dV/dt term is zero. It also has to be remembered that reverse flows can occur, in particular from the cylinder to the inlet manifold, and less significantly from the exhaust manifold into the cylinder. Also needed is an equation for mass conservation, and this is solved simultaneously with equation (10.36), so that the mass, $m(i)$, and temperature, $T(i)$, can be found on a step-by-step basis. The modelling of the gas exchange process is discussed comprehensively by Wallace, F. J. (1986a, b), and it also forms a significant part of the discussion of modelling by Watson and Janota (1982).

10.2.6 Engine friction

So far no account has been taken of engine frictional losses; the results from modelling will be in terms of indicated performance parameters. To convert the indicated performance to brake performance, it is necessary to predict the frictional losses. The difficulties in determining the frictional losses have been explained elsewhere, it is thus necessary to use the following predictions with caution.

Heywood (1988) suggests the use of the following correlation from Barnes-Moss (1975), for the frictional mean effective pressure (fmep) for automotive four-stroke spark ignition engines:

$$\text{fmep} = 0.97 + 0.15 \left(\frac{N}{1000}\right) + 0.05 \left(\frac{N}{1000}\right)^2 \qquad (10.37)$$

with fmep (bar) and N (rpm).
Millington and Hartles developed a similar correlation from motoring tests on Diesel engines:

$$\text{fmep} = (r_v - 4)/14.5 + 0.475 \left(\frac{N}{1000}\right) + 3.95 \times 10^{-3} \, \bar{v}_p^2 \qquad (10.38)$$

However, these correlations were for motored engines in which the pumping work and frictional losses will both be underestimated. This means that the earlier correlation proposed by Chen and Flynn (1965), which accounts for the pressure loading, has merit:

$$\text{fmep} = 0.137 + \frac{p_{max}}{200} + 0.162 \, \bar{v}_p \qquad (10.39)$$

with fmep (bar), p_{max} (bar), $\bar{v}_p$ (m/s).

Winterbone (1986) discusses a correlation that has the same form as equation (10.39):

$$\text{fmep} = 0.061 + \frac{p_{max}}{60} + 0.294 \frac{N}{1000} \qquad (10.40)$$

Winterbone also discusses the way in which the frictional losses increase during transients.

10.3 Application of modelling to a turbocharged medium-speed Diesel engine

10.3.1 Introduction

With naturally aspirated engines, it is quite easy to envisage the trade-offs associated with varying parameters such as ignition timing or valve timing. For example, earlier opening of the exhaust valve will reduce the expansion work produced by the engine, but it will also lead to a reduction in the pumping work. With a turbocharged engine the interactions are more complex. Again considering the effect of exhaust valve opening, earlier opening will increase the work done by the turbine and thus the pressure ratio across the compressor. The higher inlet pressure increases the work done on the piston during induction, and for many operating points the pumping work will be a positive quantity. (That is, the work done by the engine during the exhaust stroke is less than the work done by the incoming charge on the engine.) But what will the effect on a turbocharger operating point be? The turbocharger will be operating faster, but will the compressor be closer to surge, will the turbine inlet temperature become too high, will the response to transients be improved?

So, for turbocharged engines it is difficult to predict what happens when parameters are changed, unless a modelling technique is used. Even when predictions have been made once, it can be dangerous to generalise, as the efficiencies of the compressor and turbine are highly dependent on their operating points, and the turbocharger efficiencies will in turn affect the engine operating point.

To illustrate the use of a parametric investigation, a medium-speed turbocharged Diesel engine has been simulated using SPICE (Simulation Program for Internal Combustion Engines). SPICE is a filling-and-emptying type of model, and its use and background theory are documented by Charlton (1986).

10.3.2 Building and validating the model

The engine to be modelled is described in table 10.2, and an essential part of the modelling process is the validation of the model results against test data. The purpose is to use the model for a parametric investigation of valve timing.

The engine has a compact water-cooled exhaust manifold, and is turbocharged and inter-cooled. A pulse turbocharging approach is employed, with cylinders 1, 2, 3 and 4, 5, 6 exhausting into separate manifolds. In this particular case the ignition delay was a user input, but it was found

Table 10.2   Medium-speed Diesel engine specification

Configuration	6 cylinder in-line
Bore	200 mm
Stroke	215 mm
Swept volume	40 litres
Compression ratio	12.93
Firing order	153624
Rated output	840 kW at 1500 rpm
Compressor pressure ratio	2.6
Standard valve timing:	
inlet opens	60° btdc
inlet closes	45° abdc
exhaust opens	65° bbdc
exhaust closes	60° atdc
Weight	4363 kg

subsequently that the Watson correlation (equation 10.18) gave an accurate prediction. The combustion was modelled by the Watson correlation (equations 10.21 and 10.22), and the empirical constants were varied with engine load. Heat transfer within the cylinder was predicted by the Woschni (1967) correlation, and the heat transfer within the water-cooled exhaust manifold was predicted using a heat transfer coefficient to tally with experimental data. The Chen and Flynn (1965) model was used for predicting the friction levels, as it has a dependence on cylinder pressure; but under some low load conditions friction was predicted by the Millington and Hartles (1968) correlation.

In SPICE, the turbocharger is modelled from compressor and turbocharger maps that are constructed from tabulated data. The compressor map is shown in figure 10.7, and various operating points have been identified on it. The compressor data were from steady-flow tests, while the turbine data were one-dimensional steady-flow predictions using the Ainley and Mathieson method, as modified by Dunham and Came. The inter-cooler was modelled from manufacturer's data. To predict the flow through the valves, the effective flow areas were tabulated as a function of crank angle.

Model validation is an important yet difficult part of any simulation. There is never as much experimental data as might be desired, and some measurements are notoriously difficult or even impossible to make (for example, measuring the trapped mass in the cylinder). However, by conducting systematic variations of parameters in the sub-model correlations (for heat transfer, combustion etc.) it is possible to obtain a good match with experimental data.

There is also inevitably some concern over ignoring wave action effects, and in using steady-flow turbomachinery performance measurements or predictions. Watson and Janota (1982) suggest that wave action effects can be neglected if the pressure wave takes less than 15° to 20° crank angle to

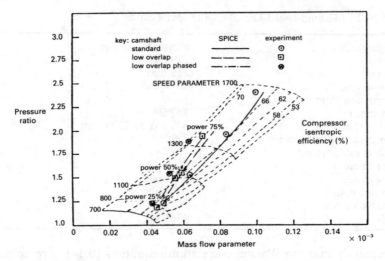

Figure 10.7   A comparison between the experimental and simulated operating
points on the compressor map

travel the length of the manifold and back. In the case of this medium speed engine the exhaust manifold is compact, and a typical wave travel time is about 40° crank angle. It was none the less thought to be acceptable to ignore the wave action effects, as this is still a small duration compared with the valve period (305° crank angle).

Another difficulty is the modelling of the turbocharger. Firstly, the compressor performance computed from engine test data was less efficient than that predicted by the map. This was accounted for in part by the map data not including the performance of the compressor air filter and inlet. The remaining differences were attributed to the pressure drop in the inter-cooler. The compressor map was thus modified by reducing all efficiencies by a 90 per cent scaling factor to agree with the performance when measured on the engine. Secondly, the turbine is subject to a pulsating flow, which both reduces the turbine efficiency, and leads to a non-stationary operating locus. Again the turbine map was modified so as to give a closer agreement with the turbine performance data obtained from the engine.

Figure 10.8 shows a comparison of the model predictions with the test data. The matching of the bmep and bsfc could have been improved by modifying the frictional losses. This has not been done as the purpose of this model was to predict the effects of valve timing variations. Figure 10.8 shows a comparison between experimental and predicted engine and turbocharger performance for a range of loads. The model was matched for the 120° overlap camshaft data, and then not changed when used to predict the performance with the other camshaft timings.

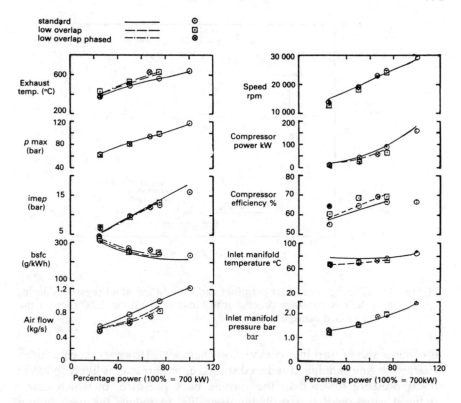

Figure 10.8 Comparison between experimental and simulated operating points with three different camshaft timings for the engine operating at 1500 rpm for a range of loads

10.3.3 The effect of valve overlap on engine operation

In general, highly turbocharged compression ignition engines have lower compression ratios in order to limit the peak pressures and temperatures. The lower compression ratios increase the clearances at top dead centre, thus permitting greater valve overlaps. In the case of large engines with quiescent combustion systems, cut-outs in the piston to provide a valve clearance have a less serious effect on the combustion chamber performance. Thus large turbocharged engines designed for specific operating conditions, can have a valve overlap of 150° or so. At a full load operating condition, the boost pressure from the compressor will be greater than the back pressure from the turbine. Consequently the large valve overlap allows a positive flow of air through the engine; this ensures excellent scavenging and cooling of the components with high thermal loadings (that

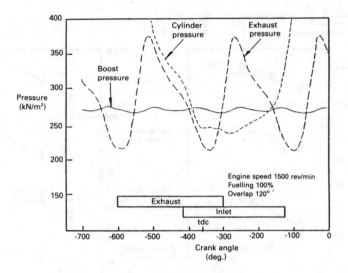

Figure 10.9   The cylinder, inlet manifold and exhaust manifold pressures during
the gas exchange process at full load (750 kW) at 1500 rpm for the
standard camshaft

is exhaust valves, turbine, combustion chamber). However, such a turbocharged engine running at reduced speeds or at part load is likely to have a
boost pressure lower than the turbine back pressure, in which case a
reduced valve overlap is probably desirable, to reduce the back flow of
exhaust into the induction system. Too much reverse flow can lead to
combustion deposits fouling the inlet port and thereby throttling the air
flow. The purpose of this type of investigation is to decide whether or not a
system that could control the valve timing might improve the engine
performance.

Figure 10.9 shows the cylinder, exhaust and inlet pressures, along with
the valve events for the standard camshaft (ivo 60° btdc) at full load. It can
be seen that initially the cylinder pressure is greater than the inlet manifold
pressure, but as the inlet valve opens slowly there is only a small amount of
reverse flow for about 25° crank angle (some 0.4 per cent of the ultimate
trapped mass). Figure 10.10 is equivalent to figure 10.9, except that the
fuelling has been reduced to 25 per cent. The compressor boost pressure
level has fallen significantly (from 2.7 to 1.4 bar) and, although the cylinder
pressure is lower, it does not equal the inlet manifold pressure until about
tdc. An immediate consequence is that there is reverse flow of exhaust
gases into the inlet manifold for about 60° crank angle, and this amounts to
about 3.3 per cent of the ultimate trapped mass. It is this flow of exhaust
gases into the inlet system that can cause inlet port fouling problems.

The effect of delaying the inlet valve opening is shown in figure 10.11 for

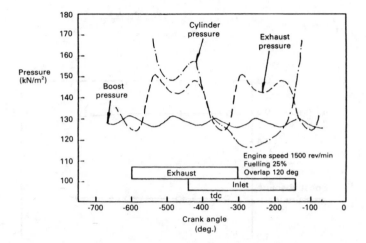

Figure 10.10   The cylinder, inlet manifold and exhaust manifold pressures
during the gas exchange process at 1500 rpm for the standard
camshaft at the 25 per cent load level

the 25 per cent fuelling case. There are negligible changes in the mean
boost pressure, and the mean exhaust pressure, but there is a slight
increase in the pumping work. Of greater importance than the mean
pressure levels in the manifolds is the instantaneous pressure difference
across the inlet valve; this is shown in figure 10.12. for a range of inlet valve
opening angles.

Figure 10.12 shows that the back flow into the inlet system would only be
eliminated if the inlet valve opening was delayed to around top dead
centre. The reduction in the inlet valve period also reduces the inflow into
the cylinder, and the overall result is the slight reduction in volumetric
efficiency shown in figure 10.11. If the inlet valve opening is delayed to 20°
btdc, then the reverse flow is reduced to 0.33 per cent of the ultimate
trapped mass at 25 per cent fuelling (compared with 3.3 per cent with the
standard inlet valve opening of 60° btdc).

Figure 10.13 summarises the effects of the different inlet valve timings at
the 25 per cent fuelling level. The substantial reduction in the reverse flow
at inlet valve opening can be seen when the inlet valve opening is delayed
by 40° crank angle. When the inlet valve closure is also delayed by 40°
crank angle, there is an even larger reverse flow, but as this is essentially
air, it should not lead to any inlet port fouling. This suggests that if the inlet
valve opening is to be delayed, then it is probably preferable (and certainly
easier) to vary the phasing of the inlet valve events. The results from the
simulation are discussed further, along with a discussion of how the
variable valve timing might be achieved, by Charlton *et al.* (1990).

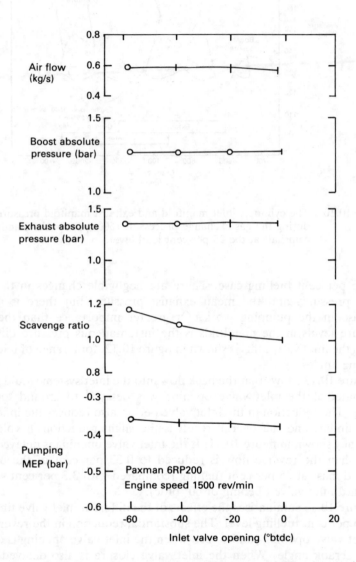

Figure 10.11   The effect on overall engine performance of delay in opening the
inlet valve at the 25 per cent load level, 1500 rpm

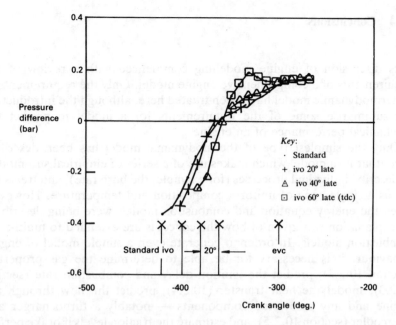

Figure 10.12   The effect on the instantaneous pressure difference across the
                inlet valve, of delaying the inlet valve opening at the 25 per cent
                load level, 1500 rpm

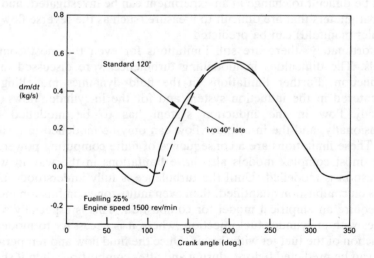

Figure 10.13   The air flow rate through the inlet valve, with a fuelling level for
                25 per cent load and a speed of 1500 rpm, for: the standard
                camshaft timing, and the inlet valve opening delayed by 40° ca

10.4 Conclusions

This discussion of engine modelling commenced with a review of the requirements of a comprehensive engine model. Only the requirements of a thermodynamic model have been treated here, although the introduction did summarise some of the requirements for a model to predict the mechanical performance of an engine.

Only the simplest type of thermodynamic model has been described here, that is, a model which makes use of a series of empirical sub-models to describe individual processes (for example, the burn rate), and treats the gas as a single zone of uniform composition and temperature. However, when the energy equation and combustion models were being described, an explanation was given of how the concepts are extended to multi-zone combustion models. In order to generate even a simple model of engine behaviour, it is necessary to: be able to determine the gas properties (section 10.2.2); predict the ignition delay and combustion rate (section 10.2.3); model the heat transfer (10.2.4); predict the flow through the engine and any associated components — notably a turbocharger and intercooler (section 10.2.5); and estimate the friction levels if brake performance parameters are required (section 10.2.6).

The application of this type of model has been illustrated by investigating the effect of valve overlap, on the part load performance of a highly rated medium-speed Diesel engine. This has illustrated how variables that might be difficult to change in an experiment can be investigated, and also how parameters that are difficult to measure (such as the reverse flow into the inlet manifold) can be predicted.

Unfortunately, there are still limitations for even the most complex models. The difficulties in modelling turbulence were discussed in the introduction. Further limitations in the fluid dynamics modelling are encountered in the induction system and for the in-cylinder flows. The unsteady flow in the induction system has to be modelled one-dimensionally, and the in-cylinder flow can only be modelled in a steady flow. These limitations are a consequence of finite computing power.

The most complex models also have limitations in the way in which combustion is modelled. Until the turbulence is fully understood and its effects on combustion quantified, then even multi-zone combustion models will require an empirical model for combustion. This is especially so for engines with in-cylinder fuel injection, where it is necessary to model the interaction of the fuel jet with the air. Once the fluid flow and temperature fields can be predicted before, during and after combustion, then it should be possible to predict heat transfer on a more fundamental basis.

Clearly there is scope for much further work in the area of thermodynamic and fluid mechanic engine modelling.

11 Mechanical Design Considerations

11.1 Introduction

Once the type and size of engine have been determined, the number and disposition of the cylinders have to be decided. Very often the decision will be influenced by marketing and packaging considerations, as well as whether or not the engine needs to be manufactured with existing machinery.

The engine block and cylinder head are invariably cast, the main exception being the fabrications used for some large marine Diesel engines. The material is usually cast iron or an aluminium alloy. Cast iron is widely used since it is cheap and easy to cast; once the quenched outer surfaces have been penetrated it is also easy to machine. Aluminium alloys are more expensive but lighter, and are thus likely to gain in importance as designers seek to reduce vehicle weight.

Pistons are invariably made from an aluminium alloy, but in higher-output compression ignition engines the piston crown needs to be protected by either a cast iron or ceramic top. The piston rings are cast iron, sometimes with a chromium-plated finish. The valves are made from one or more alloy steels to ensure adequate life under their extreme operating conditions.

Engine bearings are invariably of the journal type with a forced lubrication system. To economise on the expensive bearing alloys, thin-wall or shell-type bearings are used; these have a thin layer of bearing metal on a strip steel backing. These bearings can easily be produced in two halves, making assembly and replacement of all the crankshaft bearings (main and big-end) very simple. For the more lightly loaded bearings the need for separate bearing materials can be eliminated by careful design. The use of roller or ball bearings in crankshafts is limited because of the ensuing need for a built-up crankshaft; the only common application is in some motor-cycle engines.

397

The role of the lubricant is not just confined to lubrication. The oil also acts as a coolant (especially in some air-cooled engines), as well as neutralising the effects of the corrosive combustion products.

Only an outline of the main mechanical design considerations can be given here. Further information can be found in the SAE publications, and books such as those detailed in Baker (1979), Newton *et al.* (1983) and Taylor (1985b).

11.2 The disposition and number of the cylinders

The main constraints influencing the number and disposition of the cylinders are:

(1) the number of cylinders needed to produce a steady output
(2) the minimum swept volume for efficient combustion (say 400 cm³)
(3) the number and disposition of cylinders for satisfactory balancing.

The most common engine types are: the straight or in-line, the 'V' (with various included angles), and the horizontally opposed — see figure 11.1.

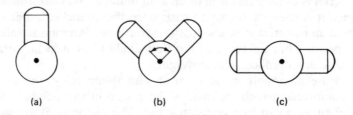

(a)   (b)   (c)

Figure 11.1   Common engine arrangements. (a) In-line; (b) 'V'; (c) horizontally opposed

'V' engines form a very compact power unit; a more compact arrangement is the 'H' configuration (in effect two horizontally opposed engines with the crankshafts geared together), but this is an expensive and complicated arrangement that has had limited use. Whatever the arrangement, it is unusual to have more than six or eight cylinders in a row because torsional vibrations in the crankshaft then become much more troublesome. In multi-cylinder engine configurations other than the in-line format, it is advantageous if a crankpin can be used for a connecting-rod to each bank of cylinders. This makes the crankshaft simpler, reduces the

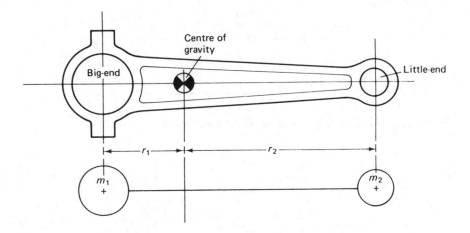

Figure 11.2   Connecting-rod and its equivalent

number of main bearings, and facilitates a short crankshaft that will be less prone to torsional vibrations. None the less, the final decision on the engine configuration will also be influenced by marketing, packaging and manufacturing constraints.

In deciding on an engine layout there are the two inter-related aspects; the engine balance and the firing interval between cylinders. The following discussions will relate to four-stroke engines, since these only have a single firing stroke in each cylinder once every two revolutions. An increase in the number of cylinders leads to smaller firing intervals and smoother running, but above six cylinders the improvements are less noticeable. Normally the crankshaft is arranged to give equal firing intervals, but this is not always the case. Sometimes a compromise is made for better balance or simplicity of construction; for example, consider a twin cylinder horizontally opposed four-stroke engine with a single throw crankshaft — the engine is reasonably balanced but the firing intervals are 180°, 540°, 180°, 540° etc.

The subject of engine balance is treated very thoroughly by Taylor (1985b) with the results tabulated for the more common engine arrangements.

When calculating the engine balance, the connecting-rod is treated as two masses concentrated at the centre of the big-end and the centre of the little-end — see figure 11.2. For equivalence

$$m_1 = m_1 + m_2$$

and
$$m_1 r_1 = m_2 r_2 \qquad (11.1)$$

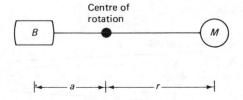

Figure 11.3   Geometry of the crank slider mechanism

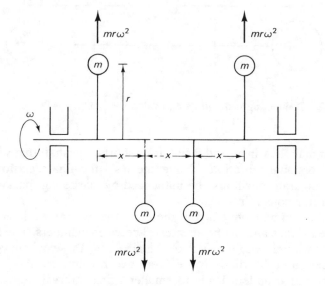

Figure 11.4   Balancing arrangements

The mass m_2 can be considered as part of the mass of the piston assembly (piston, rings, gudgeon pin etc.) and be denoted by m_r, the reciprocating mass. The crankshaft is assumed to be in static and dynamic balance, figure 11.3. For static balance $Mr = Ba$, where B is the balance mass. For dynamic balance the inertia force from the centripetal acceleration should act in the same plane; this is of importance for crankshafts since they are relatively long and flexible. As a simple example, consider a planar crankshaft for an in-line four-cylinder engine, as shown diagramatically in figure 11.4. By taking moments and resolving at any point on the shaft, it can be seen that there is no resultant moment or force from the individual centipetal forces $mr\omega^2$.

The treatment of the reciprocating mass is more involved. If the connecting-rod were infinitely long the reciprocating mass would follow simple harmonic motion, producing a primary out-of-balance force. However, the finite length of the connecting-rod introduces higher harmonic forces.

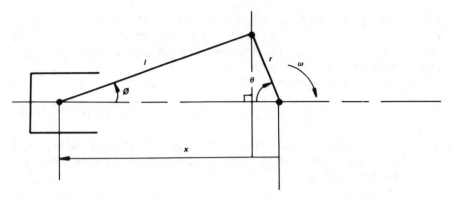

Figure 11.5   Diagrammatic representation of a four-cylinder in-line engine crankshaft

Figure 11.5 shows the geometry of the crank–slider mechanism, when there is no offset between the little-end (or gudgeon-pin or piston-pin) axis and the cylinder axis. The little-end position is given by:

$$x = r \cos \theta + l \cos \phi \qquad (11.2)$$

Inspection of figure 11.5 indicates that

$$r \sin \theta = l \sin \phi$$

and recalling that $\cos \phi = \sqrt{(1 - \sin^2 \phi)}$, then

$$x = r(\cos \theta + l/r \sqrt{\{1 - (r/l)^2 \sin^2 \theta\}}) \qquad (11.3)$$

The binomial theorem can be used to expand the square root term:

$$x = r\{\cos \theta + l/r [1 - \tfrac{1}{2}(r/l)^2 \sin^2\theta - \tfrac{1}{8}(r/l)^4 \sin^4\theta + \ldots]\} \quad (11.4)$$

The powers of $\sin \theta$ can be expressed as equivalent multiple angles:

$$\sin^2\theta = \tfrac{1}{2} - \tfrac{1}{2} \cos 2\theta$$

$$\sin^4\theta = \tfrac{3}{8} - \tfrac{1}{2} \cos 2\theta + \tfrac{1}{8} \cos 4\theta \qquad (11.5)$$

Substituting the results from equation (11.5) into equation (11.4) gives

$$x = r\{\cos \theta + l/r [1 - \tfrac{1}{2}(r/l)^2 (\tfrac{1}{2} - \tfrac{1}{2} \cos 2\theta)$$

$$- \tfrac{1}{8}(r/l)^4 (\tfrac{3}{8} - \tfrac{1}{2} \cos 2\theta + \tfrac{1}{8} \cos 4\theta) + \ldots]\} \quad (11.6)$$

The geometry of engines is such that $(r/l)^2$ is invariably less than 0.1, in which case it is acceptable to neglect the $(r/l)^4$ terms, as inspection of equation (11.6) shows that these terms will be at least an order of magnitude smaller than the $(r/l)^2$ terms.

The approximate position of the little-end is thus:

$$x \approx r\{\cos \theta + l/r \left[1 - \tfrac{1}{2}(r/l)^2 \left(\tfrac{1}{2} - \tfrac{1}{2} \cos 2\theta\right)\right]\} \qquad (11.7)$$

Equation (11.7) can be differentiated once to give the piston velocity, and a second time to give the piston acceleration (in both cases the line of action is the cylinder axis):

$$\dot{x} \approx -r\omega \left(\sin \theta + \tfrac{1}{2}r/l \sin 2\theta\right) \qquad (11.8)$$

$$\ddot{x} \approx -r\omega^2 \left(\cos \theta + r/l \cos 2\theta\right) \qquad (11.9)$$

This leads to an axial force

$$F_r \approx m_r\omega^2 r \left(\cos \theta + \frac{r}{l} \cos 2\theta\right) \qquad (11.10)$$

where
F_r = axial force due to the reciprocating mass
m_r = equivalent reciprocating mass
ω = angular velocity, $d\theta/dt$
r = crankshaft throw
l = connecting-rod length
$\cos \theta$ = primary term
$\cos 2\theta$ = secondary term

In other words there is a primary force varying in amplitude with crankshaft rotation and a secondary force varying at twice the crankshaft speed; these forces act along the cylinder axis. Referring to figure 11.4 for a four-cylinder in-line engine it can be seen that the primary forces will have no resultant force or moment.

By referring to figure 11.6, it can be seen that the primary forces for this four-cylinder engine are 180° out of phase and thus cancel. However, the secondary forces will be in phase, and this causes a resultant secondary force on the bearings. Since the resultant secondary forces have the same magnitude and direction there is no secondary moment, but a resultant force of

$$4m_r\omega^2 \frac{r^2}{l} \cos 2\theta \qquad (11.11)$$

For multi-cylinder engines in general, the phase relationship between cylinders will be more complex than the four-cylinder in-line engine. For cylinder n in a multi-cylinder engine:

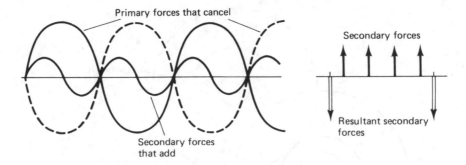

Figure 11.6   Secondary forces for the crankshaft shown in figure 11.4

$$F_{r,\,n} \approx m_{r,\,n}\omega^2 r[\cos\,(\theta + a_n) + r/l\,\cos(2\theta + 2a_n)] \qquad (11.12)$$

where a_n is the phase separation between cylinder n and the reference cylinder.

It is then necessary to evaluate all the primary and secondary, forces and moments, for all cylinders relative to the reference cylinder, to find the resultant forces and moments. A comprehensive discussion on the balancing and firing orders of multi-cylinder in-line and 'V' engines can be found in Taylor (1985b).

In multi-cylinder engines the cylinders and their disposition are arranged to eliminate as many of the primary and secondary forces and moments as possible. Complete elimination is possible for: in-line 6-cylinder or 8-cylinder engines, horizontally opposed 8-cylinder or 12-cylinder engines, and 12-cylinder or 16-cylinder 'V' engines. Primary forces and moments can be balanced by masses running on two contra-rotating countershafts at the engine speed, while secondary out-of-balance forces and moments can be balanced by two contra-rotating countershafts running at twice the engine speed — see figure 11.7. Such systems are rarely used because of the extra cost and mechanical losses involved, but examples can be found on engines with inherently poor balance such as in-line 3-cylinder engines or 4-cylinder 'V' engines.

In vehicular applications the transmission of vibrations from the engine to the vehicle structure is minimised by the careful choice and placing of flexible mounts.

Two additional forms of engine that were very popular as aircraft engines were the radial engine and the rotary engine. The radial engine had stationary cylinders radiating from the crankshaft. For a four-stroke cycle an odd number of cylinders is used in order to give equal firing intervals. If there are more than three rows of cylinders there are no unbalanced primary or secondary forces or moments. In the rotary engine

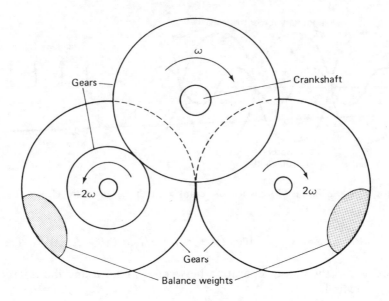

Figure 11.7   Countershafts for balancing secondary forces

the cylinders were again radiating from the crankshaft, but the cylinders rotated about a stationary crankshaft.

11.3   Cylinder block and head materials

Originally the cylinder head and block were often an integral iron casting. By eliminating the cylinder head gasket, problems with distortion, thermal conduction between the block and head, and gasket failure were avoided. However, manufacture and maintenance were more difficult. The most widely used materials are currently cast iron and aluminium alloys. Typical properties are shown in table 11.1.

There are several advantages associated with using an aluminium alloy for the cylinder head. Aluminium alloys have the advantage of lightness in weight and ease of production to close tolerances by casting — very important considerations for the combustion chambers. The high thermal conductivity also allows higher compression ratios to be used, because of the reduced problems associated with hot spots. The main disadvantages are the greater material costs, the greater susceptibility to damage (chemical and mechanical), and the need for valve seat inserts and valve guides (see figure 6.2). Furthermore, the mechanical properties of aluminium

Table 11.1   Properties of cast iron and aluminium alloy

	Cast iron	Aluminium alloy
Density (kg/m³)	7270	2700
Thermal conductivity (W/m K)	52	150
Thermal expansion coefficient (10^{-6}/K)	12	23
Young's Modulus (kN/mm²)	115	70

alloys are poorer than cast iron. The greater coefficient of thermal expansion and the lower Young's Modulus make the alloy cylinder head more susceptible to distortion. None the less, aluminium alloy is increasingly being used for cylinder heads.

When aluminium alloy is used for the cylinder block, cast iron cylinder liners are invariably used because of their excellent wear characteristics. The principal advantage of aluminium alloy is its low weight, the disadvantages being the greater cost and lower stiffness (Young's Modulus). The reduced stiffness makes aluminium alloy cylinder blocks more susceptible to torsional flexing and vibration (and thus noisier). Furthermore, it is essential for the main bearing housings to remain in accurate alignment if excessive wear and friction are to be eliminated. These problems are overcome by careful design, with ribs and flanges increasing the stiffness.

In order to facilitate design, much use is now made of finite element methods. These enable the design to be optimised by carrying out stress analysis and vibration analysis with different arrangements and thicknesses of ribs and flanges. In addition, the finite element method can be applied to heat transfer problems, and the thermal stresses can be deduced to complete the model.

An interesting example of an aluminium cylinder block is the Chevrolet Vega 2.3 litre engine. The open-deck design reduces the torsional stiffness of the block but enables the block to be diecast, thus greatly easing manufacture. The aluminium alloy contains 16–18 per cent silicon, 4–5 per cent copper and 0.45–0.65 per cent magnesium. Cast iron cylinder liners are not used; instead the cylinder bore is treated to form a wear-resistant and oil-retaining surface by electrochemical etching to expose the hard silicon particles. To provide a compatible bearing surface the piston skirts are electroplated successively with zinc, copper, iron and tin. The zinc bonds well to the piston alloy, and the copper protects the zinc; the iron provides the bearing material, while the tin protects the iron and facilitates the running-in.

In larger (non-automotive) engines, steel liners are often used because of their greater strength compared to that of cast iron. To provide an inert, oil-retaining, wear-resistant surface a carefully etched chromium-plated finish is often used.

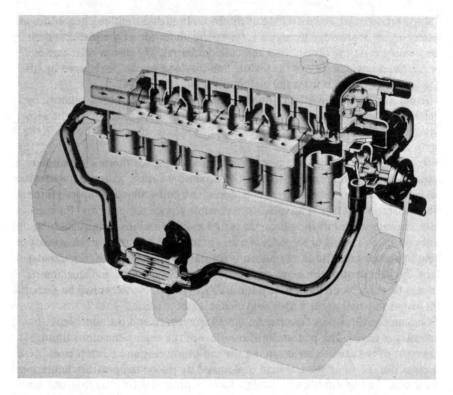

Figure 11.8   Cooling-water system; see also figures 1.10 and 11.17 (courtesy of Ford)

The majority of engines use a water-based coolant, and the coolant passages are formed by sand cores during casting. Water is a very effective heat transfer medium, not least because, if there are areas of high heat flux, nucleate boiling can occur locally, thereby removing large quantities of energy without an excessive temperature rise. Because of the many different materials in the cooling system it is always advisable to use a corrosion inhibitor, such as those that are added to ethylene glycol in antifreeze mixtures. Coolants and cooling systems are discussed extensively in chapter 12.

A typical cooling-water arrangement is shown in figure 11.8. This is for the Ford direct injection compression ignition engine that is discussed in chapter 1, section 1.3, and shown in figure 1.10. At the front of the engine is a water pump which enhances the natural convection flow in the cooling system. The pump is driven by a V-belt from the crankshaft pulley. The outflow from the pump is divided into two flows which enter opposite ends of the cylinder block. On four-cylinder engines the flow would not normally be divided in this way. The flow to the far end of the engine first passes through the oil cooler. Once the water enters the cylinder block it passes around the cylinders, and rises up into the cylinder head. While the engine is reaching its working temperature the flow passes straight to the pump inlet since the thermostats are closed.

When the engine approaches its working temperature first one thermostat opens and then, at a slightly higher temperature (say 5 K), the second thermostat opens. Once the second thermostat has opened about 50 per cent of the flow passes through the radiator. The use of two thermostats prevents a sudden surge of cold water from the radiator, and also provides a safety margin should one thermostat fail. The optimum water flow pattern is often obtained from experiments with clear plastic models.

Air-cooling is also used for reasons of lightness, compactness and simplicity. However, the higher operating temperatures require a more expensive construction, and there is more noise from the combustion, pistons and fan.

11.4 The piston and rings

The design and production of pistons and rings is a complicated job, which is invariably carried out by specialist manufacturers; a piston assembly is shown in figure 11.9.

Pistons are mostly made from aluminium alloy, a typical composition being 10–12 per cent silicon to give a relatively low coefficient of thermal expansion of 19.5×10^{-6} K^{-1}. The low density reduces the reciprocating mass, and the good thermal conductivity avoids hot spots. The temperature of the piston at the upper ring groove should be limited to about 200°C, to avoid decomposition of the lubricating oil and softening of the alloy. In high-output engines, additional piston cooling is provided by an oil spray to the underside of the piston; otherwise cooling is via the piston rings and cylinder barrel. The piston skirt carries the inertial side loading from the piston, and this loading can be reduced by offsetting the gudgeon pin (piston pin) from the piston diameter.

The gudgeon pin is usually of hollow, case hardened steel, either retained by circlips or by an accurate diametral fit. The centre hole reduces the weight without significantly reducing the strength. The piston is reinforced by bosses in the region of the gudgeon pin.

One of the key problems in piston design is allowing for thermal expansion and distortion. The thermal coefficient of expansion for the piston is greater than that of the bore, so that sufficient clearances have to be allowed to prevent the piston seizing when it is at its maximum possible service temperature. Furthermore, the asymmetry of the piston leads to

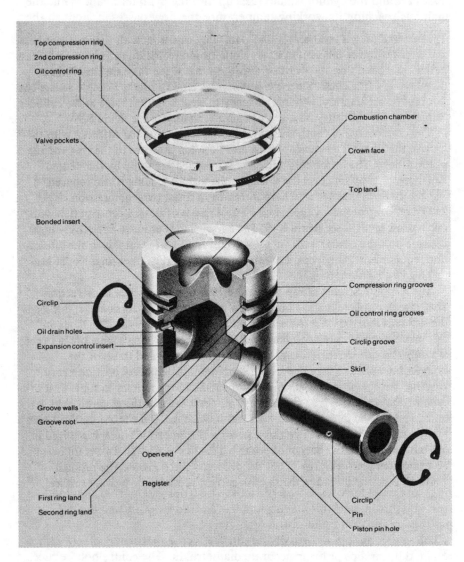

Figure 11.9   Piston assembly (with acknowledgement to GKN Engine Parts
          Division)

non-uniform temperature distributions and asymmetrical expansions. To ensure minimal but uniform clearances under operating conditions, the piston is accurately machined to a non-circular shape. To help control the expansion, carefully machined slots and steel inserts can also be used. None the less, it is inevitable that the clearances will be such that piston slap will occur with a cold engine.

Combustion chambers are often in the piston crown, and the additional machining is trivial. The piston can also influence the engine emissions through the extent of the quench areas around the top piston ring and the top land. However the extent of the top land is governed by piston-temperature limitations.

More complex pistons include those with heat-resistant crowns, articulated skirts, and raised pads on the skirt to reduce the frictional losses.

Very high output Diesel engines sometimes use cast steel or cast iron pistons. Such engines can also use an integral annular cooling gallery that surrounds the piston bowl. The cooling gallery is fed with oil through drillings in the connecting rod and gudgeon pin. Cast iron or steel pistons have a greater tolerance of higher temperatures than aluminium alloy pistons, and they also have a lower coefficient of thermal expansion. With their low expansion, iron and steel pistons can be designed to have lower clearances with the piston bore. This leads to reduced exhaust emissions and piston slap; piston slap is a significant source of noise in engines during warm-up.

The three main roles of the piston rings are:

(1) to seal the combustion chamber
(2) to transfer heat from the piston to the cylinder walls
(3) to control the flow of oil.

The material used is invariably a fine grain alloy cast iron, with the excellent heat and wear resistance inherent in its graphitic structure. Piston rings are usually cast in the open condition and profile-finished, so that when they are closed their periphery is a true circle. Since the piston rings tend to rotate a simple square-cut slot is quite satisfactory, with no tendency to wear a vertical ridge in the cylinder. Numerous different ring cross-sections have been used — see figure 11.10.

The cross-sectional depth is dictated by the required radial stiffness, with the proviso that there is adequate bearing area between the sides of the ring and the piston groove. The piston ring thickness is governed primarily by the desired radial pressure; by reducing the thickness the inertial loading is also reduced.

Conventional practice is to have three piston rings: two compression rings, and an oil control ring. A typical oil control ring is of slotted construction with two narrow lands — see figure 11.11. The narrow lands

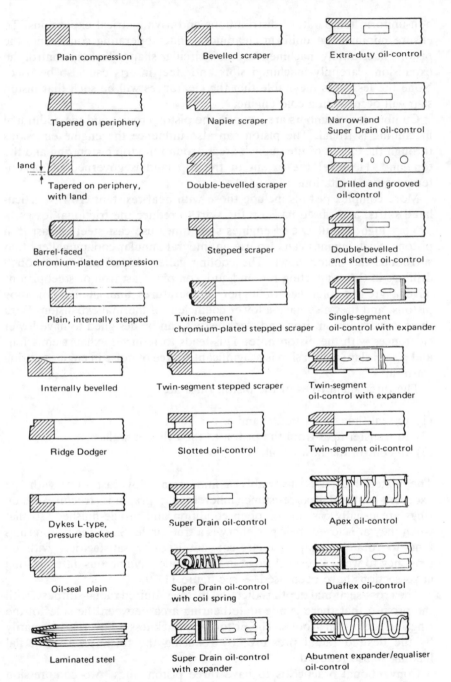

Figure 11.10   Different types of piston ring (with acknowledgement to Baker (1979); and the source of data: AE Group)

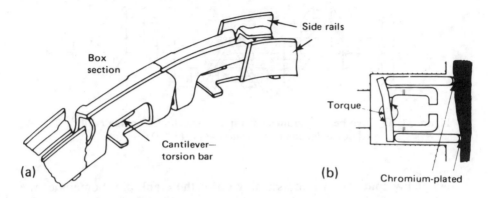

Figure 11.11   Oil control ring construction (with acknowledgement to Newton *et al.* (1983))

produce a relatively high pressure on the cylinder walls, and this removes oil that is surplus to the lubrication requirements. Otherwise the pumping action of the upper rings would lead to a high oil consumption.

The Hepolite SE ring shown in figure 11.11 is of a three-piece construction. There are two side rails and a box section with a cantilever torsion bar; this loads the side rails in a way that can accommodate uneven bore wear. The lands are chromium-plated to provide a hard wearing surface. For long engine life between reboring and piston replacement, the material selection and finish of the cylinder bore are also very important. Chromium-plated bores provide a very long life, but the process is expensive and entails a long running-in time. Cast iron cylinder bores are also very satisfactory, not least because the graphite particles act as a good solid lubricant. The bore finish is critical, since too smooth a finish would fail to hold any oil. Typically, a coarse silicon carbide hone is plunged in and out to produce two sets of opposite hand spiral markings. This is followed by a fine hone that removes all but the deepest scratch marks. The residual scratches hold the oil, while the smooth surface provides the bearing surface.

11.5   The connecting-rod, crankshaft, camshaft and valves

Connecting-rods are invariably steel stampings, with an 'H' cross-section centre section to provide high bending strength. Titanium, aluminium alloys and cast irons have all been used for particular applications, with the manufacture being by forging and machining. The big-end bearing is

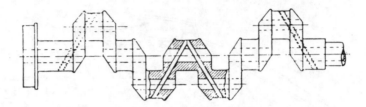

Figure 11.12   Five-bearing crankshaft for a four-cylinder engine (with
acknowledgement to Newton *et al.* (1983))

invariably split for ease of assembly on to the crank pin. Sometimes the
split is on a diagonal to allow the largest possible bearing diameter. The
big-end cap bolts are very highly loaded and careful design, manufacture
and assembly are necessary to minimise the risk of fatigue failure. The
little-end bearing is usually a force-fit bronze bush.

Connecting-rods should be checked for the correct length, the correct
weight distribution, straightness and freedom from twist.

Crankshafts for many automotive applications are now made from SG
(spheroidal graphite) or nodular cast iron as opposed to forged steel. The
cast iron is cheaper to manufacture and has excellent wear properties, yet
the lower stiffness makes the shaft more flexible and the superior internal
damping properties reduce the dangers from torsional vibrations. In normal cast iron the graphite is in flakes which are liable to be the sources of
cracks and thus reduce the material's strength. In SG cast iron, the copper,
chromium and silicon alloying elements make the graphite particles occur
as spheres or nodules; these are less likely to introduce cracks than are
flakes of graphite with their smaller radii of curvature.

A five-bearing crankshaft for a four-cylinder engine is shown in figure
11.12. The drilled oil passages allow oil to flow from the main bearings to
the big-end bearings. The journals (bearing surfaces) are usually hardened
and it is common practice to fillet-roll the radii to the webs. This process
puts a compressive stress in the surface which inhibits the growth of cracks,
thereby improving the fatigue life of the crankshaft. The number of main
bearings is reduced in some instances. If the crankshaft in figure 11.12 had
larger journal diameters or a smaller throw, it might be sufficiently stiff in a
small engine to need only three main bearings. Whatever the bearing
arrangement, as the number of main bearings is increased it becomes
increasingly important for the journals and main bearings to be accurately
in-line.

To reduce torsional vibration a damper can be mounted at the front end
of the crankshaft. A typical vibration damper is shown in figure 11.13,
where a V-belt drive has also been incorporated. An annulus is bonded by
rubber to the hub, and the inertia of the annulus and the properties of the

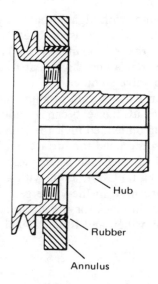

Figure 11.13   Torsional vibration damper (with acknowledgement to Newton
              et al. (1983))

rubber insert are chosen for the particular application. The torsional
energy is dissipated as heat by the hysteresis losses in the rubber. The
annulus also acts as a 'vibration absorber' by changing the crankshaft
vibration characteristics.

Camshafts are typically made from hardened steel, hardened alloy cast
iron, nodular cast iron or chilled cast iron. Chill casting is when suitably
shaped iron 'chills' are inserted into the mould to cause rapid cooling of
certain parts. The rapid cooling prevents some of the iron carbide dissociating, and thus forms a very hard surface. A variety of surface hardening
techniques are used, including induction hardening, flame hardening,
nitriding, Tufftriding and carburising. The material of the cam follower has
to be carefully selected since the components are very highly loaded, and
the risk of surface pick-up or cold welding must be minimised.

The inlet valves and in particular the exhaust valves have to operate
under arduous conditions with temperatures rising above 500°C and 800°C,
respectively. To economise on the exhaust valve materials a composite
construction can be used; a Nimonic head with a stellite facing may be
friction-welded to a cheaper stem. This also allows a material with a low
coefficient of thermal expansion to be used for the stem. The valve guide
not only guides the valve, but also helps to conduct heat from the valve to
the cylinder head. In cast iron cylinder heads the guide is often an integral
part of the cylinder head, but with aluminium alloys a ferrous insert is
used.

In spark ignition engines with high outputs (say over 60 kW/litre), then sodium-cooled exhaust valves are used. These valves have a hollow exhaust valve stem, and as the sodium melts (98°C) the liquid is shaken inside the chamber. This provides a very high heat transfer coefficient between the valve head and the valve stem.

Valve seat inserts have to be used in aluminium cylinder heads, while with cast iron cylinder heads the seats can be induction-hardened. In spark ignition engines running on leaded fuel, the lead compounds lubricate the valve seat, so obviating the need for surface hardening.

The factors which affect valve gear friction and wear have been reviewed by Narasimhan and Larson (1985), along with a comprehensive overview of the materials that are used in the valve gear. Measurements, and the associated measuring techniques for the strain and temperature distribution are discussed by Worthen and Rauen (1986).

11.6 Lubrication and bearings

11.6.1 Lubrication

The frictional energy losses inside an engine arise from the shearing of oil films between the working surfaces. By motoring an engine (the engine is driven without firing) and sequentially dismantling the components it is possible to estimate each frictional component. The results are not truly representative of a firing engine since the cylinder pressures and temperatures are much reduced. None the less, the results in figure 11.14 derived from Whitehouse, J. A. and Metcalfe (1956) indicate the trends. A useful rule of thumb is that two-thirds of engine friction occurs in the piston and rings, and two-thirds of this is friction at the piston rings. The frictional losses increase markedly with speed, and at all speeds the frictional losses become increasingly significant at part load operation. Currently much work is being conducted with highly instrumented engines and special test rigs in order to evaluate frictional losses over a range of operating conditions. For example, valve train losses can be derived from strain gauges measuring the torque in the camshaft drive, and piston friction can be deduced from the axial force on a cylinder liner.

The scope for improving engine efficiency by reducing the oil viscosity is limited, since low lubricant viscosity can lead to lubrication problems, high oil consumption and engine wear. The SAE oil viscosity classifications are widely used: there are four categories (5W, 10W, 15W, 20W) defined by viscosity measurements at — 18°C (0°F), and a further four categories (20, 30, 40, 50) defined by measurements at 99°C (210°F). Multigrade oils have

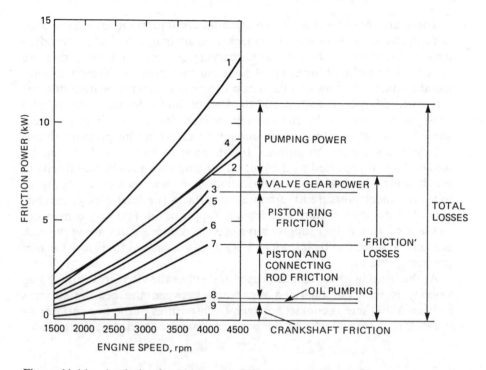

Figure 11.14   Analysis of engine power loss for a 1.5 litre engine with oil
viscosity of SAE 30 and jacket water temperature of 80° C.
Curve 1, complete engine; curve 2, complete engine with push
rods removed; curve 3, cylinder head raised with push rods
removed; curve 4, as for curve 3 but with push rods in operation;
curve 5, as for curve 3 but with top piston rings also removed;
curve 6, as for curve 5 but with second piston rings also removed;
curve 7, as for curve 6 but with oil control ring also removed;
curve 8, engine as for curve 3 but with all pistons and
connecting-rods removed; curve 9, crankshaft only (adapted from
Blackmore and Thomas (1977))

been developed to satisfy both requirements by adding polymeric additives
(viscosity index improvers) that thicken the oil at high temperatures; the
designations are thus SAE 10W/30, SAE 20W/40 etc. Multigrade oils give
better cold start fuel economy because the viscosity of an SAE 20W/40 oil
will be less than that of an SAE 40 oil at ambient conditions.

A disadvantage of the SAE classification is that viscosity measurements
are not made under conditions of high shear. When multigrade oils are
subject to high shear rates the thickening effects of the additives are
temporarily reduced. Thus, any fuel economy results quoted for different
oils need a more detailed specification of the oil; a fuller discussion with
results can be found in Blackmore and Thomas (1977).

There are three lubrication regimes that are important for engine components; these are shown on a Stribeck diagram in figure 11.15. Hydrodynamic lubrication is when the load-carrying surfaces of the bearing are separated by a film of lubricant of sufficient thickness to prevent metal-to-metal contact. The flow of oil and its pressure between the bearing surfaces are governed by their motion and the laws of fluid mechanics. The oil film pressure is produced by the moving surface drawing oil into a wedge-shaped zone, at a velocity high enough to create a film pressure that is sufficient to separate the surfaces. In the case of a journal and bearing the wedge shape is provided by the journal running with a slight eccentricity in the bearing. Hydrodynamic lubrication does not require a supply of lubricant under pressure to separate the surfaces (unlike hydrostatic lubrication), but it does require an adequate supply of oil. It is very convenient to use a pressurised oil supply, but since the film pressures are very much greater the oil has to be introduced in such a way as not to disturb the film pressure.

As the bearing pressure is increased and either the viscosity or the sliding velocity is reduced, then the separation between the bearing surfaces reduces until there is contact between the asperities of the two surfaces — point A on figure 11.15. As the bearing separation reduces, the solid-to-

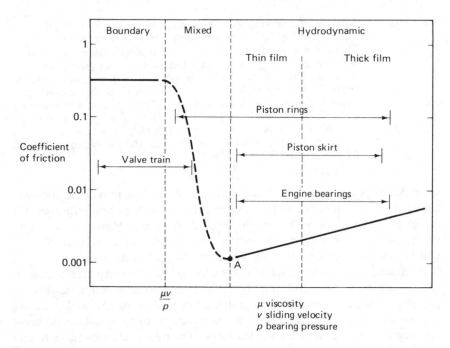

Figure 11.15   Engine lubrication regimes

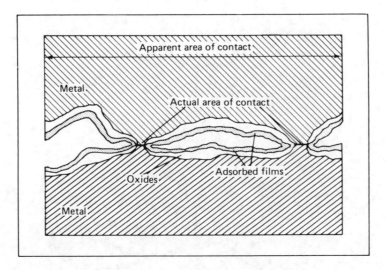

Figure 11.16   Contact between surfaces with boundary lubrication

solid contact increases and the coefficient of friction rises rapidly, so leading ultimately to the boundary lubrication mode shown in figure 11.16. The transition to boundary lubrication is controlled by the surface finish of the bearing surfaces, and the chemical composition of the lubricant becomes more important than its viscosity. The real area of contact is governed by the geometry of the asperities and the strength of the contacting surfaces. In choosing bearing materials that have boundary lubrication, it is essential to choose combinations of material that will not cold-weld or 'pick-up' when the solid-to-solid contact occurs. The lubricant also convects heat from the bearing surfaces, and there will be additives to neutralise the effect of acidic combustion products.

Part of an engine lubrication system is shown in figure 11.17; this is the same engine as is shown in figures 1.10 and 11.8. Oil is drawn from the sump through a mesh filter and into a multi-lobe positive displacement pump. The pump is driven by spur gears from the crankshaft. The pump delivery passes through the oil cooler, and into a filter element. Oil filters should permit the full pump flow, and incorporate a pressure-relief system, should the filter become blocked. Filters can remove particles down to 5 μm and smaller.

The main flow from the filter goes to the main oil gallery, and thence to the seven main bearings. Oil passes to the main bearings through drillings in the crankshaft; these have been omitted from this diagram for clarity. The oil from the main bearings passes to the camshaft bearings, and is also sprayed into the cylinders to assist piston cooling. A flow is also taken from the centre camshaft bearing to lubricate the valve gear in the cylinder head.

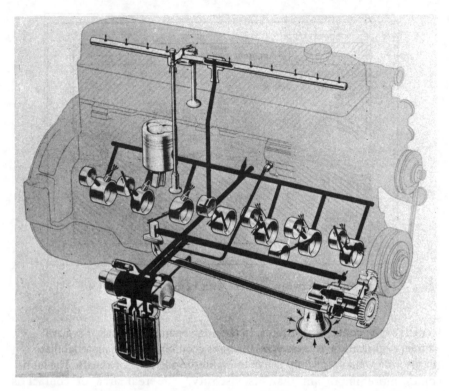

Figure 11.17   Engine lubrication system; see also figures 1.10 and 11.8 (courtesy of Ford)

11.6.2 Bearing materials

There are two conflicting sets of requirements for a good bearing material:

(1) the material should have a satisfactory compressive and fatigue strength
(2) the material should be soft, with a low modulus of elasticity and a low melting point.

Soft materials allow foreign particles to be absorbed without damaging the journal. A low modulus of elasticity enables the bearing to conform readily to the journal. The low melting point reduces the risk of seizure that could occur in the boundary regime — all bearings at some stage during start-up will operate in the boundary regime. These conflicting requirements can be met by the steel-backed bearings discussed below.

Initially Babbit metal, a tin-antimony-copper alloy, was widely used as a bearing material in engines. The original composition of Babbit metal or white metal was 83 per cent tin, 11 per cent antimony and 6 per cent copper. The hard copper-antimony particles were suspended in a soft copper-tin matrix, to provide good wear resistance plus conformability and the ability to embed foreign particles. The disadvantages were the expense of the tin and the poor high-temperature performance, consequently lead was substituted for tin. The white metal bearings were originally cast in their housings, or made as thick shells sometimes with a thick bronze or steel backing. In all cases it was necessary to fit the bearings to the engine and then to hand-scrape the bearing surfaces. This technique made the manufacture and repair of engines a difficult and skilled task.

These problems were overcome by the development of thin-wall or shell-type bearings in the 1930s. Thin-wall bearings are made by casting a thin layer of white metal, typically 0.4 mm thick, on to a steel strip backing about 1.5 mm thick. The manufacture is precise enough to allow the strip to be formed into bearings that are then placed in accurately machined housings. These bearings kept all the good properties of white metal, but gained in strength and fatigue life from the steel backing. To provide bearings for higher loads a lead-bronze alloy was used, but this required hardening of the journals — an expensive process. The overcome this difficulty a three-layer bearing was developed. A thin layer of white metal was cast on top of the lead-bronze lining. To prevent diffusion of this tin into the lead-bronze layer, a plated-nickel barrier was necessary. The expense of three-layer bearings led to the development of single-layer aluminium-tin bearings with up to 20 per cent tin. More recently an 11 per cent silicon-aluminium alloy has been developed for heavily loaded bearings.

The manufacture of such bearings is a specialist task carried out by firms such as GKN Vandervell. Table 11.2 lists some of their bearing materials.

11.7 Advanced design concepts

The Steyr–Daimler–Puch M1 engine will be used here to illustrate various advanced concepts (Freudenschuss (1988)). This engine is a turbocharged, high-speed, direct injection Diesel engine, with many innovative features; these include:

(a) monoblock construction
(b) acoustic mounting and enclosure
(c) unit injectors.

Table 11.2   Properties and composition of bearing materials

Designation and composition		Load-carrying capacity (MN/m^2)	Comments
Lead-bronze VP1			
Lead	14–20%	62	
Tin	4–6%		
Iron	0.5% max.		
Other impurities	0.75% max.		
Copper	Remainder		
Lead-bronze VP2			
Lead	20–26%	48	High-strength bearings
Tin	1–2%		with good fatigue life.
Iron	0.5% max.		Often used with a plated
Other impurities	0.75% max.		lead-indium overlay
Copper	Remainder		
Lead-bronze VP10 (equivalent to SAE 792)			
Lead	9–11%	82	
Tin	9–11%		
Iron	0.6% max.		
Zinc	0.75% max.		
Antimony	0.3%		
Phosphorus	0.05% max.		
Nickel	0.5%		
Copper	Remainder		
Tin-based Babbitt (also Micro-Babbitt) VP17 (equivalent to SAE 12)			
Tin	88.25% min.	15	Excellent conformability,
Antimony	7.25–7.75%		embedability and
Copper	3.0–3.5%		corrosion resistance
Tellurium	0.10–0.14%		
Lead	0.25% max.		
Iron	0.08% max.		
Arsenic	0.10% max.		
Aluminium	0.005% max.		
Other impurities	0.16% max.		
Lead-based Babbitt (also Micro-Babbitt) VP18 (equivalent to SAE 15)			
Antimony	14.5–15.5%	15	Superior strength to VP17
Tin	0.9–1.25%		at high temperatures.
Arsenic	0.85–1.15%		Excellent conformability,
Copper	0.5% max.		embedability and
Aluminium	0.005% max.		corrosion resistance
Zinc	0.005% max.		
Other impurities	0.25% max.		
Lead	Remainder		

Table 11.2   (continued)

Designation and composition		Load-carrying capacity (MN/m²)	Comments
Aluminium–tin VP19			
Tin	18–22%	31	For use in medium-loaded
Copper	0.7–1.3%		half-shell bearings.
Iron	0.4% max.		Excellent corrosion
Silicon	0.4% max.		resistance
Manganese	0.2% max.		
Other impurities	0.35% max.		
Aluminium	Remainder		

The engine is available as a four- or six-cylinder unit, with a range of ratings for automotive and marine applications. A cross-section of the engine is shown in figure 11.18, and the specification of the marinised engine (the version with the highest rating) is given in table 11.3.

The details of the monoblock construction and the installation of the unit injector can be seen more clearly in figure 11.19. Monoblock construction means that the cylinder head is integral with the cylinder liner, and it is a type of construction that has been used occasionally for many years. Monoblock construction has several advantages:

(a) The absence of cylinder head bolts gives greater freedom in optimising the positions of the inlet and exhaust ports, and the fuel injector.
(b) Some machining operations are eliminated, and there can be no cylinder head gasket problems.
(c) The absence of cylinder head clamping forces (which cannot be uniform circumferentially) means that the cylinder bore is distorted less, and expansion due to thermal and pressure effects is also uniform.

The reduction of bore distortion is important, since this can be a major cause of piston blow-by and oil consumption. Figure 11.19 also shows drilled coolant passages around the combustion chamber. These ensure that the coolant flow can be directed into regions with high thermal loadings, namely around the injector and the valve bridge area.

The unit injector is inclined at about 15° to the cylinder axis, and the fuel feed and return are by drilled holes in the cylinder head. This eliminates much pipework, and simplifies installation of the injectors. The fuelling is regulated by a 'rack' that is controlled by a computer-based engine management system. A single overhead camshaft operates the valve gear through finger followers, and the same camshaft operates the pump within the unit injector by a rocker arm with a roller follower. The roller follower

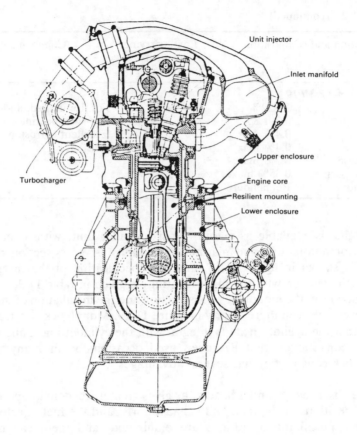

Figure 11.18   Cross-section of the Daimler–Steyr–Puch M1 monoblock DI
Diesel engine, showing the acoustic enclosures (adapted from
Freudenschuss (1988))

Table 11.3   Specification of the Daimler–Steyr–Puch M1 turbocharged and
aftercooled Diesel engine

Arrangement	6-cylinder in-line	
Swept volume	3.2	litres
Bore	85	mm
Stroke	94	mm
Maximum output	136; 4300	kW; rpm
Maximum torque	390; 2800	Nm; rpm
Minimum bsfc	218	g/kWh

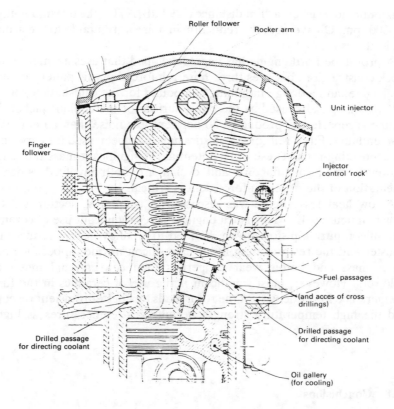

Figure 11.19   Cross-section of the unit injector, its fuel supply and cooling
details of the Daimler–Steyr–Puch M1 monoblock DI Diesel
engine (adapted from Freudenschuss (1988))

is used to minimise frictional losses on what is the most heavily loaded cam.
The injector uses a two-spring system that gives two stages to the injection
(also known as pilot injection); this helps to reduce the ignition delay and
the combustion noise (chapter 5, section 5.6.2).

The acoustic mounting and enclosure can be seen in figure 11.18. The
main structure of the engine (the cylinder block and crankshaft bearings) is
made from cast iron, and this is encased by an acoustic enclosure in two
parts. The lower part of the enclosure is diecast from aluminium alloy in
two parts, and it acts as the sump, and the mounting for auxiliary equipment and any gearbox. This lower enclosure is mounted through elastic
support rings to provide acoustic isolation. The upper housing is made
from sheet metal, and it is attached to the lower enclosure. All apertures in
the housing are carefully sealed, to minimise the transmission of noise. The

maximum noise level (at 1 m distance) is 94 dB(A) for the 6 cylinder engine at 4300 rpm. On average, the reduction in noise attributable to the acoustic enclosure is 8 dB(A).

A prototype Ford engine has been developed that uses a similar mono-block construction, but uses three fibre-reinforced plastic panels to encase the engine, and a dough moulding compound for the camshaft cover. The side panels are also used as the outer surface of the coolant jacket. The engine is reported as producing 30 per cent less noise than an equivalent all-metal unit. Fibre-reinforced plastics have also been used for experimental connecting-rods. These offer potential for significant mass reductions. However, their manufacture and design are both involved, since the orientation of the fibre reinforcement is critical.

A low heat loss engine was discussed in chapter 9, section 9.3 (and shown in figure 9.13). This Isuzu engine makes significant use of ceramics. The silicon nitride piston crown, bonded to an alloy piston, reduces heat transfer and the reciprocating mass. The silicon nitride tappet face has a lower mass and better wear properties than conventional metal cam followers. However, most significant is the use of ceramics in the turbocharger. The lower density of the rotor leads to a better transient response, and the high temperature strength permits gas temperatures as high as 900°C.

11.8 Conclusions

By careful design and choice of materials, modern internal combustion engines are manufactured cheaply with a long reliable life. Many ancillary components such as carburettors, radiators, ignition systems, fuel injection equipment etc. are made by specialist manufacturers. Many engine components such as valves, pistons, bearings etc. are also manufactured by specialists. Consequently, engineers also seek advice from specialist manufacturers during the design stage.

12 Heat Transfer in Internal Combustion Engines

12.1 Introduction

There are two aspects to heat transfer within internal combustion engines. Firstly there is heat transfer from within the combustion chamber to its boundaries (discussed in chapter 10, section 10.2.4), and secondly there is heat transfer from the combustion chamber to its cooling media — this aspect is discussed here.

First of all, the cooling requirements are considered on a global basis, as a function of: engine type, load and speed, and cooling system type (that is, liquid or air-cooled). Also included here is a discussion of the heat flow in low heat loss Diesel engines (so-called 'adiabatic' engines) and the use of ceramic components or coatings as insulators. Such engines have already been discussed in chapter 9, section 9.3 and chapter 11, section 11.7.

Secondly, a conventional liquid coolant system is described, and this leads to a review of alternative coolants and cooling systems that might cause a lower part load fuel consumption. There is also a discussion of engine performance during warm-up.

12.2 Engine cooling

12.2.1 Background

There are three reasons for cooling engines: firstly to promote a high volumetric efficiency, secondly to ensure proper combustion, and thirdly to ensure mechanical operation and reliability. The cooler the surfaces of the combustion chamber, then the higher the mass of air (and fuel) that can be

425

trapped in the cylinder. In general, the higher the volumetric efficiency, then the higher the output of the engine.

In the case of spark ignition engines, cooling of the combustion chamber inhibits the spontaneous ignition of the air–fuel mixture. Since spark ignition engines have an essentially homogeneous mixture of fuel and air, then spontaneous ignition can affect a significant quantity of mixture, and the subsequent rapid pressure rise or so-called detonation, generates the characteristic 'knocking' sound. This process destroys the thermal boundary layer, and can lead to overheating of components and ensuing damage.

There are three ways in which overheating can affect the mechanical performance of an engine. Firstly, overheating can lead to a loss of strength. For example, aluminium alloys soften at temperatures over about 200°C, and the piston ring grooves can be deformed by a creep mechanism. Furthermore, if the spontaneous ignition is sufficiently severe, then the piston can be eroded in the top-land region. This is usually the hottest region of the piston, and it also coincides with the end-gas region, the region where spontaneous ignition most frequently occurs. Figure 12.1 shows that such 'detonation' damage can also occur with Diesel engines, if the rapid combustion following the ignition delay period is too severe. Secondly, the top piston ring groove temperature must also be limited to about 200°C if the lubrication is to remain satisfactory. Above this temperature lubricants can degrade, leading to both a loss of lubrication, and packing of the piston ring groove with products from the decomposed oil. Finally, failure can result through thermal strain. Data for material failure are usually expressed in terms of stress, either for a single load application, or alternatively as fatigue data where the number of load applications also has to be specified. Such data can also be considered in terms of the strain that would cause failure: strain can be caused by either mechanical or thermal loading. The thermal strain is directly proportional to the temperature gradient. Failure is not likely from a single occurrence, but as a consequence of thermal fatigue. The regions most likely to suffer from thermal fatigue are those within the combustion chamber that have both a high temperature, and a high temperature gradient. This is exemplified by the valve bridge region, the area between the inlet and exhaust valve seats, (figure 12.1).

12.2.2 Spark ignition engines

The heat rejected to the coolant in spark ignition engines is a function of the: speed, load, ignition timing and air/fuel ratio. The large number of dependent variables means that even when comprehensive energy balance data are published, not all the variables are likely to have been investigated systematically, nor will the test conditions have necessarily been fully defined.

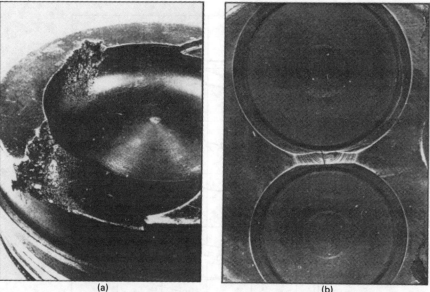

(a)   (b)

Figure 12.1   Examples of thermal loading. (a) Detonation damage on the edge
of a Diesel engine piston; (b) crack across the valve bridge,
French (1974). [Reproduced by kind permission of Butterworth–
Heinemann Ltd, Oxford]

Gruden and Kuper (1987) present a series of contour plots for the
different energy flows (fuel in, brake power, coolant, oil, exhaust) as
functions of bmep and engine speed for a 2.5 litre spark ignition engine.
They also present contour plots of the brake, mechanical and indicated
efficiency. Figure 12.2 shows the brake and mechanical efficiency as a
function of the engine load and speed. The brake efficiency results imply
that the engine has been tuned for maximum economy at part load, while
at full throttle the mixture has been richened to give the maximum power.
The mechanical efficiency is directly affected by the load (with zero
mechanical efficiency by definition at no load). Figure 12.2 also shows that
the mechanical efficiency at full load falls from about 90 per cent at 1000
rpm to 70 per cent at 6000 rpm; at 6000 rpm the frictional losses represent
about 34 kW. Friction dissipates useful work as heat, some of which
appears in the coolant and some in the oil. The heat loss recorded to the oil
is almost solely a function of speed, with about 5 kW dissipated at 3000
rpm, and 15 kW dissipated at 6000 rpm.

Figure 12.3 shows the contours of the energy flow to the coolant as a
function of the load and speed. For convenience, the brake power output
hyperbolae (calculated from the bmep and speed) have also been added.
At a bmep of about 1 bar, the energy flow to the coolant is about twice the
brake power output, while at a load of 3 bar bmep, the energy flow to the

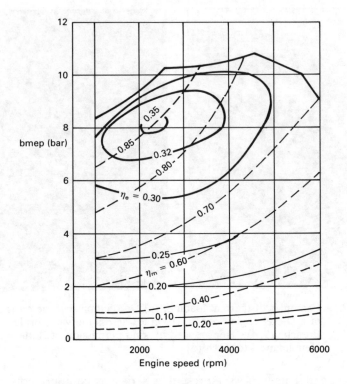

Figure 12.2   The brake efficiency and mechanical efficiency contours of the four
cylinder 2.5 litre spark ignition engine, whose thermal performance
is illustrated by figures 12.3 and 12.4 (adapted from Gruden and
Kuper (1987))

coolant is comparable to the brake power output. In the load range of 8–10
bar bmep, the energy flow to the coolant is about half the brake power
output. However, of greater importance to the vehicle cooling system are
the absolute values of the heat rejection, and figure 12.3 shows that heat
rejected to the coolant is a stronger function of speed than load.

Figure 12.4 is the counterpart to figure 12.3, as it shows contours of the
exhaust gas energy superimposed on the brake power output contours.
Almost without exception, the energy flow in the exhaust is greater than
the brake power output. However, the exhaust energy comprises in part
the chemical energy of the partial combustion products: at full load the
chemical energy is comparable to the exhaust thermal energy. The use that
can be made of the exhaust energy will be discussed later in the context of
engine warm-up.

The effect of the air/fuel ratio or equivalence ratio on the energy flow to
the coolant is illustrated by some data obtained from a gas engine (Brown,

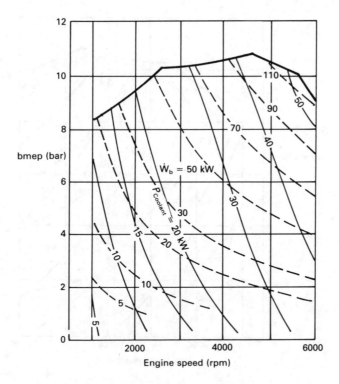

Figure 12.3   The brake power hyperbolae and the contours of heat flow
dissipated in the coolant system for the engine also described by
figures 12.2 and 12.4 (adapted from Gruden and Kuper (1987))

C. N. (1987)). The gas engine has an entirely homogeneous air–fuel
mixture, so that it is possible to run with a very wide range of equivalence
ratios. The equivalence ratio of 0.75 corresponds to an air/fuel ratio of
about 20:1 for a gasoline fuelled engine. There are several ways of
expressing the energy flow to the coolant as a function of the equivalence
ratio, and these are illustrated by figure 12.5. Firstly, in terms of the total
fuel energy supplied, the heat flow to the coolant is nearly constant at 28
per cent of the supplied energy, with a slight fall with rich mixtures to 25
per cent at an equivalence ratio of 1.2. Next, the energy flow to the coolant
can be considered as a fraction of the brake power output, but as this
remains close to unity (within experimental tolerances), it has not been
plotted here. Finally, the absolute values of the energy flow to the coolant
have been plotted. From the foregoing discussion, it can be seen that the
energy flow to the coolant is a reflection of the way in which the engine
brake output responds to the variation of the equivalence ratio. The

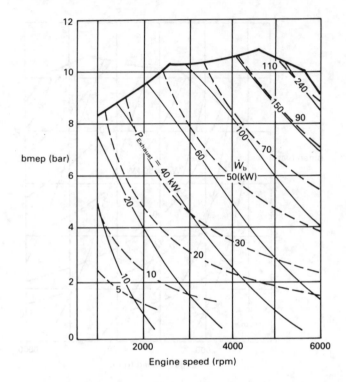

Figure 12.4   The brake power hyperbolae and the contours of thermal energy
in the exhaust, for the engine already described by figures 12.2 and
12.3 (adapted from Gruden and Kuper (1987))

explanation is that both the work output and the heat rejected are functions of the combustion temperature.

Advancing the ignition timing leads to an increased absolute value of
heat rejected to the coolant, if the throttle, speed and air/fuel ratio are
fixed. Earlier ignition causes higher temperatures in both the burned and
unburned gas, and this leads to higher levels of heat rejection from the
combustion chamber.

Raising the compression ratio also increases the in-cylinder gas temperatures, but this does not necessarily lead to an increase in the heat flow to
the coolant. The higher compression ratio increases the work output from
each charge that is ignited, and thus the exhaust temperature is lowered.
Consequently, the heat rejected to the exhaust valve and exhaust port is
reduced, and this can offset any increase in heat flow from the combustion
chamber. As the compression ratio continues to be increased, the gains in
work output reduce and the surface-to-volume ratio of the combustion
chamber deteriorates, so there will be a compression ratio above which the

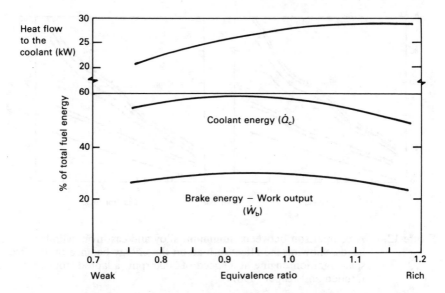

Figure 12.5   A comparison of the energy flow to the coolant and the brake
output for a Waukesha VRG220 gas engine, operating at 1500 rpm
full throttle with varying air/fuel ratio and MBT ignition timing
(Brown, C. N. (1987))

heat flow to the coolant increases. However, this is not necessarily a
compression ratio that is likely to be encountered in practice, for the
reasons discussed in chapter 4, section 4.2.1.

The material of the engine, and in particular the cylinder head, can affect
the engine performance. Tests reported by Gruden and Kuper (1987)
included comparisons of the energy balance with cast iron and aluminium
alloy cylinder heads that were otherwise identical. Some results for a speed
of 2000 rpm are presented in figure 12.6, showing: the brake power ($\dot{W}_b$);
the power flow to the coolant ($\dot{Q}_c$), oil ($\dot{Q}_{oil}$), exhaust ($\dot{Q}_{ex}$); the chemical
energy flow in the exhaust ($\dot{Q}_{chem}$); and the kinetic or dynamic energy flow
in the exhaust (KE_{ex}); the balance was assigned to extraneous heat transfer
by radiation and convection ($\dot{Q}_{ext}$). Gruden and Kuper (1987) concluded
that the energy flows were identical for a wide range of loads. However, at
full load the aluminium cylinder is less susceptible to self-ignition, so that
ignition timing is not knock limited. The explanation for this is given by
Finlay et al. (1985), who also made direct comparisons between cast iron
and aluminium alloy cylinder heads. Figure 12.7 shows that for identical
full load operation at 2000 rpm, the aluminium alloy cylinder head had
metal temperatures that were in the range 17–60 K lower than the cast iron
cylinder head. Aluminium alloys have a thermal conductivity that is
typically three times that of cast iron, and the lower surface temperatures

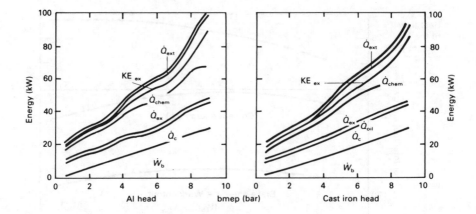

Figure 12.6   A comparison between aluminium alloy and cast iron cylinder
heads, showing the effect of load on the energy balance of a 2 litre
spark ignition engine at a speed of 2000 rpm (adapted from
Gruden and Kuper (1987))

reduce the likelihood of knock in the end-gas. However, the overall energy
flow is not affected significantly, as the highest thermal resistance is in the
thermal boundary layer of the combustion chamber.

Air cooling is most frequently encountered on engines of low output,
where simplicity in the absence of an additional cooling medium is an
advantage. Small two-stroke engines are frequently air cooled. Air-cooled
engines tend to be noisier for several reasons:

  (i) the combustion chamber walls can radiate sound
 (ii) the engine structure tends to be less rigid, as liquid cooling passages
       result in a box type construction
(iii) the cooling fan is generating significant air motion.

The inefficiency of simple cooling fans also means that the cooling power
requirements of air-cooled engines are greater than those of liquid-cooled
engines. When spark ignition engines were the common form of aircraft
propulsion, it was accepted that air-cooled engines had a slightly lower
output and efficiency. However, the weight saving associated with an
air-cooled engine meant that for journeys of up to five or six hours'
duration, the overall 'engine plus fuel' weight of an air-cooled engine was
lower than the weight of a liquid-cooled engine and its fuel (Judge (1967)).
A direct comparison has been made by Gruden and Kuper (1987) between
liquid- and air-cooled engines; they concluded that there were no significant differences. Figure 12.8 shows an energy balance for 2000 rpm, in
which the following energy flows were evaluated: the brake power ($\dot{W}_b$);

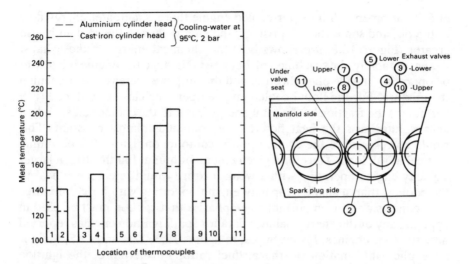

Figure 12.7   Illustrates the lower metal temperatures that occur in aluminium alloy cylinder heads for full load operation at 2000 rpm (adapted from Finlay *et al.* (1985))

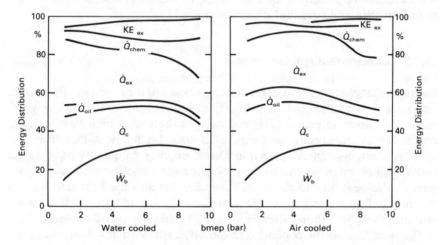

Figure 12.8   A comparison between air and liquid cooled engines, showing the effect of load on the energy balance of a 2 litre spark ignition engine at a speed of 2000 rpm (adapted from Gruden and Kuper (1987))

the power flow to the coolant ($\dot{Q}_c$), oil ($\dot{Q}_{oil}$), exhaust ($\dot{Q}_{ex}$); the chemical energy flow in the exhaust ($\dot{Q}_{chem}$); and the kinetic or dynamic energy flow in the exhaust (KE_{ex}); the balance was assigned to extraneous heat transfer by radiation and convection ($\dot{Q}_{ext}$). Both the air-cooled and the liquid-cooled engine have a maximum brake efficiency of 32 per cent in the range of 6–7 bar bmep. With the air-cooled engine there is a greater energy flow to the oil, and somewhat surprisingly the energy flow to the coolant is also greater. Figure 12.8 also shows how the chemical energy in the exhaust increases rapidly above a bmep of 7 bar. At this stage the engine is likely to be operating at wide open throttle, and the output is increased by richening the mixture. For the liquid-cooled engine operating close to full load (8–9 bar bmep) the fraction of the total energy input to the coolant falls, and on an absolute basis the heat flow to the coolant is almost constant. This would correspond to the coolant power contours on figure 12.3 becoming vertical at full load. Thus for a given engine output at full throttle, a larger capacity engine operating with a weak mixture would reject more heat to the coolant than a smaller capacity engine operating with a rich mixture.

In conclusion, it is important to emphasise that caution must be used in applying any of the energy balance data for spark ignition engines reported here to other engines. For example, in figures 12.3, 12.4 and 12.5, there is no explicit information on the air/fuel ratio. Furthermore, the ignition timing and the compression ratio both influence the energy balance. It is unfortunate that comprehensive energy balance data are not readily available in the literature. However, so far as the engine designer is concerned it is the maximum thermal loading that is the main interest, and this occurs at full throttle and (usually) maximum speed.

12.2.3   Compression ignition engines

Energy balance data for compression ignition engines are equally sparse. The heat flow to the coolant is higher in indirect injection (IDI) engines than with direct injection (DI) engines, as there is a high heat transfer coefficient in the throat, and this leads to a higher heat loss from the cylinder contents. Direct injection Diesel engines frequently have an oil cooler, which employs the engine coolant as the cooling medium; consequently the heat flow to the coolant usually includes the heat removed by the oil. Indirect injection engines are usually limited to smaller power outputs, for which there often is no need to employ an oil cooler.

The heat flow to the coolant is frequently expressed as a function of the fuel flow rate. Alcock *et al.* (1958) present results from a single cylinder direct injection engine with a swept volume of 1.78 litres. They concluded that

$$Q_c \propto (m_f)^{0.64} \qquad (12.1)$$

However, within the data used for deriving this correlation, there are clearly additional dependencies: the constant of proportionality increases with increasing speed and reducing boost pressure. The heat flow to the coolant was always within the range of 18–34 per cent of the fuel energy,

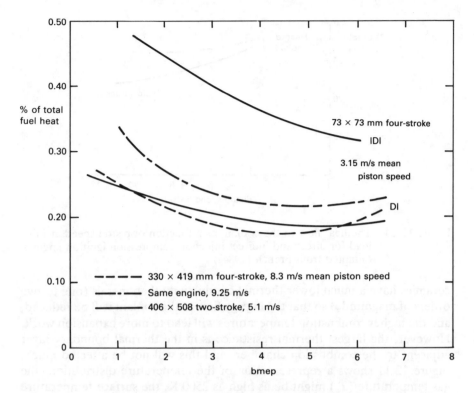

Figure 12.9   The heat flow to the coolant as a function of load for direct and indirect injection compression ignition engines; data from Taylor (1985a) and Brunel University

with the highest percentage heat rejection occurring at no load with atmospheric induction.

Watanabe *et al.* (1987) also present results from a turbocharged direct injection engine with a cylinder capacity of 1.945 litres; their data suggest a linear dependence on load with about 14 per cent of the fuel energy entering the coolant.

Taylor (1985a) presents cooling heat flow data from large two-stroke and four-stroke engines, which are shown here in figure 12.9. The heat flow to the coolant is shown as a function of the engine load, and is in the range 17–32 per cent of the fuel energy, and this is consistent with the Alcock *et al.* data.

French (1984) presents some data that compares the heat rejected to the coolant for both direct and indirect injection engines as a function of the piston speed: this is presumably full load data and is shown in figure 12.10.

In recent years, there has been much interest in the application of ceramics to compression ignition engines. The simple argument is that

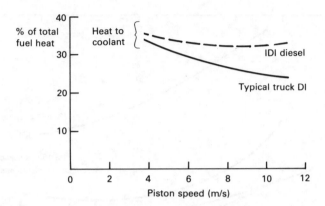

Figure 12.10   The heat flow to the coolant as a function of piston speed at full
load for direct and indirect injection compression ignition engines
(adapted from French (1984))

ceramics have a much lower thermal conductivity than metals (one to two
orders of magnitude) so that the energy flow to the coolant will be reduced,
and the higher combustion temperatures will lead to more expansion work.
However, the largest thermal resistance is in the thermal boundary layer
adjacent to the combustion chamber, and this will not be affected much.
Figure 12.11 shows a representation of the temperature distribution: the
gas temperature (T_g) might be as high as 2500 K, the surface temperature
on the gas side ($T_{w,g}$) might be 600 K, the surface temperature on the
coolant side ($T_{w,c}$) might be 400 K with a coolant temperature (T_c) of 360 K.
This system can be modelled as a series of thermal resistances to represent
the thermal boundary layers on the gas side and the coolant side, and a
thermal resistance to control the heat flow through the combustion chamber wall. If the heat flow is considered to be steady, then the thermal
resistances are proportional to the temperature differences (cf. voltage
differences). If the heat flow (cf. current) is to be changed, then the largest
effect will be obtained by changing the largest thermal resistance, that is
the gas side heat transfer coefficient. Thus an order of magnitude change to
the thermal conductivity of the combustion chamber wall does not lead to
an order of magnitude change in the heat flow. The thermal resistances
used by Moore and Hoehne (1986) to simulate the effect of a 1.25 mm
polystabilised zirconia (PSZ) coating on a cylinder head are summarised in
table 12.1.

It should be noted that the gas side heat transfer coefficient is increased,
so that some of the gain in cylinder head insulation is offset. However,
because of the time-dependent variation of gas temperature within the
combustion chamber, there are other factors controlling the heat flow
which will be considered later.

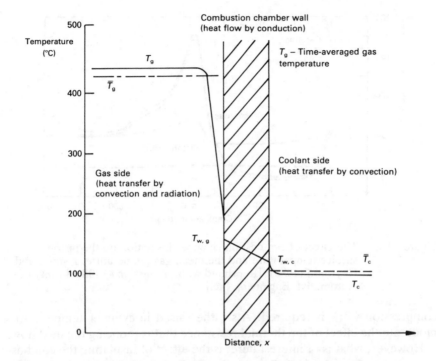

Figure 12.11   Schematic of the temperature distribution, showing the large
temperature drop in the gaseous thermal boundary layer

Table 12.1   Thermal resistances (K/kw) for a conventional and insulated (1.25
mm of PSZ) cylinder head (Moore and Hoehne (1986))

Region	Uninsulated	Insulated
Coolant side	7.9	7.9
Cylinder head	13.6	12.1
Insulation	—	80.3
Gas side	67.3	54.4
TOTAL	88.5	155

Some predictions of surface and gas temperatures have been made by
Assanis and Heywood (1986) for a partially insulated turbocompounded
direct injection engine. Figure 12.12 shows that the insulation causes a
significant increase in the piston surface temperature. This leads to heat
transfer to the induction gas (thus lowering the volumetric efficiency) and
heat transfer to the gas much further into the compression stroke. This
raises the compression and combustion temperatures, so that more

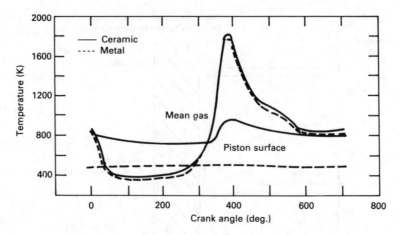

Figure 12.12   The effect of combustion chamber insulation on the piston
surface temperature and the mean gas temperature, Assanis and
Heywood (1986). [Reprinted with permission © 1986 Society of
Automotive Engineers, Inc.]

compression work is required. Also, the raised in-cylinder temperatures
moderate the effect of the thermal resistance that is reducing the heat flow.

However, what is of interest here, is the effect of insulating the combustion chamber on the heat flow to the coolant, and this is influenced by
several factors. Firstly, the cylinder head face represents a small proportion of the bore area (especially with four valve-per-cylinder arrangements). Secondly, heat transfer from the exhaust valve and port are not
affected.

When Moore and Hoehne (1986) applied a 1.25 mm PSZ coating to a
cylinder head, there was no net change in the fuel energy rejected to the
coolant. The reduction of 0.2 percentage points in the heat rejected to the
cylinder head coolant (to 7.4 per cent of the fuel energy) was exactly
balanced by the increase in the heat rejected to the oil (to 6.9 per cent of
the fuel energy). When a similar coating was applied to the piston crown
and top land, the results were as summarised in table 12.2.

Table 12.2   The effect of a 1.25 mm PSZ piston coating on fuel energy
distribution at full load (Moore and Hoehne (1986))

	Percentage of fuel energy	
	Uninsulated pistons	Insulated pistons
Cylinder head	8.4	8.2
Oil	6.3	5.4

In absolute terms there was no significant change in the heat flow to the coolant, as the engine brake efficiency fell. The piston insulation has a direct effect on the heat flow to the oil, as the undersides of the piston were cooled by an oil spray. These results should be sufficient to demonstrate that the use of ceramics does not produce a significant change in the heat flow to the coolant.

Some compression ignition engines are air cooled, and their use is most prevalent in off-highway applications. The development of an engine family that comprises both air- and liquid-cooled options is presented by Stevens and Tawil (1989).

12.3 Liquid coolant systems

12.3.1 Conventional coolant systems

The cooling circuit from a typical automotive application is shown in figure 12.13. Originally a coolant pump was not used, instead natural convection led to a thermosyphon effect. Initially with a cold engine, the thermostat is

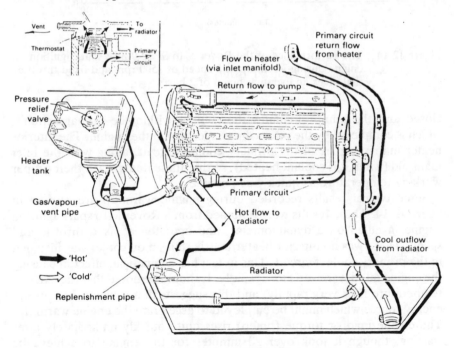

Figure 12.13   Schematic diagram of the Rover M16 cooling system

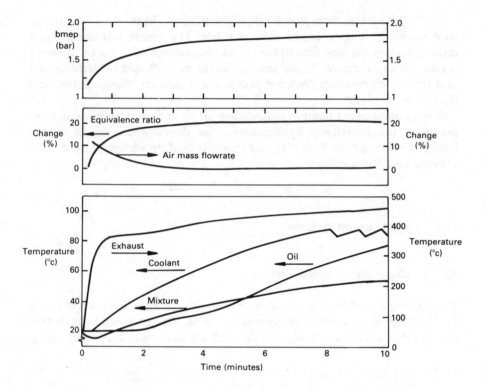

Figure 12.14   The performance of the Rover M16 SPi 2.0 litre spark ignition engine during warm-up at a speed of 1500 rpm and fixed throttle, that gives an eventual bmep of 2 bar

closed and the pump circulates the coolant within the primary circuit, which is completed by the internal passages within the engine. The interior heater matrix is part of the primary coolant circuit, along with the inlet manifold (when it is coolant heated), so that these items reach their proper working temperature as quickly as possible.

Some typical results recorded during engine warm-up are plotted in figure 12.14. These results were obtained from a Rover M16 spark ignition engine installed on a dynamometer, and operated at fixed throttle and speed. There was no interior heater matrix, but an oil cooler was installed in the primary coolant circuit. This in fact has a beneficial effect in causing the engine oil to warm-up more rapidly. It should be noted that the exhaust temperature rises very rapidly, and this suggests that it would be a useful energy source which might be employed to accelerate the engine warm-up. The engine load or torque (bmep) rises quite quickly immediately after starting, though it took over 20 minutes for the engine to achieve its steady-state output under these conditions with a bmep of 2 bar.

This particular version of the engine had a single point injection system with a coolant heated inlet manifold: the mixture temperature is seen to track the coolant temperature, but with an initial fall that is attributable to the evaporative cooling of the fuel. The engine management system was arranged to give a fixed fuel flow rate into the engine. Soon after starting there is a 10 per cent fall in the air flow rate, but this only partly accounts for the 24 per cent richening of the burnt air/fuel ratio. The remaining difference is probably due to changes in the liquid fuel film on the wall of the inlet manifold.

Once the engine has reached its operating temperature (usually around 90°C) the thermostat opens so that some coolant flows through the main circuit, and is cooled by the radiator. Figure 12.14 shows that once the thermostat opens, there are fluctuations in the temperature of the coolant in the primary circuit. This is due to cold coolant entering from the main circuit; these oscillations would be smaller if the thermostat opened more slowly, but there would then be a risk of overheating if the thermostat did not open quickly enough at a high engine load. Even when the thermostat is fully open, there will still be a significant flow through the primary circuit.

The only remaining element to discuss in figure 12.13 is the expansion tank, which is at the highest elevation. The vent pipe allows any gas to leave the cooling system and to be separated from the coolant. Such gas comes from the degassing of dissolved gases in the coolant, and under some circumstances there might be a leakage of combustion gases through the cylinder head gasket. (This can be identified by using an exhaust gas analyser.) The expansion tank has a pressure relief valve to limit the system pressure (usually about 1 bar gauge pressure), and there is also an outlet into the main circuit so that make-up liquid can enter the engine. The expansion tank should be large enough to accommodate expansion of the coolant, and have a large enough gas/vapour space, so that the gas/vapour is not compressed to a high enough pressure to open the pressure relief valve. Otherwise, when the system cools down the pressure would fall below atmospheric pressure.

The coolant pump is usually located at a low part of the cooling system with a cooled inlet flow; this is to minimise the risk of cavitation. Frequently the pump is partly formed by the cylinder block, as this simplifies assembly and reduces the component costs. Traditionally, the coolant pump is of a very simple design with a correspondingly low efficiency. The rotor may be stamped from sheet metal, or be a simple casting. Fisher (1989) argues that the coolant pump can be made much more efficient, and that it can then be driven by an electric motor. This gives independent control of the coolant flow, and the overall power requirement can be less, even allowing for the low efficiency of the motor and alternator.

The coolant flow pattern within a Ford Dover direct injection compression ignition engine has already been seen in figure 11.8.

The outflow from the pump is divided into two flows which enter at opposite ends of the cylinder block. On four-cylinder engines the flow would not normally be divided in this way. The flow to the far end of the engine first passes through the oil cooler. Once the water enters the cylinder block it passes around the cylinders, and rises up into the cylinder head. The flow rate to different parts of the cylinder head is frequently controlled by adjusting the size of the coolant holes in the cylinder head gasket. The coolant distribution is often arrived at with the help of clear plastic models. The flow can be visualised by means of particles or small gas bubbles, and photographic techniques can be used to record the flow pattern, for example continuous ciné or single frame but with illumination from controlled duration flash(es). Laser Doppler anemometry and hot wire anemometry techniques can be utilised for generating more quantitative information.

The traditional coolant passage in the cylinder head is a void, which has been defined by: the walls of the combustion chambers, the inlet and exhaust ports, the top deck and other features. Some control of the flow is achieved by regulating the flow from around the cylinder liners, and in selecting the outflow from the cylinder head. This somewhat haphazard arrangement is satisfactory in many cases, because overheating of the combustion chamber is prevented by nucleate boiling. If boiling occurs, then the heat transfer coefficient can increase by an order of magnitude, and the metal temperatures are then linked to the saturation temperature of the coolant.

With compression ignition engines, it is quite common to have a drilled passage for coolant passing through the valve bridge area or some other means of directing the coolant. Drilled passages are preferable to relying on holes made by cores, as such cores have a very high aspect ratio, and are thus susceptible to damage or movement during the casting process. Drilled coolant passages have already been seen in figure 11.19.

12.3.2 Cooling media performance

Water is a very effective cooling media, with a high enthalpy of vaporisation, high specific heat capacity and a high thermal conductivity. Its saturation temperature of 99.6°C at 1 bar is also very convenient; less convenient are its freezing point of 0°C and its contribution towards corrosion — especially when there are different metals in the cooling system.

Removal of heat from metal into a liquid coolant can be by: forced convection, subcooled boiling (where the bubbles detach when small and collapse into the bulk fluid that is below its saturation temperature), and

Table 12.3   Characteristics and requirements of an engine coolant (Rowe
(1980))

High specific heat and good thermal conductivity
Fluidity within the temperature range of use
Low freezing point
High boiling point
Non-corrosive to metals;   minimum degradation of non-metals
Chemical stability over the temperature range and conditions of use
Non-foaming
Low flammability;   high flash point
Reasonable compatibility with other coolants or oil
Low toxicity;   no unpleasant odour
Reasonable cost;   available in large quantities

saturated boiling (which has large bubbles and no condensation in the bulk
fluid). With saturated boiling there is the risk of vapour blankets and film
boiling — this can cause overheating which leads to thermal fatigue,
fracture of the component, or some form of abnormal combustion.

Rowe (1980) provides a very comprehensive summary of the requirements of automobile engine coolants and their methods of evaluation, and
he precedes his treatment with a comprehensive historical overview. The
requirements of an engine coolant are summarised by Rowe in table 12.3.

Rowe (1980) concludes that an aqueous ethylene glycol solution provides a sound basis for an automotive engine coolant. However, to prevent
corrosion and other shortcomings, it is necessary to have a range of
additives in the ethylene glycol concentrate, and these are typically:

(a) inhibitors to prevent metal corrosion
(b) alkaline substances to provide a buffering action against acids
(c) an antifoam additive
(d) a dye for ready identification
(e) a small amount of water to dissolve certain additives.

Mercer (1980a) describes the background to the selection of corrosion
inhibitors. Sodium benzoate and sodium nitrite were found to be suitable
for engines of cast iron and steel construction. In parallel, a combination of
triethanolammonium phosphate (TEP) and sodium mercaptobenzothia-
zole (NaMBT) was found to be suitable for engines of aluminium alloy
construction with copper radiators. However, it was subsequently found
that both corrosion inhibitor systems were acceptable for either type of
engine construction.

Mercer (1980b) discusses laboratory tests for determining the heat
transfer coefficients of different coolants under nucleate boiling conditions,
with and without flow. (Nucleate boiling is where vapour bubbles are

nucleated and grow from separate sites, and liquid remains in contact with part of the heated surface.) Mercer makes comparisons between as-cast and machined surfaces for cast irons, and also investigates ageing effects.

The condition of the metal/coolant interface can have a profound effect on the heat transfer coefficient (and thereby the metal temperatures). O'Callaghan (1974) reviews some of these issues, but a very comprehensive summary is presented by French (1970). French considered a wide range of different cast steels, cast irons, and coolants (including coolant contaminated by sea water) with different corrosion inhibitors. He conducted rig tests for flow and pool boiling, as well as presenting data from Diesel engine tests. French (1970) also conducted some long-term rig tests of up to 2500 hour duration, with a fixed heat flux of 520 kW/m² and a coolant temperature of 70°C. He recorded the change in wall superheat (the temperature difference between the metal surface and the saturation temperature), and found that initially the wall superheat was in the range 6–32 K. However, after prolonged boiling, the wall superheat rose by as much as a further 68 K. For one particular additive, the wall superheat rose from 30 K to 70 K after only 300 hours of testing, and was still increasing when the tests was stopped. These results emphasise the care which is needed in selecting additives if they are not to lead to elevated metal temperatures. French (1970) also reported on a 'thermal barrier' that could be found with some cast irons. The thermal barrier leads to an increase in the wall superheat, and French found that it was a property of the surface sub-layer (of about 1 mm thickness), and this sub-layer had to be completely machined away to obtain a normal value of the wall superheat. It appears that this was not an effect of the surface finish.

As described in section 12.3.1, boiling is an important heat transfer mechanism in liquid-cooled internal combustion engines. When boiling occurs, its high heat transfer coefficient restricts the increase in the metal temperature. Boiling is a very complex phenomenon that defies a thorough analytical approach. However, for certain geometries there are correlations that have been found to give good predictions; such a geometry is a drilled hole that might occur in the valve bridge area (or in a precision cooled system that is discussed later in section 12.3.3).

Finlay et al. (1987a) found that the heat transfer coefficient for a drilled hole could be predicted, by using the Dittus–Boelter correlation for forced convection, and the Chen correlation for nucleate boiling. Heat transfer coefficients have been evaluated here using these correlations, and although the relevant geometry is not widely encountered in engines, the correlations will illustrate the increase in heat transfer coefficient that is associated with nucleate boiling.

If the heat transfer mechanism is assumed to be a combination of convection and nucleate boiling, then the heat flux is given by the following equation:

$$q = h_c(T_w - T_l) + h_b(T_w - T_s) \qquad (12.2)$$

where T_w is the surface temperature
       T_l is the liquid bulk temperature
       T_s is the saturation temperatue of the fluid.

For turbulent convection the heat transfer coefficient (h_c) is predicted with the Dittus–Boelter correlation:

$$h_c = 0.023\ Re_l^{0.8}\ Pr^{0.4} \qquad (12.3)$$

$$Re_l = \frac{\rho_l\, v_l d}{\mu_l} \quad \text{and} \quad Pr = \frac{\mu_l\, C_{p,1}}{k_l}$$

where v_l = liquid flow velocity (m/s)
       d = hole diameter (m)
       k_l = liquid thermal conductivity (W/m K).

The boiling heat transfer coefficient (h_b) in equation (12.2) is predicted using the Chen (1966) correlation:

$$h_b = 0.00122 \left(\frac{k_l^{0.79}\ C_{p,1}^{0.45}\ \rho_l^{0.49}}{\sigma^{0.5}\ \mu_l^{0.29}\ h_{fg}^{0.29}\ \rho_v^{0.24}} \right) (T_w - T_s)^{0.24}\ \Delta p_s^{0.75} S \qquad (12.4)$$

where σ = surface tension (N/m)
       h_{fg} = enthalpy of vaporisation (J/kg K)
       Δp_s = (p_{sat} at T_w) − p
       S = suppression factor.

The suppression factor has been plotted as a function of Reynolds number by Finlay et al. (1987a), and this empirical result is shown in figure 12.15.

    Some results obtained from this model (equations 12.2, 12.3 and 12.4), for a 50 per cent by volume ethylene glycol/water mixture are plotted in figure 12.16 for two flow velocities. In the convective heat transfer regime, the heat transfer coefficient rises as the surface temperature rises. This is because the dominant effect is the fall in the mixture viscosity as the temperature rises. It should be noted that for the lower flow rate (0.25 m/s), the flow is laminar until the viscosity has been reduced sufficiently by the rise in temperature. For each flow rate, raising the system pressure by 0.5 bar delays the onset of nucleate boiling by about 10 K. The nucleate boiling heat transfer coefficient is so high that once boiling occurs, then there will only be a slight difference in surface temperature (the combustion chamber can be considered as a constant heat flux source). In other words, once boiling occurs the flow velocity is of secondary importance,

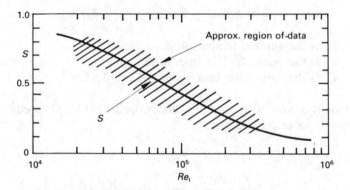

Figure 12.15   The suppression factor (*S*) used in the Chen correlation for the prediction of convective boiling heat transfer (Finlay *et al.* (1987a))

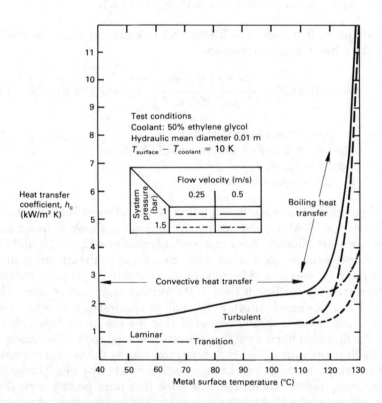

Figure 12.16   The heat transfer coefficient as a function of surface temperature, for different pressure and velocity coolant flows; note the effect on the transition to boiling heat transfer

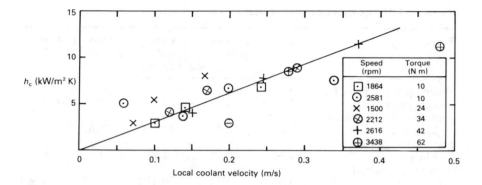

Figure 12.17   The heat transfer coefficient as a function of load coolant velocity around the combustion chamber and cylinder liner for a Ford Valencia 1.1 litre spark ignition engine with 50 per cent water/ethylene glycol coolant (derived from Sorrell (1989))

although the higher flow velocity does delay slightly the onset of nucleate boiling.

It must not be assumed that this model (equations 12.2–12.4) can be applied to conventional cooling systems. Tests were conducted by Sorrell (1989) in which the flow velocity, and the coolant side heat transfer coefficient were measured separately, in a Ford Valencia 1.1 litre spark ignition engine. Data for the heat transfer coefficients and the local flow velocity are plotted for five locations around the cylinder head and cylinder liner in figure 12.17. The data are from a range of operating points up to a torque of 62 N m at 3488 rpm, and the heat transfer coefficients are about an order of magnitude higher than those predicted by the Dittus–Boelter correlation.

Presumably the complex geometry of the flow passages defeats correlations for heat transfer coefficients, which in general assume some form of fully developed flow. Willumeit and Steinberg (1986) suggest that the heat transfer coefficient for the cylinder liner to the coolant can be predicted by correlation for tube bundles; a result not supported by the work of Sorrell (1989).

Aqueous mixtures of ethylene glycol are now well established as an engine coolant, with the percentage of ethylene glycol ranging from typically 25 to 60 per cent on a volumetric basis (that is, 26.9–61.9 per cent by weight, or 9.7–32.1 per cent on a molar basis). None the less, many alternative coolants have also been considered, but in general their heat transfer properties are inferior to those of water, or water/ethylene glycol mixtures. For example oil, which is frequently used for splash cooling the undersides of pistons, has been suggested for cooling the cylinder liner. Propylene glycol, Fluorinet (a fluid developed by the 3M company), and

Table 12.4   Properties of liquid coolants

Property	Water	Propylene glycol	Ethylene glycol water (50/50)	Ethylene glycol	Flourinet FC–77
Boiling point, 1 bar (°C)	100	187	111	197	97
Freezing point (°C)	0	−14	−37	−9	−110
Enthalpy of vaporisation (MJ/kmol)	44.0	52.5	41.2	52.6	—
Specific heat capacity (kJ/kg K)	4.25	3.10	3.74	2.38	1.05
Thermal conductivity (w/m K)	0.69	0.15	0.47	0.33	0.06
Density, 20°C (kg/m^3)	998	1038	1057	1117	1780
Viscosity, 20°C (cS, 10^{-6}m^2/s)	0.89	60	4.0	20	0.80

pure ethylene glycol have all been proposed as engine coolants. Their properties are compared with the properties of water and water/ethylene glycol in table 12.4.

Pure propylene glycol has been advocated as an engine coolant by Light (1989). Mark and Jetten (1986) also discuss the use of propylene glycol as a base fluid for automotive coolants. They conclude that aqueous solutions of propylene glycol and aqueous solutions of ethylene glycol have similar heat transfer properties, but propylene glycol has environmental advantages, and a greater resistance to cavitation. Pure propylene glycol has a high boiling point (187°C), and thus is likely to make any boiling sub-cooled, in which case the vapour bubbles tend to collapse, thereby avoiding any risk of vapour pockets 'insulating' the components that are to be cooled. Table 12.4 also shows that the molar enthalpy of vaporisation is high for propylene glycol; this implies that a small volume of vapour is formed per unit heat input. However, the high boiling point of propylene glycol might also lead to component overheating.

In conclusion, water has exceptional properties as a heat transfer medium for both convective and boiling heat transfer, these properties are moderated when water organic mixtures are used as a coolant. However, the heat transfer characteristics are still almost an order of magnitude better than those of pure organic liquids. This implies that water-based systems are likely to be needed for cooling the cylinder head. Organic coolants (such as oil or propylene glycol) might be suitable for cooling the cylinder liner.

12.3.3   Advanced cooling concepts

There are two main disadvantages with conventional cooling systems. Firstly the large volume of coolant in the primary circuit can lead to a slow engine warm-up, and a high short-journey fuel consumption. Secondly, the cooling system tends to overcool parts of the engine, especially the cylinder liners at light loads. To overcome these two shortcomings, many alternative cooling concepts have been proposed. These include:

(a) precision cooling
(b) dual circuit cooling
(c) controlled component temperature cooling
(d) evaporative cooling.

These systems all aim at making the engine temperature more uniform, and less sensitive to the operating point (in terms of the torque and speed). For example, if the cylinder liner temperature is raised, then the piston and ring pack friction are reduced. The significance of reductions in the mechanical losses increase as the load is reduced, and this leads to corresponding reductions in the fuel consumption. The tendency for the combustion to be quenched in the thermal boundary layer is reduced, as is the tendency for any hydrocarbons to be absorbed in the oil film; there is thus a reduction in the emissions of unburnt hydrocarbons (Wentworth (1971)). Unfortunately, the rise in the in-cylinder temperature might also lead to an increase in the emissions of nitrogen oxides (the formation of which is strongly temperature dependent). The emissions of NO_x (nitrogen oxides) can be reduced by either retarding the ignition timing or employing exhaust gas recirculation (EGR). However, both these remedies impose a fuel consumption penalty that might or might not be offset by the reduction in fuel consumption due to the reduced frictional losses. It should be remembered that for Diesel engines, the NO_x emissions are only likely to be problematic close to full load.

With spark ignition engines, another problem that can occur at full load is the reduction in the knock margin caused by the higher in-cylinder temperatures. Again, this can be remedied by EGR or retarding the ignition timing (usually retarding the ignition timing).

Precision cooling

The principle of precision cooling, originally developed for Diesel engines, is to provide cooling only where it is needed, at a rate proportional to the local heat flux. Figure 12.18 shows a precision cooled cylinder head (Priede and Anderton (1974)). A precision cooled system has small local passages

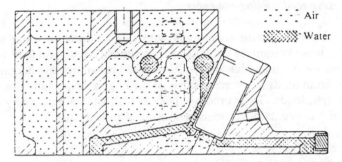

Figure 12.18   A precision cooled cylinder head (Priede and Anderton (1974), reproduced by permission of the Council of the Institution of Mechanical Engineers)

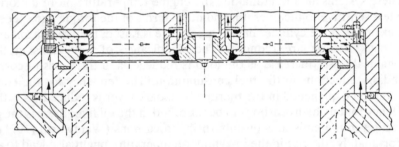

Figure 12.19   The Ruston flame plate cylinder head design (French (1970), reproduced by permission of the Council of the Institution of Mechanical Engineers)

which are used to cool critical regions such as the injector nozzle, valve bridge, exhaust valve region and valve guides, with much space in the head filled with air. A similar result is achieved by the Ruston flame plate system (figure 12.19). The application of precision cooling to an aluminium alloy cylinder head is described by Ernest (1977).

With precision cooling, many regions are not directly cooled and their temperatures rise, but the system is designed to keep the temperature within safe limits. The corresponding benefit of this is a more even temperature distribution producing less thermal strain, and the possible elimination of hot spots to allow higher compression ratios in spark ignition engines.

The design of a precision cooling system is difficult because: the heat flow patterns in the engine are complex and poorly defined, it is difficult to obtain the correct flow in the cooling passages, and it is difficult to predict the boiling heat transfer performance. In particular, a phenomenon known as dry-out can occur. If a vapour bubble grows to such a size, so as to fill the passage, then it is possible for the thin liquid layer between the bubble and surface (the microlayer) to 'dry-out'. If the surface is not re-wetted, then the vapour effectively insulates the surface, and the metal temperatures will rise in an uncontrolled manner. Also, the flow in the passages is driven by a pressure differential (it is not a fixed volume flow rate), and Finlay *et al.* (1987a) have identified an oscillatory flow caused by bubbles growing and collapsing. This leads to 'dry-out' occurring at comparatively low heat fluxes.

Dual circuit cooling

This process is characterised by separate head and block cooling circuits, with the head coolant temperature being lower than that of the block. The higher block temperature reduces the frictional losses associated with the piston and ring pack, and this leads to a reduced fuel consumption, especially at part load. The lower coolant temperature in the cylinder head tends to reduce the indicated efficiency, but as the risk of knock in spark ignition engines is also reduced, a higher compression ratio can be used. The overall result is an improved efficiency. The effect on emissions is more complex. If the combustion temperatures decrease (either through a lower compression ratio or overall cooler combustion chamber surfaces), then the NO_x emissions should decrease significantly. Consequently, the effect of dual circuit cooling on emissions is likely to be highly engine dependent.

Figure 12.20 summarises the results from tests reported by Kobayaski *et al.* (1984), in which the cylinder head coolant temperature and the compression ratio were varied. At full load (WOT), the ignition timing was limited by the onset of knock at 1200 rpm. If a cylinder head coolant temperature of 50°C could be maintained, then the compression ratio could be increased to from 9 to 12:1 yet still maintain the same output at 1200 rpm. The raised compression ratio gives about a 5 per cent reduction in the fuel consumption at 20 N m torque. At higher engine speeds the ignition timing will not be knock limited, so the raised compression ratio will also improve the full throttle output and efficiency.

Work on dual circuit cooling by Finlay *et al.* (1987b) concentrated on raising the cylinder block coolant temperature as high as 150°C. This led to reductions in the fuel consumption and hydrocarbon emissions, but an increase in the NO_x emissions.

Controlled component temperature cooling (CCTC)

This process keeps a particular component temperature constant at all speeds and loads by varying the amount of cooling. The system was first described by Willumeit *et al.* (1984), and the temperature used was that of the cylinder liner at the top ring reversal point at TDC. This philosophy could also be applied to a dual circuit cooling system.

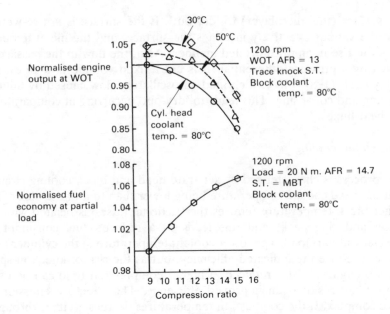

Figure 12.20   The effect of compression ratio on WOT output and part load fuel economy for various cylinder head coolant temperatures, Kobayashi *et al.* (1984). [Reprinted with permission © 1984 Society of Automotive Engineers, Inc.]

Figure 12.21 shows the cylinder liner temperature at the upper ring reversal position for varying engine speed at zero and full loads; the temperature only reaches a maximum at maximum power, and is over-cooled at other points.

Figure 12.22 shows the reductions in fuel consumption that are obtained when the controlled component temperature concept is applied. Up to 20 per cent fuel savings can be made at low loads. Other benefits of high metal temperatures (as discussed with precision cooling above) are also obtained, like reduced unburnt hydrocarbon emissions. It is also suggested that with these high temperatures, it is possible to run the engine leaner, and to reduce emissions further. However, at full load with low speeds, raising the component temperatures towards their rated load and speed values will promote knock, thereby reducing the benefits of CCTC at full load.

A not unimportant problem with the system is its response to transients; when the throttle is opened suddenly the cooling system must be able to react fast enough to prevent metal temperatures rising — otherwise knock is likely to occur (especially at low speeds). However, the role of the engine management system could be extended to control the cooling system. Coolant temperatures would be monitored, and used in conjunction with a map of the cooling requirements (analogous to mixture or

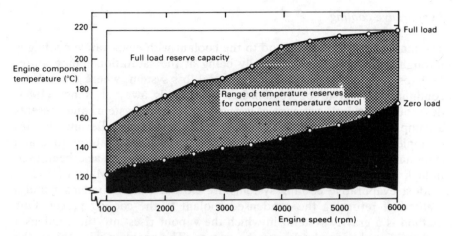

Figure 12.21   Component temperature as a function of engine speed for full
load and zero load, Willumeit *et al.* (1984). [Reprinted with
permission © 1984 Society of Automotive Engineers, Inc.]

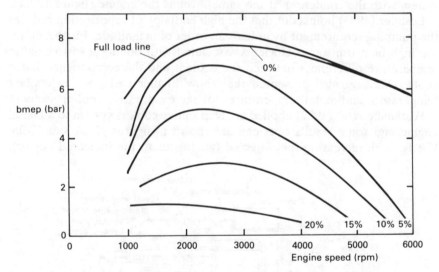

Figure 12.22   The percentage reduction in fuel consumption obtained by using
the controlled component temperature cooling concept,
Willumeit *et al.* (1984). [Reprinted with permission © 1984
Society of Automotive Engineers, Inc.]

ignition timing control). The control system could then respond to transients before any change might be detected in the metal temperature. The
metal temperature measurement would then provide a check on the
cooling system performance, and allowance could be made automatically
for the fouling of the water/metal interface.

Evaporative cooling

The main heat transfer method to the coolant with evaporative cooling is through boiling, the liquid usually being at the saturation temperature. Perhaps the most basic system is the total loss system, where the coolant continually needs to be replenished as it boils away. Leshner (1983) describes some early versions of evaporative cooling systems and presents a comprehensive overview of the system requirements. Leshner quotes examples of evaporative cooling that have been used in automotive and stationary applications. However, evaporative cooling has also been used in high performance aircraft engines by Rolls Royce (McMahon (1971)).

Most evaporative cooling systems are closed, with a condenser and a method of returning the condensed coolant to the cooling jacket. One method is a gravity system, in which the vapour rises into the condenser and the liquid drops back into the engine. The problem with this is the space required, so most systems are likely to pump the condensate back into the engine, and look very similar to the standard convective cooling system with the condenser at the same level as the engine (figure 12.23).

Leshner (1983) points out that the high enthalpy of vaporisation reduces the pumping requirement by at least an order of magnitude. Furthermore, the high heat transfer coefficients associated with boiling lead to smaller temperature differences between the coolant and the components that it cools. Data are also presented that show smaller variations in coolant temperature and metal temperatures for the evaporative cooling system.

Watanabe *et al.* (1987) applied an evaporative cooling system to a Diesel engine and some results obtained are shown in figures 12.24 and 12.25. Firstly, with increasing fuel injected (analogous to the indicated power),

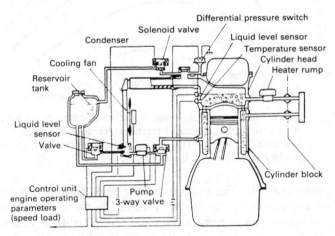

Figure 12.23   The elements of an evaporative cooling system, Leshner (1983). [Reprinted with permission © 1983 Society of Automotive Engineers, Inc.]

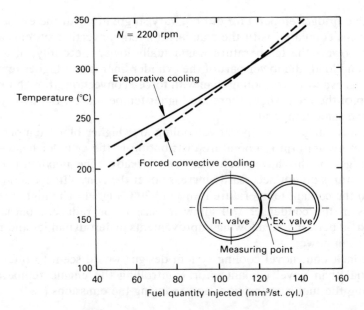

Figure 12.24   The effect of load on the valve bridge temperature in a 4-cylinder 11.7 litre Diesel engine, with evaporative and convective cooling systems, Watanabe *et al.* (1987). [Reprinted with permission © 1987 Society of Automotive Engineers, Inc.]

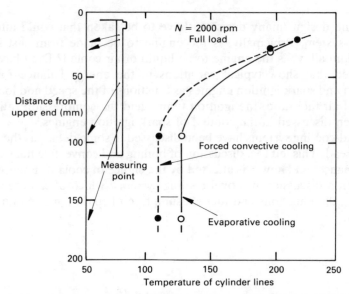

Figure 12.25   The temperature distribution in the cylinder liner at full load on a 4-cylinder 11.7 litre Diesel engine, Watanabe *et al.* (1987). [Reprinted with permission © 1987 Society of Automotive Engineers, Inc.]

the valve bridge temperature did not vary so greatly with the evaporative system, as compared with the standard forced convective system, and at higher powers the temperature was actually lower. Secondly, at a given speed and load, the lower part of the cylinder liner is at a higher temperature with evaporative cooling than with forced convection, though towards the top of the liner where there are higher temperatures, the two systems have the same temperature.

The advantages of evaporative cooling are: higher block temperatures and a more uniform temperature distribution in the cylinder head — the advantages of this have been discussed already. The maximum engine temperatures do not necessarily increase over the convective system, even though the coolant temperature is around 20°C higher. The higher temperatures in the condenser will allow its size to be reduced, perhaps by around 30 per cent, so allowing improvements in aerodynamics and reducing the fan power.

In conclusion, novel cooling system designs would seem to offer more potential than novel coolants, for: controlling the engine temperature, reducing the fuel consumption, and reducing the emissions.

12.4   Conclusions

In engine design, many decisions have to be taken that could affect the cooling system. Principally, is the engine to be made from cast iron or aluminium alloy? is the engine to be liquid or air cooled? Data have been presented that show typical variations in the energy balance for compression and spark ignition engines as functions of the speed and load. The effect of air/fuel ratio, the ignition timing and the compression ratio have also been discussed in the context of spark ignition engines.

Liquid-cooling systems have been discussed in some detail, as their use is widespread. This led to a discussion of boiling and convective heat transfer performance, and how it is affected by the choice of coolant. Finally, there has been a discussion of novel cooling systems, which offer potential for reducing the emissions and fuel consumption of spark ignition and Diesel engines.

13 Experimental Facilities

13.1 Introduction

The testing of internal combustion engines is an important part of research, development and teaching of the subject. Engine test facilities vary widely. The facilities used for research can have very comprehensive instrumentation, with computer control of the test and computer data acquisition. On the other hand, a more traditional test cell with the engine controlled manually, and the data recorded by the operator, can be better for educational purposes. This second type of test cell is covered in some detail by Greene and Lucas (1969), and is dealt with first in this chapter.

Some of the instrumentation described in this chapter is unlikely to be encountered by students. The instrumentation most likely to be met by students is introduced first. Thus, the emissions analysis equipment and combustion analysis techniques are introduced at the end of the chapter. Within the sections dealing with each measurement, the simplest instrumentation is discussed first. For example, the section on fuel flow measurement starts with control volumes, and ends with flowmeters that give a continuous reading of the instantaneous flow rate.

The chapter ends with a case study of an advanced engine test system using microprocessor engine control and data acquisition. A final class of test facility not separately discussed here is those used for acceptance tests. Most engines are tested immediately after manufacture to check power output and fuel consumption; the main requirement here is ease of installation.

Before dealing with any test facility it is important to remember that there are certain advantages in using single-cylinder engines for research and development purposes:

(1) No inter-cylinder variation. The manufacturing and assembly tolerances in multi-cylinder engines cause performance differences between cylinders. This is attributable to differences in compression ratio, valve timing etc.

457

(2) No mixture variation. With fuel injection systems it is difficult to calibrate pumps and injectors to give identical fuel distribution. In carburated engines it is difficult to design the inlet manifold to give the same air/fuel mixture to all cylinders for all operating conditions.
(3) For a given cylinder size the fuel consumption will be less and a smaller capacity (cheaper) dynamometer can be used.

13.2 Quasi-steady engine instrumentation

Figure 13.1 shows the schematic arrangement of a simple engine test rig for a Ruston Oil Engine. This single-cylinder compression ignition engine has the advantage of simplicity and ruggedness. The engine has a bore of 143 mm and a stroke of 267 mm, giving a displacement of 4.29 litres. The engine is governed to operate at 450 rpm, but fine speed adjustment is still necessary. The slow operating speed means that a particularly simple engine indicator can be used to determine the indicated work output of the engine.

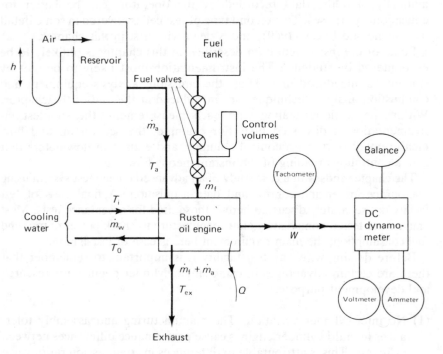

Figure 13.1   Schematic test arrangement for a Ruston Oil Engine

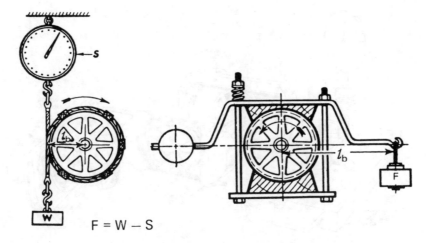

Figure 13.2   Friction brake dynamometers (courtesy of Froude Consine Ltd)

13.2.1   Dynamometers

The dynamometer is perhaps the most important item in the test cell, as it is used to measure the power output of the engine. The term 'brake horse power' (BHP) derives from the simplest form of engine dynamometer, the friction brake. Typically the engine flywheel has a band of friction material around its circumference, and the torque reaction on the friction material corresponds to the torque output of the engine — see figure 13.2.

Another type of dynamometer is the electric dynamometer which acts as a generator to absorb the power from the engine. An advantage of this type of dynamometer is that it can also be used as a motor for starting the engine, and for motoring tests (when the engine is run at operating speeds without combustion) to determine the mechanical losses in the engine. The torque output or load absorbed by the dynamometer is controlled by the dynamometer field strength. The disadvantages of this type of dynamometer are the cost and limitations on speed which may be as low as 3000 rpm. Very often these dynamometers are fitted with voltmeters and ammeters; these must not be used for calculating power unless the dynamometer efficiency is known for all operating conditions.

Another common type of dynamometer is the water brake, figure 13.3. A vaned rotor turns adjacent to a pair of vaned stators. The sluice gates separate the stators from the rotor, and these control the load absorbed by the dynamometer. Figure 13.4 shows typical absorption curves for a hydraulic dynamometer; only engine-operating points between the upper and lower solid lines can be attained.

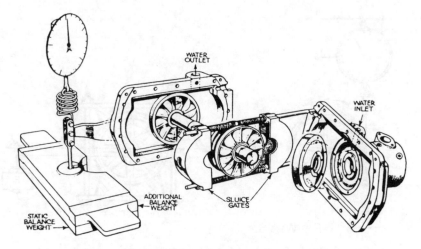

Figure 13.3   Hydraulic dynamometer (courtesy of Froude Consine Ltd)

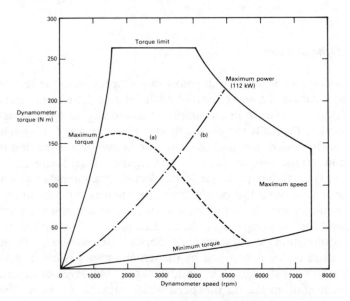

Figure 13.4   Absorption characteristics of a broad DPX2 dynamometer,
showing the operating envelope with operating lines for: (a)
constant engine throttle position, but varying the sluice gate
position; (b) constant sluice gate position but varying the engine
throttle position

Figure 13.4 also shows some dynamometer operating lines for constant
sluice gate settings and some fixed throttle operating lines. The dynamometer
operating lines show how the torque absorbed by the dynamometer increases
with speed, and the throttle operating lines show how the torque varies at a
fixed throttle setting if the speed is allowed to vary. By varying the throttle

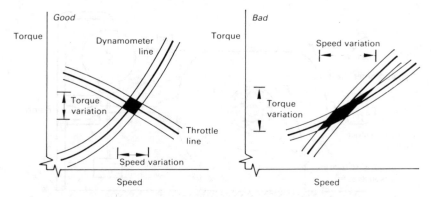

Figure 13.5   The influence of throttle and dynamometer operating line
perturbations on the stability of the operating point for well and
ill-conditioned engine and dynamometer operating lines

and sluice gate setting, any operating point should be attainable. For stable operation the dynamometer operating lines and throttle lines should intersect as close as possible to 90°. If both the operating lines are nearly parallel there will be poor stability. For example, if the operating lines are nearly parallel to the speed axis, a small change in torque will allow a large variation in speed. This is illustrated by figure 13.5 where the effect of perturbations from the throttle setting and dynamometer operating lines are considered. Such problems might occur, for example, with an engine operating at full throttle, when coupled to an electrical dynamometer that has a flat torque/speed characteristic.

Two other dynamometer types that might be encountered are the hydraulic dynamometer and the eddy current dynamometer. The hydraulic dynamometer consists of a high pressure hydraulic pump connected to the engine. The hydraulic pump will need a control system that also controls the expansion of the high pressure oil; the system is completed by an oil cooler. Since hydraulic dynamometers utilise standard components they can be made cheaply.

Eddy current dynamometers have a thin rotor that rotates within a magnetic field (the flux lines being parallel to the dynamometer axis). The torque reaction is controlled by the strength of the magnetic field, and the system is suited to electronic control. The energy is dissipated by the eddy currents (short circuited) within the rotor, and the rotor is invariably water cooled.

Most dynamometers measure the load absorbed by the torque reaction on the dynamometer casing. The dynamometer is mounted in bearings co-axial with the shaft, so that the complete dynamometer is free to rotate, but usually within a restricted range, figure 13.6. The torque reaction (T), is equal to the product of the effective lever arm length (l_b) and the net force on the lever arm, F:

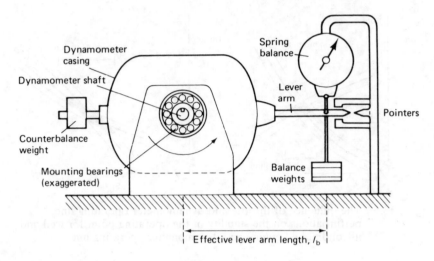

Figure 13.6   Dynamometer mounting system

$$T = F.l_b \ (\text{N m}) \tag{13.1}$$

The force F must be measured in the datum position for two reasons:

(1) the effective lever arm length becomes $l_b \cos \phi$
(2) away from the datum position a torque will be exerted by the connections for the cable or hose.

Usually a dashpot is connected to the lever arm to damp any oscillations. All these problems are avoided if a load cell is connected to the lever arm, although calibration problems are introduced instead, and some form of low pass filter will be needed.

Dynamometers usually include a tachometer for measuring engine speed; the principles of operation are the same as those for car speedometers. Alternatively, revolution counters, either mechanical or electrical, can be used. These give accurate results, especially when the engine is operating steadily and the count is timed over a long period.

Where a tachometer is not fitted a stroboscope provides a convenient means of determining the engine speed, by illuminating a marker on the flywheel or crankshaft. However, care must be taken or otherwise a submultiple ($\frac{1}{2}, \frac{1}{3}$ etc.) of the speed will be found, as the marker will be illuminated once every 2nd, 3rd etc. revolution. This problem can be avoided by starting at the highest strobe frequency, and reducing the frequency until the first steady image of the marker is obtained.

After the measurements of torque and speed the fuel-consumption measurements are next in importance.

13.2.2   Fuel-consumption measurement

A common measurement system for fuel consumption is to time the consumption of a fixed volume. This has to be converted to a gravimetric consumption by using the density as determined from a separate test, or deduced from fuel temperature for a specific fuel. A typical arrangement is shown in figure 13.7. In normal operation valve 1 is open and fuel flows directly to the engine. The calibrated volumes are filled when valve 2 is open. If the vent pipe ends below the level of the fuel tank, care is needed in filling the control volumes. To measure the fuel flow rate, valve 1 is closed and any fuel to the engine is drawn from the calibrated volumes; valve 2 must be open. Usually there are a range of volumes to give the best compromise between accuracy and speed of taking the readings. A common problem is when a vapour-bubble travels back down the fuel line; this obviously invalidates a reading. In a fuel-injection system, if the spill flow is fed back to the tank this must be measured separately; such fuel is much more prone to vapour-bubbles. Although this method gives accurate results, the readings are not instantaneous.

Figure 13.8 shows a flowmeter that does give instantaneous readings, though the accuracy is less than that of the flowmeter previously described. The float chamber provides a constant head of fuel, and the pressure drop across the orifice is proportional to the square of the volumetric flow rate. The pressure drop is measured by the difference in head between the float chamber and the sight glass; alongside is a scale calibrated directly in volumetric flow rate. Again, problems can occur with fuel injection systems.

Both these flowmeters can be adapted for automatic flow measurement. The control volume system can be automated by solenoid valves, opto-

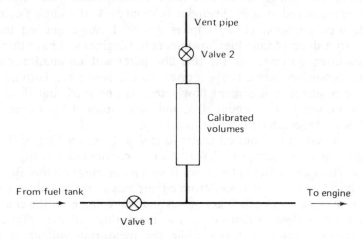

Figure 13.7   Fuel flow measurement

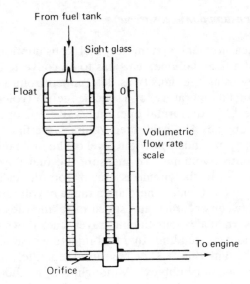

From fuel tank

Sight glass

Float

0

Volumetric
flow rate
scale

To engine

Orifice

Figure 13.8   Orifice type flowmeter

electronic liquid level sensors and a timer. The orifice flowmeter could be automated by a differential pressure transducer, but there is of course a non-linear relation between flow ($\dot{V}_f$) and pressure drop (Δp), and it is also necessary to know the fuel density (ρ_f):

$$\dot{V}_f \text{ is directly proportional to } \sqrt{(\Delta p/\rho_f)} \qquad (13.2)$$

Furthermore, the range of fuel flow rates (the turndown ratio) that might have to be measured on a single engine is about 50:1. If a single flowmeter is used, then equation (13.2) implies a 2500:1 range needed for the pressure transducer! Many fuel flowmeters for engines will have duplicate flow measuring systems, and the fuel flowmeter will automatically select the flow measuring elements appropriate to the flow rate. Turbine flowmeters or positive displacement flowmeters can be used, but their turndown ratio is usually only about 10:1, and an accuracy of 1 per cent across the full flow range can be difficult to achieve.

Ideally, a system is required that will give a direct reading of the fuel mass flow rate on a continuous basis. A quasi-continuous reading is given by systems that allow fuel to be drawn from a measuring volume. Readings of mass flow rate can be deduced from either: measurements of the change in hydrostatic pressure at the base of a cylindrical volume, or the change in buoyancy on a float immersed in the measuring volume. Obviously, measurements cannot be taken while the measuring volume is being refilled.

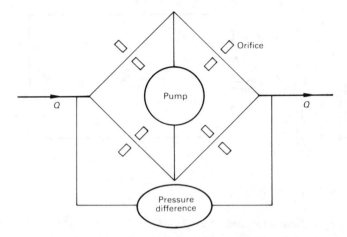

Figure 13.9   Flow system in the Flotron fuel mass flowmeter

There are two widely used systems for continuous measurement of the fuel mass flow rate. The theory of both systems is complex, but is presented in detail by Katys (1964). The first system employs a hydraulic equivalent of the Wheatstone bridge. Figure 13.9 shows the four orifices and the pump that provides a reference flow. The pressure drop is proportional to the mass flow rate, with a fast response to changes in flow, and no dependence on the fuel density or viscosity. The second system employs a vibrating 'U'-shaped tube through which the flow passes. Coriolis accelerations cause a twisting force, and the twist corresponds to the fuel mass flow rate.

With fuel injection systems, the return flow from the injection system can be greater than the fuel flow into the engine. So, even when two flowmeters are available, measuring the difference in flow would not lead to an accurate result. Instead, it is usual practice to cool the return fuel flow, and connect it to the fuel system downstream of the flow measuring system.

13.2.3   Air flow rate

A simple system to measure the air flow rate is obtained by connecting the air intake to a large rigid box with an orifice at its inlet. The box should be large enough to damp out the pulsations in flow and be free of resonances in the normal speed range of the engine. The pressure drop across the orifice can be measured by a water tube manometer, as shown in figure 13.10. For incompressible flow

$$\dot{m}_a = C_d A_o \sqrt{(2gh\rho_f\rho_a)} \qquad (13.3)$$

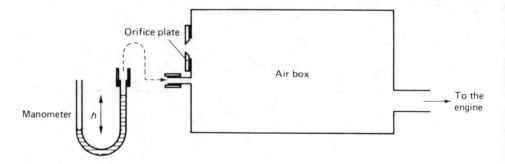

Figure 13.10   Air flow rate tank

where $\dot{m}_a$ = mass flow rate of air (kg/s)
C_d = discharge coefficient of the orifice
A_o = cross-sectional area of the orifice (m)
g = acceleration due to gravity (m/s^2)
h = height difference between liquid levels in the manometer (m)
ρ_f = density of manometer fluid (kg/m^3)
ρ_a = density of air = $(p/R_a T)$ (kg/m^3).

The accuracy depends on knowing the discharge coefficient for the orifice; this should be checked against a known standard.

In practice, it can be very difficult to make the air box work satisfactorily. Apart from making the air box act as a Helmholtz resonator (chapter 6, section 6.6.2), coupling a long tube to the inlet of an engine can have a significant effect on the engine performance. It should be remembered that even without a tuned induction system, there can be flow reversals occurring at the air inlet to the engine. A way of minimising any coupling effects and checking for their effect is shown in figure 13.11. If the flexible pipe is of large enough diameter, and the air entry protrudes far enough into the pipe, then the unsteady flow sensor will show no change to the air flow pattern when the flexible pipe is connected. The unsteady flow sensor is most likely to be a form of hot wire anemometer (see chapter 8, section 8.2.2).

An alternative approach is the viscous flowmeter. As no damping is required it is much more compact than an air box. To obtain viscous flow a matrix of passages is used in which the length is much greater than a typical diameter. Since the flow is viscous the pressure drop is proportional to the velocity or volumetric flow rate. The flowmeter has to be calibrated against a known standard.

However, in its usual form the viscous flowmeter is not well suited to pulsating flows. The instantaneous pressure drop will be a function of the

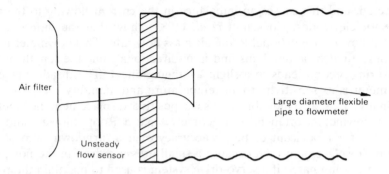

Figure 13.11   Connection of an air flow meter to minimise any effects on engine
performance

unsteady flow, and the pressure-measuring system will not necessarily
indicate a pressure drop that corresponds to the mean flow (Stone R.
(1989c)). With a pulsating flow, the acceleration and deceleration of the
flow causes an additional pressure difference term. These and other effects
have been considered by Wright (1990), who shows how the instantaneous
pressure differential can be analysed, so as to give the instantaneous
volume flow rate and the mean flow rate.

Another flowmeter that gives an instantaneous measurement is the
Lucas–Dawe air mass flowmeter. This was originally intended for use with
engine management systems, however it is better suited to laboratory use.
The principle is illustrated by figure 13.12. The central electrode is maintained at about 10 kV so that a corona discharge is formed. The exact
voltage is varied so that the sum of the currents flowing to the two collector
electrodes is constant. When air flows through the duct, the ion flow is

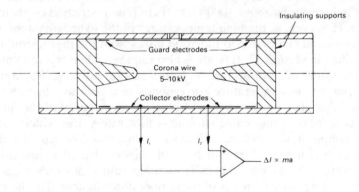

Figure 13.12   The Lucas–Dawe air mass flowmeter

deflected, and this causes an imbalance in the current flowing to the two collector electrodes. Cockshott *et al.* (1983) show that the difference in current flow is proportional to the air mass flow rate. This flowmeter has a response time of about 1 ms and is bi-directional, but the length of the measuring section leads to a slight averaging effect. The disadvantage of this meter is its sensitivity to air temperature and humidity.

Finally, it is also possible to use a positive displacement flowmeter. These flowmeters often have a geometry like a Roots blower, and are capable of a 1 per cent or better accuracy over a turndown ratio of 10. Positive displacement meters cause a small pressure drop in the flow, but this can be eliminated if a servo-drive system is used to maintain the rotor speed at the correct value. The inertia of the rotors in positive displacement flowmeters leads to a poor transient response.

In general, gas flow rate is more difficult to determine than liquid flow rate. The calibration of liquid flowmeters can always be checked gravimetrically.

Under many circumstances, the most accurate means of evaluating the air flow rate is to measure the fuel flow rate, and to calculate the air/fuel ratio from an exhaust gas analysis. The relevant techniques have already been introduced by chapter 3, example 3.2, but the methods are discussed further in section 13.4.7.

13.2.4 Temperature and pressure

Mercury-in-glass thermometers and thermocouples both provide economical means of measuring temperature, with the potential of achieving an accuracy of about ±0.1 K. However, if this level of accuracy is required, it can be cheaper to use platinum resistance thermometers. A disadvantage of thermocouples is their low output, about 40 μV/K. This can lead to the use of thermistors for temperatures up to 400°C. Thermocouples can be used at temperatures above 1000°C, so there is no restriction on their use in engines. However, at the high temperatures in the exhaust system, care is needed to avoid errors due to radiation losses from the thermocouple (this is quantified in section 13.3). Care is also required when trying to measure the temperature of a pulsating flow. Caton and Heywood (1981) argue that the thermocouple temperature corresponds to a time-averaged temperature, and this does not correspond to the average energy of the flow. For the exhaust port of an engine, the mass flow rate varies widely, and the highest temperatures occur at a time when the mass flow rate is high. The mass-averaged temperature can be 100K higher than the time-averaged temperature. This is why the exhaust temperature at entry to a turbocharger can be higher than any of the temperatures indicated in the exhaust ports.

Bourdon pressure gauges and manometers provide a cheap and accurate means of measuring steady pressures. By selecting the manometer fluid and its angle of inclination, pressures in the range of 1–100 kN/m^2 can be measured with an accuracy of about 1 per cent. When the manometer fluid is not a single compound, then care must be used in selecting the fluid, as otherwise vaporisation of the more volatile component will change the fluid density.

Most pressure transducers utilise a piezo-resistive effect, a notable exception being the piezo-electric transducers used for measuring the in-cylinder pressure (section 13.2.5). A common arrangement is to have a diaphragm etched with strain gauges in a Wheatstone bridge configuration. For low pressures a silicon diaphragm is used, while for higher pressures stainless steel is more likely to be used (for example, in measuring the injector fuel line pressure).

13.2.5 In-cylinder pressure measurement

Engine indicators record the pressure/volume history of the engine cylinder contents. The simplest form of engine indicator is the Dobbie McInnes mechanical indicator shown in figure 13.13. A piece of paper is attached to a drum, and the rotation of the drum is linked to the piston displacement by a cord wrapped around the drum. The pressure in the cylinder is recorded by a linkage attached to a spring-loaded piston in a cylinder. The

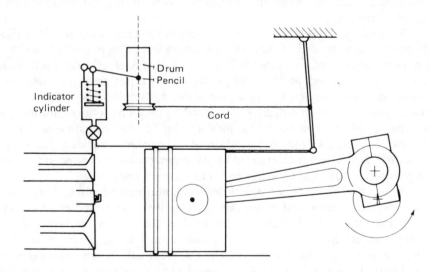

Figure 13.13   Mechanical indicator

indicator cylinder is connected to the engine cylinder by a valve. Unless the indicator cylinder has a much smaller volume than the engine cylinder, the indicator will affect the engine performance.

When the paper is unwrapped from the drum the area of the diagram can be found, and this corresponds to the indicated work per cylinder per cycle. The area can be determined by 'counting squares', by cutting the diagram out and weighing it, or by using a planimeter. The planimeter is a mechanical device that computes the area of the diagram by tracing the perimeter. Practice is necessary in order to obtain reliable results, and accuracy can be improved by tracing round the diagram several times. To convert area to work a calibration constant is needed; alternatively imep can be found more directly. The diagram area is divided by its length to give a mean height. When this height is multiplied by the indicator spring constant (bar/mm) the imep can be found directly:

$$\text{imep} = k\,h_d = k\,\frac{A_d}{l_d} \tag{13.4}$$

where A_d = diagram area
     l_d = diagram length
     h_d = mean height of diagram.

Because of the inertia effects in moving parts — friction, backlash and finite stiffness — mechanical indicators can be used only at speeds of up to about 600 rpm. Also, this simple type of mechanical indicator is not sensitive enough to record the 'pumping losses' during the induction and exhaust strokes.

Electronic systems are now very common for recording indicator diagrams. Care is needed in their interpretation since the pressure is plotted on a time instead of a volumetric basis. As with any electronic equipment, care is also needed in the calibration. The output from the pressure transducer is connected to the y-channel of an oscilloscope and the time base is triggered by an inductive or equivalent pick-up on the crankshaft — see figure 13.14. The inductive pick-up should also be connected to a second y-channel so that the position of tdc can be accurately recorded. Since tdc occurs during the period of maximum pressure change, a 1° error in position of tdc can cause a 5 per cent error in imep.

The location of tdc is not straightforward, not least because of the finite stiffness of the crank–slider mechanism. The flexibility of the crankshaft is such, that at full load there can be about 1° of twist at certain speeds. Ultrasonic techniques have been used to determine the dynamic position of tdc, but more usually the measurements are made when the engine is stationary. If the cylinder head is removed, then a dial gauge can be used to

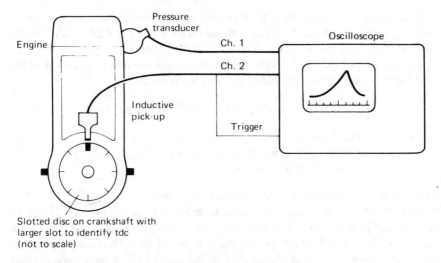

Figure 13.14   Electronic engine indicator system

measure the piston height. The dial gauge needs to act in-line with the gudgeon pin, otherwise the readings will be influenced by the piston rocking. Two angular positions are found either side of tdc with the same piston height, and the angle is then bisected to give tdc.

The included angle needs to be as large as possible to give accurate results. Some engines have valves whose axes are parallel to the cylinder axis, in which case it might be acceptable to let the valve act as a 'follower' on the piston, to obviate the need to remove the cylinder head. With the appropriate piston close to tdc, the valve spring is removed, and the valve is allowed to hit the piston.

The position of tdc has to be recorded by the data acquisition system, and the options are:

(a) use a separate marker
(b) use a wider crank angle marker
(c) add extra markers around tdc.

One of the most popular arrangements is to have markers every 10° ca, with extra markers at ±5° tdc. When an oscilloscope is being used, the time base can be adjusted so that a division on the graticule corresponds to a convenient crank angle increment (5° or 10° etc.). If there is a shortage of channels, the crank markers can be used on the z-input to modulate the trace brightness. The output from the oscilloscope can be recorded photographically. If a transient recorder is connected before the oscilloscope, the output can alternatively be directed to an x-y plotter.

To convert the time base to a piston displacement base it is usual to assume constant angular velocity throughout each revolution. Assuming that the gudgeon pin or 'little-end' is not offset (that is, assuming that the line of motion of the pivot in the piston intersects the axis of rotation of the crankshaft), the piston displacement (x) is given by

$$x = r(1 - \cos \theta) + (l - \sqrt{(l^2 - r^2 \sin^2 \theta)}) \qquad (13.5)$$

where θ = crank angle measured from tdc
         r = crank radius (half piston stroke)
         l = connecting-rod length.

When $l \gg r$ the motion becomes simple harmonic.

A voltage can be generated that corresponds to equation (13.4), by either analogue or digital electronics, Hudson *et al.* (1988). When this is correctly phased, it can be used as the x-input to an oscilloscope operating in its x–y mode. This is illustrated by figure 13.15, which shows three successive cycles superimposed to give an indication of the cycle-by-cycle variation in combustion. Figure 13.15 also shows an enlargement of the pumping loop, and this is useful when tuned induction or exhaust systems and valve timing are being investigated.

Figure 13.16 is from the same operating point as figure 13.15, except a time base has been used. The cylinder pressure trace and tdc markers are self-explanatory. When additional channels are available, on the oscilloscope or data logging system, they can be used for a variety of purposes. With spark ignition engines it is useful to record the ignition timing. This can be done by wrapping a wire around the appropriate spark plug lead, or by monitoring the coil LT voltage. The voltage should be measured at the switched side of the coil (negative terminal for negative earth systems). The signals in figure 13.16 indicate when the coil is switched on and off, and the spark duration.

For Diesel engines it is useful to measure the fuel line pressure and the injector needle lift. The needle lift can be measured by proprietary Hall effect transducer or a FM (frequency modulated) system that records the change in inductance of a coil. Figure 13.17 shows the construction of a needle lift transducer. The armature connected to the injector needle extends about half-way into the coil. The coil forms part of a tuned circuit that is resonating at close to 2 MHz. As the armature position moves the inductance of the coil changes and the resonant frequency changes. This frequency modulation is converted to an analogue voltage that is proportional to needle lift for a uniformly would coil. The construction of such a system is described by Osborne (1985). Figure 13.18 shows the output from a needle lift transducer, along with the cylinder pressure (a), the fuel line pressure (c) and crank angle markers (d).

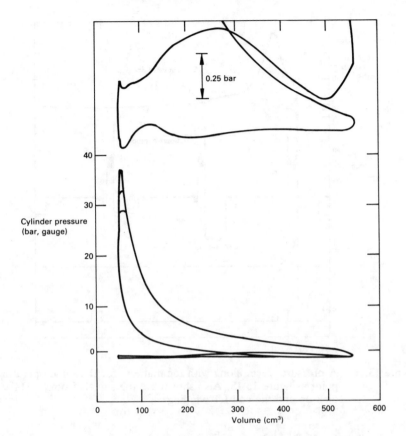

Figure 13.15   The pressure/volume or indicator diagram from a Rover M16
engine operating at 2000 rpm; with an enlargement of the
pumping loop. bmep = 3.8 bar, imep = 4.6 bar (including the
pumping work of 0.45 bar pmep)

The in-cylinder pressure transducer requirements are very demanding
because of the high temperature and pressures, and the need for a high-
frequency response. The transducers usually have a metal diaphragm
which is displaced by the pressure. The displacement can be measured,
inductively, by capacitance, by a strain gauge or by a piezo-electric crystal.
Piezo-electric transducers are common, but have the disadvantage that
they respond only to the rate of change in pressure; thus they have to be
used in conjunction with a charge amplifier that integrates the signal.

The piezo-electric transducer produces an electrical charge that is proportional to pressure (typically between 2 and 50 pC/bar). When the signal
is integrated it is necessary to define the pressure/voltage datum, this can
be done in a variety of ways:

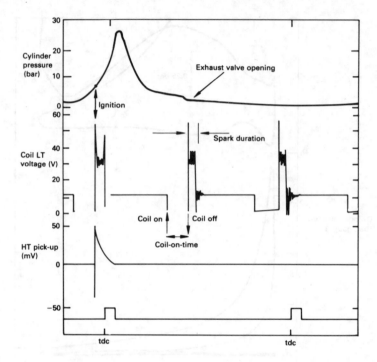

Figure 13.16   A pressure trace, along with tdc markers for the same operating point as figure 13.15. Also shown are the signals from an HT pick-up and the coil LT winding

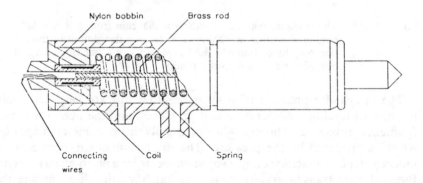

Figure 13.17   The installation of an injector needle-lift transducer

(a) By assuming that the minimum pressure recorded in the cycle corresponds to a particular value, for example the inlet manifold pressure.

(b) By adjusting the datum value so that the compression process is described by a polytropic process (pv^k = constant).

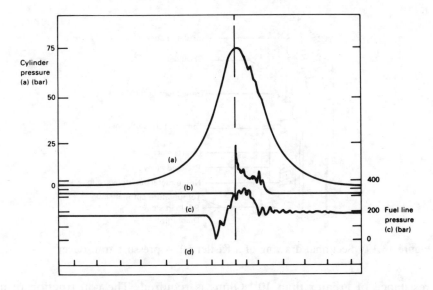

Figure 13.18   Cylinder pressure (a), injector needle lift (b), fuel line pressure (c) and crank marker (d) (10° ca with additional ±5° ca marks) from an indirect injection Diesel engine operating at 1000 rpm, no load

(c) By using a cycle simulation program to define the pressure at a particular crank angle.
(d) By using a clipper adaptor — a pressure transducer mounting designed to record a datum pressure, and pressures either above or below the datum pressures (according to the clipper adaptor design).

For calculations such as imep it is not necessary to know the absolute values of pressure, and for peak pressures (invariably above 30 bar) uncertainty of 0.3 bar or so in the datum will not be significant. However, for combustion analysis it is important to know the absolute pressure more accurately (section 13.5.2).

The electrical charge is prone to both leakage and accumulation, and this causes the voltage output from the charge amplifier to drift. By using a coupling with an appropriate time constant, the effect of the drift is eliminated. However, when the pressure transducer is calibrated by static pressures (often by means of a dead-weight pressure tester), then a long time constant is needed, so that a steady pressure corresponds to a steady voltage. This places considerable demands on the input resistance of the charge amplifier, the internal resistance of the pressure transducer, and the interconnecting cable. To prevent the electrical charge being dissipated, a

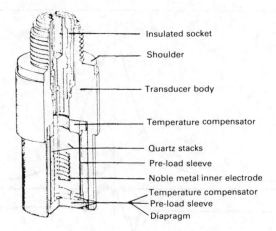

Figure 13.19   Sectional drawing of a Kistler 601A pressure transducer

resistance of greater than 10^{14} Ohms is required. The construction of a piezo-electric pressure transducer is shown in figure 13.19.

The pressure transducer response should be independent of temperature, and the calibration should be free from drift. The pressure transducer should be mounted flush with the cylinder wall or as close to the engine cylinder as possible through a small communicating passage. This minimises the lag in the pressure signal and should avoid introducing any resonances in the connecting passage.

13.2.6   Techniques for estimating indicated power

Very often a pressure transducer cannot be readily fitted to an engine, so alternative means of deducing imep are useful. The difference between indicated power and brake power is the power absorbed by friction, and this is often assumed to be dependent only on engine speed. Unfortunately, the friction power also depends on the indicated power since the increased gas pressures cause increases in piston friction etc.; this is shown by figure 13.20. When extrapolated to zero imep, fmep is about 2.25 bar. This can be compared to a 1.75 bar motoring mep — the equivalent of the power output of the electric dynamometer turning the engine at the same speed.

If the friction power is assumed to be independent of the indicated power, then the friction power can be deduced from the Morse test. This is applicable only to multi-cylinder engines (either spark or compression ignition), as each cylinder is disabled in turn. When a cylinder is disabled the load is reduced so that the engine returns to the test speed; the

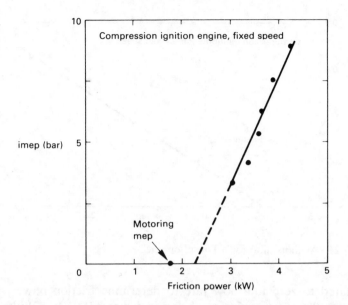

Figure 13.20   Dependence of friction power on indicated power at constant speed

reduction in power corresponds to the indicated power of that cylinder. For a n-cylinder engine

$$\sum^{n} \text{indicated power} - \sum^{n} \text{friction power} = (\text{brake power})_n \quad (13.6)$$

With one-cylinder disabled

$$\sum^{n-1} \text{indicated power} - \sum^{n} \text{friction power} = (\text{brake power})_{n-1} \quad (13.7)$$

Subtracting:

indicated power of disabled cylinder = reduction in brake power   (13.8)

This underestimates the friction power since the disabled cylinder also has reduced friction power. However, the test does check that each cylinder has the same power output.

A method for estimating the friction power of compression ignition engines is Willans' line. Again it is assumed that at constant speed the friction power is independent of indicated power, but in addition it is assumed that the indicated efficiency is constant. This is a reasonable assumption away from maximum power. Figure 13.21 shows a plot of fuel consumption against power output. Willans' line is when the graph is

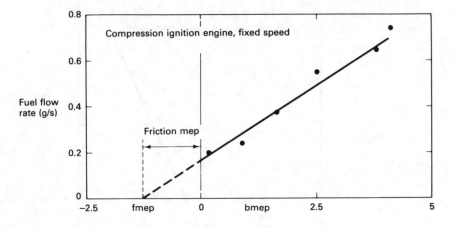

Figure 13.21   Willans' line for a Diesel engine

extrapolated to zero fuel flow rate to determine friction power. Figure 13.20 is for the same engine as shown in figure 13.21, but Willans' line suggests a friction mep of only 1.25 bar.

13.2.7 Engine test conditions

Various standards authorities (BS, DIN, SAE) are involved with specifying the test conditions for engines, and how allowance can be made for variations in ambient conditions.

In the past a wide range of performance figures could be quoted for a given engine, depending on which standard was adopted and how many of the engine ancillary components were being driven (water pump, fan, alternator etc.). Obviously it is essential to quote the standard being used, and to adhere to it.

Corrections for datum conditions vary, and in general they are more involved for compression ignition engines, whether turbocharged or naturally aspirated. Corrections for spark ignition engines in the *SAE Handbook* are as follows:

$$\left. \begin{array}{l} \text{ambient test conditions} \\ \text{should be in the range} \end{array} \right\} \begin{array}{l} 95 < p < 101 \text{ kN/m}^2 \\ 15 < T < 43°\text{C} \end{array}$$

where p is ambient pressure and T is ambient temperature. The corrections are applied to indicated power, where $(W_i)_o$ is the observed value and $(W_i)_c$ is the corrected value:

$$(\dot{W}_i)_c = (\dot{W}_i)_o \left(\frac{99}{p_d}\right)\sqrt{\left(\frac{T + 273}{298}\right)} \qquad (13.9)$$

where $p_d = p - p'_{water}$, the partial pressure of dry air (kN/m^2). Values for brake power (W_b) are found from a knowledge of the friction power (W_f):

$$\dot{W}_i = \dot{W}_b + \dot{W}_f \qquad (13.10)$$

This approach relies on knowing the friction power. If this has not been found, an alternative approach is

$$(\dot{W}_b)_c = (\dot{W}_b)_o \left[1.18 \left(\frac{99}{p_d}\right)\sqrt{\left(\frac{T + 273}{298}\right)} -0.18\right] \qquad (13.11)$$

13.2.8 Energy balance

Experiments with engines very often involve an energy balance on the engine. Energy is supplied to the engine as the chemical energy of the fuel and leaves as energy in the cooling water, exhaust, brake work and extraneous heat transfer. Extraneous heat transfer is often termed 'heat loss', but this usage is misleading as heat is energy in transit and the 1st Law of Thermodynamics states that energy is conserved.

The heat transfer to the cooling water is found from the temperature rise in the coolant as it passes through the engine and the mass flow rate of coolant. The temperature rise is most commonly measured with mercury in glass thermometers. The mass flow rate of coolant is usually derived from the volumetric flow rate. Common flow-measuring devices include tanks, weirs and variable-area flowmeters such as the Rotameter. The Rotameter has a vertical-upwards flow through a diverging graduated tube; a float rises to an equilibrium position to indicate the flow rate. To conserve water the coolant is usually pumped in a loop with some form of heat exchanger. The heat exchanger should be regulated to control the maximum engine-operating temperature. The engine coolant flow can be adjusted to make the temperature rise sufficiently large to be measured accurately without making the flow rate too small to be measured accurately.

The energy leaving in the exhaust is more difficult to determine. If the gas temperature is measured, the mean specific heat capacity can be estimated in order to calculate the enthalpy in the exhaust.

The enthalpy of the exhaust can be calculated from the polynomial functions that define the enthalpy of the constituents in the exhaust (table 10.1). This requires a knowledge of the temperature and composition.

Table 13.1   Mean specific heat capacity of exhaust products, for idealised combustion of C_nH_{2n} (datum of 25°C)

Temperature K	Equivalence ratio							
	0	0.2	0.4	0.6	0.8	1.0	1.2	1.4
400	1.010	1.024	1.037	1.050	1.063	1.075	1.100	1.124
600	1.024	1.041	1.057	1.072	1.087	1.102	1.133	1.164
800	1.045	1.064	1.082	1.100	1.117	1.132	1.150	1.173
1000	1.067	1.087	1.107	1.126	1.145	1.162	1.179	1.194
1200	1.088	1.110	1.131	1.151	1.172	1.191	1.206	1.221

Table 13.1 presents the mean specific heat capacity of exhaust products in terms of the temperature and the equivalence ratio. The fuel composition has been taken as C_nH_{2n}, and it has been assumed that no oxygen is present in rich mixtures, and no carbon monoxide is present with weak mixtures.

Sometimes a known flow rate of water is sprayed into the exhaust, and the temperature is measured after the water has evaporated. The exhaust can include partially burnt fuel, notably with spark ignition engines operating on rich mixtures; this can make a significant difference to an energy balance.

The significance of chemical energy has already been seen in chapter 12, figures 12.7 and 12.8. When emissions data are available, then the chemical energy associated with the partially burnt fuel can be evaluated. For most purposes the energy associated with unburned hydrocarbons can be neglected, so it is only necessary to consider the chemical energy associated with the carbon monoxide and hydrogen.

Rogers and Mayhew (1980a) provide molar enthalpies of reaction for carbon monoxide and hydrogen at 25°C:

$$(\Delta H_0)_{CO} = -283.0 \text{ MJ/kmol}$$

$$(\Delta H_0)_{H_2} = -241.8 \text{ MJ/kmol}$$

Consider the reaction of a generalised fuel with air, that produces 100 kmol of dry products:

$$x(CH_yO_z) + a(O_2 + 79/21 \text{ N}_2) \rightarrow b \text{ CO} + c \text{ CO}_2 + d \text{ H}_2 + \ldots$$

where $b \equiv$ percentage of CO, $c \equiv$ percentage of CO_2, $d \equiv$ percentage of H_2 etc.

For a fuel with known composition and calorific value (CV in units of MJ/kg), then a carbon balance can be used to establish the mass of fuel to produce 100 kmol of dry products:

$$\text{C balance} \quad x = b + c \qquad (13.12)$$

$$\text{mass of fuel} \quad m = x(12 + y + 16z) \qquad (13.13)$$

Thus, the percentage of the fuel energy that remains as chemical energy in the exhaust is

$$\frac{283.0 \times b + 241.8 \times d}{CV \times (b + c) \times (12 + y + 16z)} \times 100 \text{ per cent} \qquad (13.14)$$

The data in figure 3.15 suggests that for a typical hydrocarbon fuel with an equivalence ratio of 1.1, then about 15 per cent of the fuel energy remains as chemical energy in the exhaust. If the fuel is a hydrocarbon, but its composition is unknown, then the emissions data can be used to find the fuel composition; this is discussed further in section 13.4.7. The carbon balance has been used here, since it is the simplest and usually the most accurate; of course other atomic balances could be used to determine the mass of fuel needed to produce 100 kmol of dry exhaust products.

Heat transfer from the engine cannot be readily determined from temperature measurements of the engine. If the engine is totally enclosed, the temperature rise and mass flow rate of the cooling air can be used to determine the heat transfer.

Finally, brake power should be used in the energy balance, not indicated power. The power dissipated in overcoming friction degenerates to heat, and this is accounted for already.

13.3 Experimental accuracy

Whenever an experimental reading is taken there is an error associated with that reading. Indeed, it can be argued that any reading is meaningless unless it is qualified by a statement of accuracy. There are three main sources of error:

(1) the instrument is not measuring what is intended
(2) the instrument calibration is inaccurate
(3) the instrument output is incorrectly recorded by the observer.

To illustrate the different errors, consider a thermocouple measuring the exhaust gas temperature of an engine. Firstly, the temperature of the thermocouple may not be the temperature of the gas. Heat is transferred to the thermocouple by convection, and is transferred from the thermocouple by radiation, and to a lesser extent by conduction along the wires. If the gas stream has a temperature of 600°C and the pipe temperature is about 400°C, then the thermocouple could give a reading that is 25 K low because of radiation losses (Rogers and Mayhew (1980a)).

Secondly, there will be calibration errors. Thermocouple outputs are typically $40\mu V/K$ and very close to being linear; the actual outputs are tabulated as functions of temperature for different thermocouple combinations. The output is very small, so amplification is often needed; this can introduce errors of gain and offset that will vary with time. Thermocouples also require a reference or cold junction, and this adds scope for further error whether it is provided electronically, or with an additional thermocouple junction in a water/ice mixture. The output has to be indicated on some form of meter, either analogue or digital, with yet further scope for errors.

Thirdly, errors can arise through misreading the meter; this is less likely with digital meters than with analogue meters.

In serious experimental work the instrumentation has to be checked regularly against known standards to determine its accuracy. Where this is not possible, estimates have to be made of the accuracy; this is easier for analogue instruments than for digital instruments. In a well-designed analogue instrument (such as a spring balance, a mercury in glass thermometer etc.), the scale will be devised so that full benefit can be obtained from the instrument's accuracy. In other words, if the scale of a spring balance has 1 Newton divisions and these can be subdivided into quarters, then it is reasonable to do so, and to assume that the accuracy is also $\pm \frac{1}{4}$ Newton. Of course a good spring balance would have some indication of its accuracy engraved on the scale.

This approach obviously cannot be applied to digital instruments. It is very tempting, but wrong, to assume that an instrument with a four-digit display is accurate to four significant figures. The cost of providing an extra digit is much less than that of improving the accuracy of the instrument by an order of magnitude! For example, most thermocouples with a digital display will be accurate to only one degree.

There are many books, such as Adams (1975), that deal with instrumentation and the handling of results. One possible treatment of errors uses binomial approximations. If a quantity u is dependent on the quantities x, y and z such that

$$u = x^a y^b z^c \qquad (13.15)$$

then for sufficiently small errors

$$\frac{\delta u}{u} \approx a \left(\frac{\delta x}{x} \right) + b \left(\frac{\delta y}{y} \right) + c \left(\frac{\delta z}{z} \right) \qquad (13.16)$$

where δu denotes the error associated with u etc.

As an example consider equation (9.5):

$$T_{2s} = T_1 \left(\frac{p_2}{p_1} \right)^{(\gamma-1)/\gamma} \qquad (13.17)$$

If the error in T_1 is $\pm \frac{1}{2}$ per cent and the error associated with the pressure ratio (p_2/p_1) is ± 5 per cent, then the error associated with the isentropic compressor temperature T_{2s} is

$$\pm \tfrac{1}{2} \pm \frac{1.4 - 1}{1.4} \times 5 \text{ per cent} \qquad (13.18)$$

which is ± 1.9 per cent. This is a pessimistic estimate of the error, since it assumes a worst possible combination of errors that is statistically unlikely to occur.

Sometimes it is possible for the experimenter to minimise the effect of errors. Consider the heat flow to the engine coolant:

$$\dot{Q} = \dot{m} c_p (T_{out} - T_{in}) \qquad (13.19)$$

Suppose the mass flow rate $(\dot{m})$ of coolant is 4.5 kg/s and the inlet and outlet temperatures are 73.2°C and 81.4°C, respectively. If the errors associated with mass flow rate are ± 0.05 kg/s and the errors associated with the temperature are ± 0.2 K, then

$$
\begin{aligned}
Q &= (4.5 \pm 0.05)c_p \, [(81.4 \pm 0.2) - (73.2 \pm 0.2)] \\
&= (4.5 \pm 0.05)c_p \, (8.2 \pm 0.4) \\
&= (4.5 \pm 1.11 \text{ per cent})c_p \, (8.2 \pm 4.88 \text{ per cent}) \\
&= 4.5 \times c_p \times 8.2 \pm 6 \text{ per cent} \qquad (13.20)
\end{aligned}
$$

It can be readily shown by calculus that the errors would be minimised if the percentage error in each term were equal. Denoting the optimised values of mass flow rate as $\dot{m}'$ and temperature difference as $\Delta T'$, then

$$\frac{0.05}{\dot{m}'} = \frac{0.4}{\Delta T'}, \quad \Delta T' = 8\dot{m}' \qquad (13.21)$$

and for the same heat flow

$$\dot{m}'.\Delta T' = 4.5 \times 8.2$$

Combining

$$\dot{m}' = \sqrt{\left(\frac{4.5 \times 8.2}{8} \right)}$$

$$= 2.15 \text{ kg/s}$$

$$\text{and } \Delta T' = 17.18 \qquad (13.22)$$

The error is now reduced to

$$\pm \left(\frac{0.05}{2.15} \right) \pm \left(\frac{0.4}{17.18} \right) \qquad (13.23)$$

or ± 4.65 per cent.

Since the effect of errors is cumulative, always identify the weakest link in the measurement chain, and see if it is possible to make an improvement.

13.4 Measurement of exhaust emissions

Exhaust gas emissions need to be measured because of legislation, and also because of the insights the measurements provide into engine performance. The emissions governed by legislation are: carbon monoxide (measured by infra-red absorption), nitrogen oxides (measured by chemiluminescence), unburnt hydrocarbons (measured by flame ionisation detection) and particulates. If carbon dioxide (measured by infra-red absorption) and oxygen (measured by a chemical cell, or more accurately by para-magnetism) are also analysed, then it is possible to calculate the air/fuel ratio. Each of these measurement techniques will be described in the following sections, and there also follows a discussion of how to compute the air/fuel ratio, and other parameters such as exhaust gas recirculation (EGR).

13.4.1 Infra-red absorption

Infra-red radiation is absorbed by a wide range of gas molecules, each of which has a characteristic absorption spectrum. The fraction of radiation transmitted (τ_λ) at a particular wavelength (λ) is given by Beer's Law:

$$\tau_\lambda = \exp(-\rho\alpha_\lambda L) \qquad (13.24)$$

where ρ is the gas density, α_λ is the absorptivity and L is the path length.

Figure 13.22 shows the key components in a non-dispersive infra-red gas analyser. The detector cells are filled with the gas that is to be measured (for example carbon monoxide), so that they absorb the radiation in·the wavelength band associated with that gas. The energy absorbed in the 'LUFT' detector cells causes the cell pressure to rise. The reference cell is filled with air, and the gas to be analysed flows through the sample cell. If the relevant gas (in this case carbon monoxide) is present in the sample, then infra-red will be absorbed in the sample cell, and less infra-red will be absorbed in the detector cell. This leads to a differential pressure in the detector cells, which can be measured and related to the gas (carbon monoxide) concentration. The calibration is determined by passing gases of known composition through the sample cell.

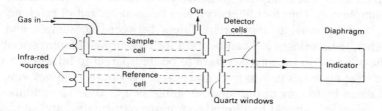

Figure 13.22   Diagram of a non-dispersive infra-red (NDIR) gas analyser

Figure 13.23 shows the absorption spectra of carbon monoxide and carbon dioxide. This shows that in the region of 4.4 microns, infra-red radiation is absorbed by both carbon dioxide and carbon monoxide. In other words, for the simple arrangement shown in figure 13.22, when carbon dioxide is present in the sample, then this will affect slightly the readings of carbon monoxide, and vice versa when carbon dioxide is being measured. This problem can be eliminated by using a 'filter' cell between the infra-red sources and the sample and reference cells. If carbon monoxide is to be measured, then the filter cell would be filled with carbon dioxide, and any carbon dioxide in the sample should not lead to any further infra-red absorption.

More recently non-dispersive infra-red analysers have been developed that use solid-state infra-red detectors, for example lead selenide. The arrangement is essentially the same as the 'LUFT' cell, except that a chopper disc is placed between the sample cell and the detector. The chopper disc has a slot, so that the infra-red detector is exposed to: the

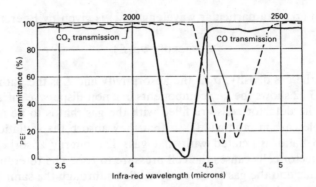

Figure 13.23   Transmittance of infra-red through gaseous carbon dioxide and carbon monoxide. Adapted from Ferguson (1986) and reprinted by permission of John Wiley and Sons Inc.

infra-red from the sample cell, the reference cell, and no direct infra-red. This enables a single detector cell to establish the background signal level, and then make a comparative measurement between the reference cell and the sample cell. Thin film filters can also be used instead of reference gas cells, and by moving appropriate filters between the sample and the detector, then a single cell can be used for measuring different species.

The windows in the analyser have to be transparent to infra-red, so are made from materials such as mica and quartz. Obviously readings would be invalidated by fouling of the windows in the sample cell. To minimise this risk, the sample should be filtered to remove particulates, and condensation (of water vapour or high molecular weight hydrocarbons) is avoided by either:

 (i) heating the sample lines and analyser, or
(ii) cooling and removing the condensate, then warming the sample to ambient temperature.

Non-dispersive infra-red absorption (NDIR) can be used for measuring the unburned hydrocarbons. However, this is not entirely satisfactory, as different hydrocarbon species have different absorption spectra. Ideally, when quantitative measurements of hydrocarbons are required a flame ionisation detection system should be used.

13.4.2 Flame ionisation detection (FID)

When hydrocarbons are burned, electrons and positive ions are formed. If the unburned hydrocarbons are burned in an electric field, then the current flow corresponds very closely to the number of carbon atoms present.

Table 13.2   Typical responses of a flame ionisation detector to different
molecular structures, normalised with respect to methane

Molecular structure	Relative response
Alkanes	1
Aromatics (benzene rings)	1
Alkynes	0.95
Alkenes	1.3
Carbonyl radical (CO^-)	0
Oxygen in primary alcohol	−0.6

Table 13.2 shows that the response is slightly dependent on the molecular structure, and furthermore the presence of some atoms in a molecule can suppress the ionisation current from the carbon atoms.

A flame ionisation detector is shown in figure 13.24. The sample is mixed with the fuel and burned in air. The fuel should not cause any ionisation, so a hydrogen or hydrogen/helium mixture is used. The air should be of high purity, again to reduce the risk of introducing hydrocarbons or other species. The fuel and sample flows have to be regulated, as the instrument response is directly related to the sample flow rate, and dependent on the fuel flow rate as this influences the burner temperature (and thence the sensitivity). The flows are regulated by maintaining fixed pressure differences across devices such as:

(a) capillary tubes
(b) porous sintered metallic elements
(c) critical flow orifices — when the pressure ratio is above about 2, the flow through the orifice is supersonic, and dependent only on upstream conditions.

In figure 13.24, the burner jet and the annular collector form the electrodes, and a potential of about 100 volts is applied between them. The signals have to be amplified, and calibration is achieved by:

(a) zeroing the instrument with a sample containing no hydrocarbons (such as pure nitrogen),
(b) using calibration gases of known hydrocarbon concentration (such as 0.1 per cent C_3H_8 in N_2).

As with NDIR analysers it is necessary to minimise the risk of sample deposition occurring in the sample line; the usual arrangement is to have a heated sample line.

A particularly ingenious FID system has been developed by Collings and Willey (1987). This system has a very fast response (about 1 ms), since the

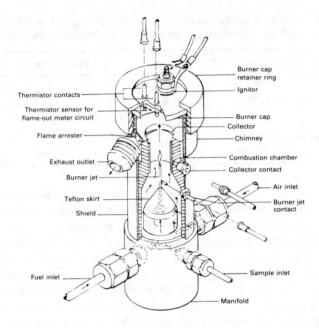

Figure 13.24   The burner and electrode system for a flame ionisation detector
(Beckman)

flame ionisation detector has been miniaturised and can be sited remotely
from the analyser, and adjacent to the sample source. The system is
capable of making cycle-resolved measurements, and has been used for:

(a) In-cylinder sampling to look at cycle-by-cycle variations in unburned
    hydrocarbons, before and after combustion, (Cheng *et al*. (1990)).
(b) Sampling of the pre-combustion chamber gases in a lean-burn gas
    engine to establish the air/fuel ratio distribution (Tawfig *et al*.
    (1990)).
(c) Measurement of the unburned hydrocarbons in the exhaust port of a
    gasoline engine (Finlay *et al*. (1990)).
(d) Measurement of in-cylinder hydrocarbons to deduce the effects of
    mixture preparation and trapped residuals on the performance of a
    port-injected spark ignition engine (Brown, C. N. and Ladommatos
    (1991)).

The flame ionisation detector cannot distinguish between different
hydrocarbon species. However gas chromatography can be used for identifying particular hydrocarbon species in the exhaust gas. The gas sample

and carrier gas are passed into the column — a long tube which contains a medium (liquid or solid) that tends to absorb the constituents in the sample. Since different molecules pass through the column at different rates, when the carrier gas leaves the column it contains the different molecules in discrete groups, and the different molecules can be identified by their residence time. The exit of the species from the column is detected by measuring the ionisation in a flame. The chromatograph is calibrated by injecting samples of known gases into the carrier gas.

13.4.3 Chemiluminescence

The chemiluminescence technique depends on the emission of light. Nitric oxide (NO) reacts with ozone (O_3) to produce nitrogen dioxide in an activated state (NO_2^*), which in due course can emit light as it converts to its normal state:

$$NO + O_3 \rightarrow NO_2^* + O_2 \rightarrow NO_2 + O_2 + photon$$

The nitrogen dioxide can also be deactivated by a collision with another molecule. Ferguson (1986) shows that if

(a) the reactor is sufficiently large
(b) the ozone flow rate is steady and high compared with a steady sample flow
(c) the reactor is at a fixed temperature

then the light emitted is proportional to the concentration of nitric oxide in the sample stream.

Both nitric oxide and nitrogen dioxide can exist in the exhaust of an engine, and NO_x is used to denote the sum of the nitrogen oxides. Nitrogen dioxide can be measured by passing the sample through a catalyst that converts the nitrogen dioxide to nitric oxide. By switching the converter in and out of the sample line, then the concentrations of NO and (NO + NO_2) can be found in the exhaust sample.

Figure 13.25 shows the arrangement of an NO_x analyser. The vacuum pump controls the pressure in the reaction chamber, and is responsible for drawing in the ozone and exhaust sample. The ozone is generated by an electrical discharge in oxygen at low pressure, and the flow of ozone is controlled by the oxygen supply pressure and the critical flow orifice (a short length of capillary tube). The sample can either by-pass or flow through the nitrogen dioxide converter. The sample flow rate is regulated by two critical flow orifices. The by-pass flow is drawn through by a sample pump. This arrangement ensures a high flow rate of sample gas, so as to

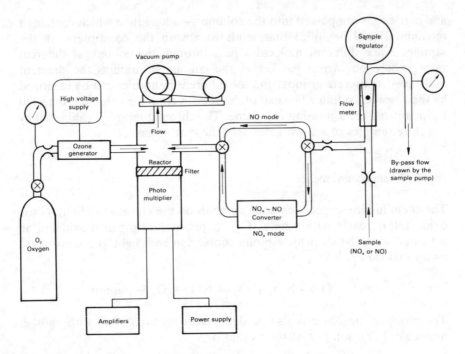

Figure 13.25   Key elements in a chemiluminescence NO_x analyser

minimise the instrument response to a change in NO_x concentration in the sample. The flow of sample into the reactor is controlled by the pressure differential across the critical flow orifice upstream of the NO_x converter; this pressure differential is controlled by a differential pressure regulator.

The light emission in the reactor is measured by a photomultiplier, and then amplified. In view of variations in gain that might occur with the photomultiplier, and in the other parameters that affect the light emission, then it is essential to calibrate the NO_x analyser regularly. This is similar to calibrating a NDIR analyser, since calibration gases are used to set the zero and check the span.

13.4.4 Oxygen and air/fuel ratio analysers

Oxygen is obviously not an undesirable exhaust emission, but its measurement is very useful when evaluating the air/fuel ratio.

The lowest cost oxygen analysers are usually based around a galvanic cell. A typical galvanic cell comprises a PTFE membrane with a gold coating that acts as the cathode. Also immersed in the electrolyte (potassium chloride gel) is a silver or lead anode. A potential is applied across the electrodes. When the oxygen diffuses through the membrane it is reduced electrochemically, and a current flows that is proportional to the partial pressure of the oxygen in the sample. The galvanic cell also responds to other gases, and the gas of greatest significance with combustion is carbon dioxide. However, a 12 per cent carbon dioxide concentration would only give an output equivalent to that of 0.1 per cent oxygen.

Paramagnetic oxygen analysers are probably the most accurate. Paramagnetism occurs in oxygen because two of the electrons in the outer shell of the oxygen molecule are unpaired, and in consequence the molecule is attracted by a magnetic field. Other gas molecules can also be attracted (and conversely some are repelled) and most significant are NO_x. Nitric oxide has 43 per cent of the magnetic susceptibility of oxygen, and nitrogen dioxide has 28 per cent of the magnetic susceptibility. However, the NO_x concentration is usually an order of magnitude lower than the oxygen concentration so its effect can be ignored. (Furthermore, the effect on any atomic balance would be even smaller.) There are two main types of oxygen analyser using the paramagnetic principle: thermomagnetic analysers and magneto-dynamic analysers.

The principle of the thermomagnetic oxygen analyser is illustrated by figure 13.26. The heated filament in the cross-tube forms part of a Wheatstone bridge. Oxygen is attracted by the magnetic field, but when it is heated by the filament its paramagnetism is reduced, so the oxygen flows away from the magnetic field. The flow of gas cools the filament, thereby changing its resistance and leading to an imbalance in the Wheatstone bridge, which can be related to the oxygen level. Care must be taken with this type of analyser for several reasons:

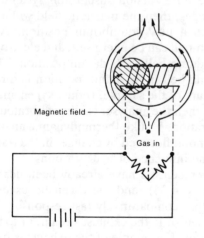

Figure 13.26   Basis of the thermomagnetic paramagnetic oxygen analyser

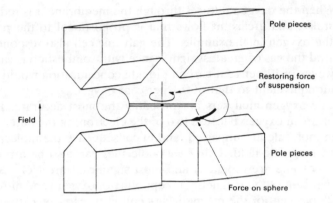

Figure 13.27   Basis of the magneto-dynamic paramagnetic oxygen analyser

(a) The heated filament is affected by changes in the transport properties of the gases in the sample.
(b) Hydrocarbons and other combustible gases can react on the filament causing changes in the filament temperature.
(c) The cross-tube must be mounted horizontally to avoid the occurrence of natural convection.

Magneto-dynamic analysers use a diamagnetic body (usually dumb-bell shaped) located in a strong non-uniform magnetic field; such an arrangement is shown in figure 13.27. The spheres are repelled by the magnetic field, and will reach equilibrium when the repulsion force is balanced by the torque from the fibre torsion suspension system. When the oxygen level in the cell changes, then the magnetic field will also change, and the dumb-bell will reach a new equilibrium position. Alternatively, a coil attached to the dumb-bell can be energised, and electro-magnetic feedback used to restore the dumb-bell to its datum position. The current required to maintain the dumb-bell in its datum position is directly related to the oxygen partial pressure of the sample. If the oxygen analyser is operated at a constant pressure, then the analyser can be calibrated directly in terms of the oxygen concentration. Unlike thermodynamic analysers, the magneto-dynamic analyser is not influenced by changes in the transport properties of the sample gas or the influence of hydrocarbons.

Exhaust gas oxygen sensors have already been described in chapter 4, section 4.7 (see figure 4.36), and these can be used as the basis of an oxygen analyser with a comparatively fast response. The two options are either to mount the sensor in the exhaust stream, or to take a sample of the exhaust to the analyser. If a sample of the exhaust is fully oxidised, then it is also possible to deduce the air/fuel ratio.

In the Cussons Lamdascan, a sample of the exhaust is taken to an analyser. If a rich mixture is being burned, then a controlled amount of oxygen is blended into the exhaust. This way, there is always oxygen present in the sample being analysed. The sample is passed through a heated catalyst to fully oxidise any partial products of combustion. For rich mixtures, one of several levels of dope air is added, so that there is a low level of oxygen in the diluted sample. If the hydrogen/carbon/oxygen ratio of the fuel is known, and also the quantity of any dope air, then the equivalence ratio or air/fuel ratio can be deduced. Consider first the combustion of such a fuel as a weak mixture with an equivalence ratio ϕ:

$$CH_xO_y + \frac{(1 + x/4 - y/2)}{\phi} \left(O_2 + \frac{79}{21} N_2 \right) \rightarrow$$

$$CO_2 \quad + (x/2) H_2O + (1 - 1/\phi) (1 + x/4 - y/2) O_2$$

$$+ 1/\phi (1 + x/4 - y/2) \frac{79}{21} N_2 \qquad (13.25)$$

Examination of the right-hand side of equation (13.25) shows that the oxygen level in the exhaust gases enables the equivalence ratio to be found. Since the system is maintained at a temperature for which the water is vaporised:

$$\% \ O_2 = \frac{21 (\phi - 1) (1 + x/4 - y/2)}{21\phi + 21x\phi/2 + (21\phi + 58) (1 + x/4 - y/2)} \times 100 \quad (13.26)$$

for which ϕ is the only unknown variable.

When a rich mixture is burned, dope air is added, denoted here by $d(O_2 + 79/21 \ N_2)$:

$$CH_xO_y + \frac{(1 + x/4 - y/2)}{\phi} (O_2 + 79/21 \ N_2) \rightarrow$$

partial combustion products $+ d(O_2 + 79/21 \ N_2)$

$$\rightarrow CO_2 + (x/2) H_2O + \{(1-1/\phi) (1 + x/4-y/2) + d\} O_2$$

$$+ \{ 1 + x/4-y/2) + d\} 79/21 \ N_2 \qquad (13.27)$$

As with equation (13.26), the equivalence ratio can be expressed in terms of: the oxygen level in the exhaust, and the fuel composition, but it is now also in terms of the dope air level (d):

$$\% \ O_2 = \frac{(1 - 1/\phi) \ (1 + x/4 - y/2) + d}{1 + x/2 + (1 + 58/21\phi) \ (1 + x/4 - y/2) + 100 \ d/21} \times 100$$

(13.28)

from which ϕ can be evaluated if d is known. The flows of exhaust and dope air are controlled by flows through critical orifices, such that the ratio of molar flows is constant. The ratio of flows is evaluated by heating nitrogen to the same temperature as the exhaust stream, and measuring the level of the oxygen, in the mixture of dope air and nitrogen:

$$N_2 + e(O_2 + 79/21 \ N_2) \rightarrow e \ O_2 + (1 + 79e/21)N_2 \qquad (13.29)$$

The percentage of oxygen in this mixture can be measured, and is denoted here by f:

$$f \% = 100 \ e/(1 + 100 \ e/21) \qquad (13.30)$$

Equation (13.30) can be evaluated to give e, and this leads to a determination of d in equation (13.27):

$$d = e \times \text{(number of kmols of partial combustion products)} \qquad (13.31)$$

The partial combustion products will include nitrogen, water vapour, carbon dioxide, carbon monoxide, hydrogen and a small amount of oxygen. The oxygen level will be small, especially with regard to the other constituents, and it has to be assigned a value which is here taken to be zero. The partial products of combustion (from equation 13.27) are thus:

$$\rightarrow (a + b)CO_2 + (1-a-b)CO + b \ H_2 + (x/2-b)H_2O$$
$$+ \ 79/21\phi(1 + x/4 - y/2) \qquad (13.32)$$

Fortunately the temporary variables (a and b) in equation (13.32) cancel when the number of kmols are being evaluated. Equations (13.29), (13.30) and (13.31) enable d to be found in terms of known variables and ϕ. This result can then be substituted into equation (13.28) to give a solution for the equivalence ratio ϕ.

The oxygen sensor is calibrated by means of atmospheric air, and bottled nitrogen. The calibration is then checked by means of a nitrogen/oxygen calibration gas, with a low level of oxygen (as might be encountered in the exhaust). The performance of the catalyst and the dope air system can be checked by a calibration gas that represents the combustion products of a rich mixture, with for example: 10 per cent carbon dioxide, 1 per cent carbon monoxide and 0.1 per cent propane in nitrogen. Such a reference gas is also useful in checking other analysers.

The Cussons Lamdascan is a unit that has a fast response (significantly less than a second) and is able to analyse combustion over a wide range of equivalence ratios (from about 0.3 to 3).

When the exhaust gas oxygen sensor is fitted in the exhaust system, there is a faster response (about 100 ms) but with a more limited range of equivalence ratios. The importance of operating the sensor at the correct temperature has already been discussed in chapter 4, section 4.7. Also described here, was how the platinum acts as an electrode and as a catalyst to fully oxidise the exhaust products. It is self-evident how such an exhaust gas oxygen sensor operates with weak or stoichiometric air/fuel ratios. However, the sensor can also be contrived for use with mixtures slightly rich of stoichiometric (up to about an equivalence ratio of 1.4). By using a second zirconia sensor, oxygen can be either 'pumped' into or out of the exhaust gas sample. The detecting cell is kept at stoichiometric, and the amount of oxygen that has to be pumped can be measured, since it is proportional to the current flow in the pumping cell.

13.4.5 Exhaust smoke and particulates

A variety of systems has been developed for measuring the smoke level in Diesel exhaust. Unfortunately, the systems respond to the particle size distributions in different ways, so that it is not possible to make comparisons between different measuring systems when either the operating condition or the engine is changed. Commercially available systems either measure the obscuration of a light beam (for either the whole flow or part of the flow) or the fouling of a filter paper. An alternative approach described by Kittelson and Collings (1987) involves measuring the electrical charge associated with the smoke particles.

However, the most widely used system is the Bosch Smokemeter, in which a controlled volume of exhaust is drawn through a filter paper. The change in the reflectance of the paper then corresponds to the smoke level. A value of zero is assigned to a clean filter paper, and a value of ten is assigned to a piece of paper that reflects no light. The calibration of intermediate values can be checked, by placing a perforated piece of non-reflecting paper over a filter paper.

Exhaust particulates are defined as material that can be collected on a filter paper maintained at 325 K. Since it is impractical to pass the whole of the exhaust stream through a filter, a sample of the exhaust is drawn off and cooled by dilution with air. This leads to a complex system, such as illustrated by figure 13.28 which enables evaluation of the fraction of the exhaust being filtered. By weighing the filter before and after use, the mass

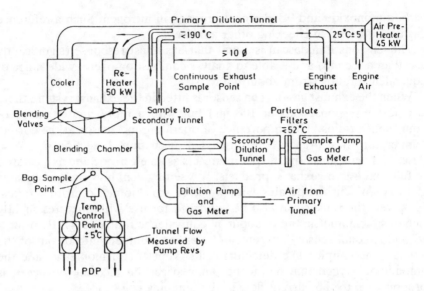

Figure 13.28   Schematic of a Federal Test Procedure (FTP) emissions sampling system

of the particulates is evaluated. The particulates consist of particles, and high molecular weight hydrocarbons.

A simpler way of estimating the mass of particulates has been proposed by Greeves and Wang (1981). They argued that as particulates comprise soot and unburned hydrocarbons, then it should be possible to deduce the mass of particulates from a smoke reading and unburned hydrocarbon reading. Greeves and Wang tested direct and indirect injection engines, and proposed the following correlation:

$$\text{Particulates (g/m}^3) = 1.024 \times \text{smoke (g/m}^3)$$
$$+ 0.505 \times \text{HC (g/m}^3) \qquad (13.33)$$

Conversion of a FID reading of unburned hydrocarbons to g/m³ is straightforward, but the smoke reading cannot be calculated from a smoke meter reading. Greeves and Wang (1981) used the results of Fosberry and Gee (1961), whose correlation for the Bosch Smoke Units is presented in figure 13.29.

13.4.6 Determination of EGR and exhaust residual (ER) levels

The exhaust gas recirculation (EGR) level can be defined volumetrically, on a molar basis, or gravimetrically. The volumetric definition is simpler,

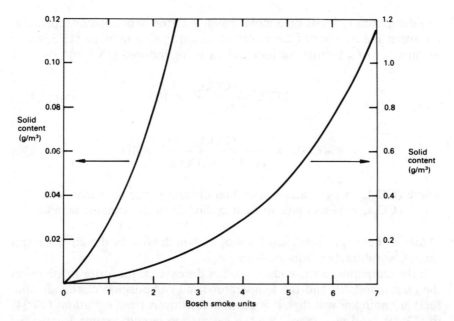

Figure 13.29   Relation between the Bosch Smoke number and the solid content
in the exhaust flow (from Fosberry and Gee (1961))

as the EGR level is defined as the percentage reduction in volume flow rate
of air at a fixed operating point:

$$\% \text{ EGR} = \frac{\dot{V}_o - \dot{V}_e}{\dot{V}_o} \times 100 \tag{13.34}$$

where $\dot{V}_o$ = volume flow rate of air with no EGR.
$\dot{V}_e$ = volume flow rate of air with EGR.

With a Diesel engine the fixed operating point will correspond to the
same speed, and either the same fuelling level or the same torque. With
spark ignition engines the speed should be fixed, and either the torque is
allowed to change or changes will have to be made to the throttling and/or
fuelling level.

Alternatively the EGR level can be deduced, by comparing the concentration of a particular species in the exhaust with its concentration in the
inlet manifold. It is usual to measure the carbon dioxide, as it is present in
the most significant quantities, and it can be measured accurately. The
mixture in the manifold is assumed to be well mixed, and a vacuum pump is
needed to draw the sample from the inlet manifold (since spark ignition
engine manifolds can be operating at less than 0.2 bar absolute). There are

two definitions of EGR on a molar basis, with the denominator representing either the number of kilomoles of mixture being induced (13.35a), or the number of kilomoles of fuel and air being induced (13.35b):

$$\% \, EGR = \frac{(CO_2)_m}{(CO_2)_e} \times 100 \qquad (13.35a)$$

$$\% \, EGR = \frac{(CO_2)_m}{(CO_2)_e - (CO_2)_m} \times 100 \qquad (13.35b)$$

where $(CO_2)_m$ = percentage of carbon dioxide in the inlet manifold
$(CO_2)_e$ = percentage of carbon dioxide in the exhaust manifold.

(*Note*: if there is a significant level of carbon dioxide in the air, then this should be subtracted from both values.)

If the temperature and composition of the exhaust just prior to mixing in the inlet manifold, and the temperature and composition of the air (and fuel) are both known, then it is possible to convert from equations (13.34) to (13.35), and vice versa. Such a calculation would assume isenthalpic mixing of the gas streams (that is, with no external heat transfer).

The volumetric definition of EGR is probably best suited to Diesel engines, since at light loads the concentrations of carbon dioxide would be low and equation (13.35) would become ill-conditioned. An alternative approach that can be suitable for Diesel engines is to measure the oxygen level in the inlet manifold and the exhaust manifold. In contrast, for spark ignition engines in which the throttle position might be varied and the levels of carbon dioxide in the exhaust will invariably be about 10 per cent, then equation (13.35) is more useful.

Since a gravimetric definition of EGR could also be used, then it is essential to state or establish which method of measuring EGR has been used, and how it has been defined.

The exhaust residuals are those products of combustion that are not displaced from the cylinder during the gas exchange processes. They are a consequence of: the clearance volume at tdc; the relative pressures in the inlet manifold, cylinder and exhaust manifold; and the valve timing. High levels of exhaust residuals will occur when there is: a large clearance volume, or a wide valve overlap, or a low inlet manifold pressure (Toda *et al.* (1976)). This has also been referred to in chapter 6, section 6.4.1.

The exhaust residual fraction in the trapped charge can be deduced by assuming perfect mixing of the residuals with the incoming charge, and then extracting a sample during the compression process. Many researchers have developed their own high speed gas sampling valve, and a typical example has been described by Yates (1988). Sampling valves are also available commercially, and the minimum sampling duration is about 1 ms.

In the absence of exhaust gas recirculation, the residual fraction on a molar basis (rf_M) is given by

$$rf_M = \frac{(CO_2)_c}{(CO_2)_e} \qquad (13.36)$$

where $(CO_2)_c$ = percentage of carbon dioxide from the sampling valve.

To convert equation (13.36) to a mass fraction, it is necessary to know the mean molecular mass of the cylinder contents and the exhaust gas. The mean molecular mass can be calculated once the composition is known, and this can be found either by measurement, or, if the air/fuel ratio is known, by assuming idealised combustion. Idealised combustion is illustrated in chapter 3 by examples 3.1, 3.4 and 3.6.

When EGR is being used, some of the carbon dioxide trapped in the cylinder prior to combustion will be a consequence of the exhaust gas recirculation. As the cylinder contents are comprised of residuals and the gas drawn in from the inlet manifold, and the carbon dioxide level in each of these is known, then it can be shown that

$$rf_M = \frac{(CO_2)_c - (CO_2)_m}{(CO_2)_e - (CO_2)_m} \qquad (13.37)$$

As with equation (13.36), so equation (13.37) could be converted to a mass fraction if the molar composition of the exhaust and induction gases are known.

Sampling valves can also be used to extract samples during or after combustion from discrete sites within the clearance volume. Such measurements are particularly useful in gaining insights into the Diesel combustion process (Whitehouse, N. D. (1987)).

13.4.7 Determination of the air/fuel ratio from exhaust emissions

The methods used in calculating the air/fuel ratio from the exhaust emissions have already been introduced in chapter 3 by example 3.2. The generalised combustion of an oxygenate fuel is given by

$$C_xH_yO_z + \frac{(x + y/4 - z/2)}{\phi}\left(O_2 + \frac{79}{21}N_2\right) \rightarrow a\,CO + b\,CO_2$$

$$+ c\,H_2O + d\,H_2 + e\,O_2 + f\,C_mH_n + g\,NO_q + h\,N_2 \quad (13.38)$$

In general the emissions measurements do not include hydrogen or nitro-gen, and are made on a dry basis. Equation (13.38) is most conveniently solved if it is assumed that there are sufficient reactants to produce 100 kmol of *dry products*, since a will correspond to the measured percentage of carbon monoxide and so on (that is why there are assumed to be x kmol of carbon atoms rather than 1 kmol).

When the fuel composition is known (as the ratio of $x{:}y$ and $x{:}z$), then there are four atomic balances and four unknowns, x and ϕ, and the concentration of the hydrogen and water vapour in the products.

As the nitrogen is not usually measured, but found by difference, there are only in fact three independent equations. However, as the ratio of fuel to air is being evaluated (the variable x could be eliminated in the ratio of fuel to air) there are only three unknowns: ϕ, c and d. Unfortunately, the solution of these equations is ill-conditioned, and the usual approach is to make assumptions about the amount of hydrogen present.

This can be illustrated by considering a carbon balance and the nitrogen balance:

$$\text{C balance:} \qquad\qquad x = a + b + f \times m \qquad\qquad (13.39)$$

where $f \times m = (\text{ppm HC}) \times 10^{-4}$.

$$\text{N}_2\text{ balance:} \qquad\qquad \frac{79}{21\phi}(x + y/4 - z/2) = h \qquad\qquad (13.40)$$

where a is the percentage of carbon monoxide etc.

Equation (13.39) allows x to be evaluated, so this can be substituted into equation (13.40).

The number of kmols of nitrogen can be found by difference, if assumptions are made about the number of kmols of hydrogen (c). The options are

(i) To neglect the hydrogen — acceptable for weak mixtures.
(ii) To make assumptions about the relative quantities of CO and H$_2$ (for example, from data such as those in figure 3.15).
(iii) To consider the water/gas equilibrium.

The normal approach is to consider the water/gas equilibrium, and to assume a typical value. Spindt (1965) adopted a value of 3.5, and this has been used by others subsequently:

$$b \text{ CO}_2 + d \text{ H}_2 \rightleftharpoons a \text{ CO} + c \text{ H}_2\text{O}$$

$$3.5 = \frac{a \times c}{b \times d} \qquad\qquad (13.41)$$

The other atomic balances (hydrogen, oxygen and nitrogen) can also be used to determine the equivalence ratio, and this provides a means of checking the quality of the emissions measurements, and the assumptions about the hydrogen level. The solution of these equations is straightforward, but lengthy and tedious, so it is best suited to the use of charts or a computer program.

A series of charts for different composition hydrocarbons has been prepared by Eltinge (1968), and the chart for a fuel with a 1.9 H:C ratio (typical of gasoline) is shown in figure 13.30.

The use of the chart is illustrated by the inset in figure 13.30 which considers an exhaust comprising: 1 per cent CO, 2 per cent O_2 and 13.0 per cent CO_2. If the emissions measurements were self-consistent (not necessarily the same as being correct) and the combustion was complete, then the triangle formed by the iso-composition lines would collapse into a point. In other words, the larger the triangle, the greater the inconsistency in the emissions measurements.

The parameter S_x is meant to indicate the level of fuel mal-distribution in a multi-cylinder engine. Consider a two-cylinder engine which is operating on a stoichiometric mixture overall, but one cylinder is operating with an equivalence ratio of 0.9, and the other with 1.11. The result of averaging these two exhaust streams will be a higher level of carbon monoxide and oxygen (but a lower level of carbon dioxide) than if both cylinders were operating with a stoichiometric mixture. Examination of figure 13.30 shows that this will lead to a higher value of S_x.

The alternative to using charts is to employ a computer program, and this is particularly convenient when the exhaust gas analysers are linked directly to a computer to undertake both data analysis and logging. Such a program can also be used to conduct a sensitivity analysis, in which the effect of possible measurement errors on the computed air/fuel ratio can be analysed. The results of a sensitivity analysis are shown in figure 13.31 (Stone R. (1989c)).

Figure 13.31 has used hypothetical emissions data that have been evaluated for a hydrocarbon fuel (H:C ratio of 1.8) at a specified air/fuel ratio. The effect of a +1 percentage point error in each of the emissions (CO_2, CO and O_2) has been examined in turn, and the error in the computed air/fuel ratio has then been evaluated and plotted against the specified air/fuel ratio. AFR1 has been evaluated by the oxygen/hydrogen atomic balance, while AFR2 has been evaluated by the carbon balance (equations 13.39 and 13.40). Figure 13.31 shows that AFR2, the carbon balance, is more likely to give reliable results, and that the computed air/fuel ratio is much more sensitive to measurement errors when the air/fuel ratio is significantly weak — such as might occur with Diesel engines. The data used in preparing figure 13.31 show that in the equivalence ratio range of $0.7 < \phi < 1.5$, a +1 percentage point error in any of the emissions measurements leads to less than a 1 per cent error in the equivalence ratio.

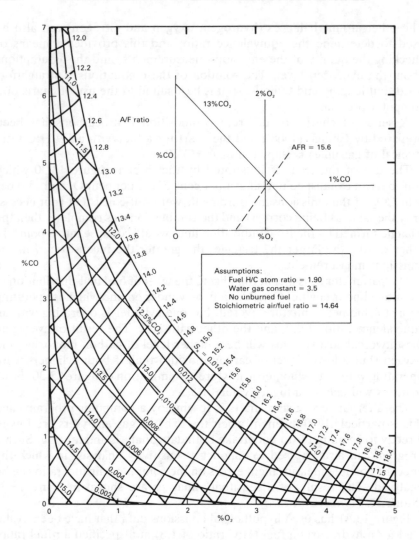

Figure 13.30   Exhaust emissions chart for evaluating the air/fuel ratio and maldistribution parameter (S_x) for a typical gasoline (adapted for Eltinge (1968))

The computer program can also be used to investigate the significance of uncertainty in the fuel composition. Such an exercise has been summarised in table 13.3 for a fuel with a H:C ratio of 1.8 for a range of air/fuel ratios.

Inspection of the data in table 13.3 shows that in the region of stoichiometry (as would be encountered with homogeneous charge spark-ignition engines), the effect of uncertainty in fuel composition on the

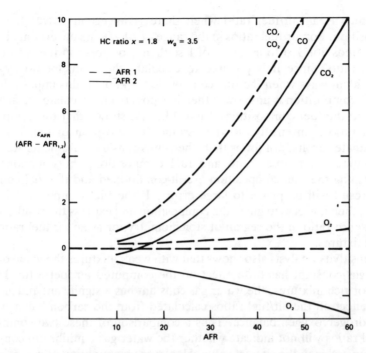

Figure 13.31 The errors in the computed air/fuel ratios caused by 0.01 errors on each of the principal emissions (CO_2, CO and O_2)

Table 13.3 Sensitivity of computed air/fuel ratio and equivalence ratio (using the carbon balance) to uncertainty in the fuel composition (actual composition 1.8 H:C)

AFR	ϕ	$AFR_{1.6}$ (assuming 1.6 H:C)	$\phi_{1.6}$	$AFR_{2.0}$ (assuming 2 H:C)	$\phi_{2.0}$
11.05	1.313	10.83	1.310	11.26	1.314
20.02	0.725	19.80	0.718	20.22	0.732
30.02	0.483	29.95	0.475	30.08	0.492
40.03	0.362	40.09	0.355	39.92	0.371
50.03	0.290	50.28	0.283	49.83	0.297
60.04	0.242	60.37	0.236	59.63	0.248
80.07	0.181	80.64	0.176	79.32	0.187

equivalence ratio is negligible. For the H:C ratio range of 1.6–2.0, there would be an error of less than 1 per cent in the computed equivalence ratio. Thus, figure 13.30 can be used for a wide range of gasolines, and the results will be most accurate if expressed as an equivalence ratio.

For the weaker mixture, as might be encountered with a Diesel engine, then table 13.3 demonstrates that the results would be expressed more accurately as an air/fuel ratio when there is uncertainty over the fuel composition. For air/fuel ratios in the range of 20–80, the uncertainty in the fuel composition led to an error of less than 1 per cent in the calculated air/fuel ratio. Since it is possible to calculate the air/fuel ratio by two independent ways when the fuel composition is known, this implies that if the fuel composition is unknown, then it is possible to calculate the air/fuel ratio and the fuel composition. Table 13.3 has shown that the computed air/fuel ratio is insensitive to the assumed composition of the fuel. An unfortunate corollary of this is that when emissions measurements are used to evaluate the air/fuel ratio and fuel composition, the equations for determining the fuel composition are ill-conditioned and the fuel composition result will be prone to large errors. If the fuel composition is not known, then it is best to guess the composition and express the results as an equivalence ratio in the region of stoichiometry, or as an air/fuel ratio for weak mixtures.

A sensitivity analysis also shows that with weak mixtures the value of the water gas constant has little effect on the computed air/fuel ratio. However, for rich mixtures, the water gas constant has a significant and almost equal effect on the air/fuel ratios calculated from the carbon balance and the hydrogen/oxygen balance. Thus a comparison of these two computed air/fuel ratios will not indicate whether the water gas equilibrium constant has been assigned the correct value. However, examination of wet exhaust emissions data presented by Heywood (1988) shows that the water gas constant remains within the range of 3–4, and this uncertainty would cause an error of less than ± 0.25 per cent ϕ.

13.5 Computer-based combustion analysis

13.5.1 Introduction

The increasing power and falling costs of personal computer systems has meant that very satisfactory systems can be bought for a modest amount. There is, of course, a need for a high-speed data acquisition card that allows for the conversion of analogue voltages to digital signals. The cost of such a computer-based data acquisition system is now comparable to the cost of a piezo-electric pressure transducer and its amplifier.

There are essentially two types of combustion analysis undertaken

(i) burn rate analysis — usually associated with spark ignition engines, and
(ii) heat release analysis — usually associated with Diesel engines.

Table 13.4   The specification of a computer-based data acquisition system comprising a Computerscope data acquisition card, and a Compaq 386/25e personal computer

Channels	1, 2, 4, 8 or 16
Multiplexing overhead	1 μs
Resolution	12 bits
Input voltage range	±10 V

	Max. sampling rate	Available memory
Computerscope on-board memory	1 MHz	256 k
System RAM	200 kHz	10 MB
Hard Disk	100 kHz	60 MB

These will be described shortly, but first it will be useful to review a computer-based data acquisition system. This will be done by means of an example — a Computerscope data acquisition card installed in a Compaq 386/25e personal computer; the resulting specification is in table 13.4.

When selecting a data acquisition card, it is necessary to decide on:

(i)   the resolution and accuracy (for example, 10 bit)
(ii) the number of channels
(iii) the maximum sampling rate
(iv) how much data are to be collected.

Firstly, the lower the resolution of the card, then the lower the cost for a given sampling rate. An analogue-to-digital converter (ADC) with 8 bits might appear adequate, as this will give a resolution of 1 part in 256. In other words, the ADC will cause an uncertainty of about 0.4 per cent, and this is comparable with the accuracy of a piezo-electric pressure transducer. However, to achieve this ADC accuracy it is necessary to use the full dynamic voltage range of the input, and this may not be convenient. Furthermore, it is desirable to minimise any sources of additional errors. In practice, 10- or 12-bit resolution ADC cards are most likely to be used.

Secondly, it is necessary to identify how many channels might need to be logged. It is possible to superimpose a marker voltage (at say bdc) on the cylinder pressure signal, and then use software later to separate the marker voltage. However, it is simpler to use a separate channel for the reference flag. Other channels might be wanted for measuring fuel line pressure or injector needle lift. Furthermore, measurements might be wanted from one or more cylinders. For this type of application it is usual to have a single ADC, and to multiplex the inputs so that each is read in turn. With multiplexing, many systems introduce a multiplexing overhead (the time to switch from one channel to the next). Thirdly, the maximum sampling rate has to be identified. This is determined by the number of channels being sampled, and the frequency of the signals that are to be recorded. Sampling theory determines that the sampling rate should be twice the highest frequency that is to be recorded from the signal. Consider the cylinder pressure transducer which might have a response that is within 1 per cent of linear to a frequency of 10 kHz. This might suggest that the maximum useful sampling rate is 20 kHz. However, if a phenomenon such as knock is being investigated, then a higher sampling rate will be wanted, even though the non-linearity of the transducer is becoming significant. In general, the sampling will be controlled by a shaft encoder on the engine that presents a signal to the 'external clock' input of the data acquisition system. This arrangement has the advantage that the angular position of each reading will be known. Thus the sampling rate will be influenced by the engine speed and the crank angle resolution that is required. For example, the injector needle lift might be wanted with a ¼° resolution to determine the start of injection. Thus

$$
\begin{aligned}
\text{Sampling rate (sample/s)} = 6 &\times \text{engine speed (rpm)} \\
&\times \text{number of readings/} \\
&\quad\text{degree} \\
&\times \text{number of channels} \\
&\quad\text{enabled}
\end{aligned}
\tag{13.42}
$$

For example, with a Diesel engine operating at 4000 rpm, with four channels enabled and readings taken every ¼°, there would be 384k sample/s.

Finally, the amount of data to be collected has to be identified, as this influences the maximum sampling rate. Large quantities of data need to be written to some form of disk, and this is slower than writing to RAM (random access memory). Consider measurements of cycle-by-cycle variation in a four-stroke spark ignition engine. The requirement for memory is

$$
\begin{aligned}
\text{Memory} = &\text{ number of cycles} \times \text{number of channels} \times \\
&\text{number of readings/degrees} \times (180 \times \text{number of} \\
&\text{strokes}) \times \text{number of bytes/sample}
\end{aligned}
\tag{13.43}
$$

For example, if 300 cycles of a four-stroke engine are to be recorded from two channels, with readings every ½° and 2 bytes are required for each reading (for a 12-bit reading), then about 3.4 MB of memory will be required.

The trade-offs between the sampling rate and the storage capacity have been illustrated by table 13.4 for a particular system. In this case, the 3.4 MB of memory would require the use of the RAM or the hard disk, and the engine speeds corresponding to the maximum sampling rates would be 8333 or 4166 rpm. These two examples illustrate that the system described in table 13.4 is capable of acquiring large amounts of data very quickly. This can then lead to analysis and archiving problems unless the combustion analysis software is quick to run.

There are other issues to be considered with computer-based data acquisition systems. Firstly, when channels are being multiplexed, then the channels are not all being read at the same time. For example, table 13.4 shows that the Computerscope system has a delay of 1 microsecond between reading successive channels. This becomes significant with high sampling rates, but it can be corrected for in software that either assigns the correct angle to the reading or interpolates between readings to give the value at a specific crank angle. Secondly, it is essential that the ADC is coupled to a sample/hold circuit. If this is not the case, then slight changes in the signal during the analogue-to-digital conversion process can lead to large errors.

When the data are being analysed, it must be remembered that they were originally a continuous analogue signal that has been assigned to a progression of digital levels. For example, when such a signal is differentiated numerically by looking at the difference in successive values, then the result is very noisy. A more satisfactory result can be obtained by using a higher-order finite difference approach and smoothing the signal.

Taylor series can be used to represent the signal, and the following example calculates the first derivative, but includes the terms up to the fourth derivative. Consider a function, $f(a)$ in which the interval between values is h:

$$f(a_{n+2}) = f(a_n) + \frac{2hf'(a_n)}{1!} + \frac{2^2h^2f''(a_n)}{2!} + \frac{2^3h^3f'''(a_n)}{3!} + \frac{2^4h^4f''''(a_n)}{4!} + \ldots$$

$$f(a_{n+1}) = f(a_n) + \frac{hf'(a_n)}{1!} + \frac{h^2f''(a_n)}{2!} + \frac{h^3f'''(a_n)}{3!} + \frac{h^4f''''(a_n)}{4!} + \ldots$$

$$f(a_n) = f(a_n) \qquad\qquad\qquad (13.44)$$

$$f(a_{n-1}) = f(a_n) - \frac{hf'(a_n)}{1!} + \frac{h^2f''(a_n)}{2!} - \frac{h^3f'''(a_n)}{3!} + \frac{h^4f''''(a_n)}{4!} + \ldots$$

$$f(a_{n-2}) = f(a_n) - \frac{2hf'(a_n)}{1!} + \frac{2^2h^2f''(a_n)}{2!} - \frac{2^3h^3f'''(a_n)}{3!} + \frac{2^4h^4f''''(a_n)}{4!} + \ldots$$

These equations can be combined to eliminate the second, third and fourth derivatives ($f''(a_n)$, $f'''(a_n)$ and $f''''(a_n)$), to give

$$f'(a_n) = (f(a_{n-2}) - 8f(a_{n-1}) + 8f(a_{n+1}) - f(a_{n+2}))/12h \qquad (13.45)$$

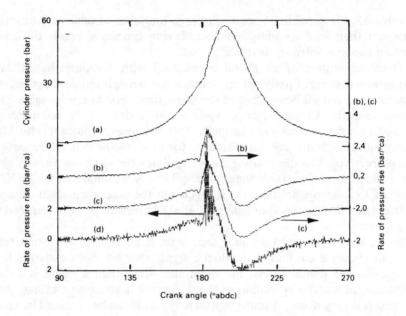

Figure 13.32   Cylinder pressure data recorded every ½° ca from a direct
injection Diesel engine at 1000 rpm with 60 per cent load. Trace
(d) shows the first derivative evaluated by equation (13.45).
Trace (b) illustrates the consequence of applying equation (13.47)
twice to smooth trace (d). Trace (c) illustrates when equations
(13.47) is used twice to smooth trace (a), prior to differentiation
with equation (13.45)

However, figure 13.32 shows that the derivative will still be noisy, and it is
advantageous to apply smoothing. The following smoothing algorithm for
$(2b + 1)$ values is widely used:

$$a_n = \frac{1}{b^2} [a_{n-(b-1)} + 2a_{n-(b-2)} + 3a_{n-(b-3)} \ldots + ba_n$$

$$+ \ldots 3a_{n+(b-3)} + 2a_{n+(b-2)} + a_{n+(b-1)}] \qquad (13.46)$$

Note: the terms in equation (13.46) are only evaluated when the part of
the subscript in brackets is not negative.

This is illustrated by the simplest case when $b = 2$:

$$a_n = \frac{(a_{n-1}) + 2(a_n) + (a_{n+1})}{4} \qquad (13.47)$$

The smoothing equation can be applied recursively (that is, more than once) and this is illustrated by figure 13.32. Figure 13.32 also shows that the order of smoothing and differentiation is not important, a result that can be shown algebraically. It is also possible to combine the smoothing and differentiation equations. The smoothing equation would be applied to each of the terms in equation (13.45) in turn, and the combined equation will reduce the amount of arithmetic to be undertaken by the computer. Temporary variables can also be used, to minimise the access of variables stored in the arrays. The theory behind these smoothing equations is beyond the scope of this book, but a thorough treatment is provided by Anon (1979), which also describes digital filtering techniques. However, it must be remembered that the smoothing process loses information, for example, any maxima will be reduced in magnitude. Care must be taken to avoid over-smoothing.

13.5.2 Burn rate analysis

A burn rate analysis is usually applied to the combustion data from spark ignition engines to calculate the mass fraction burnt (mfb). A widely used technique is the approach devised by Rassweiler and Withrow (1938). After the start of combustion, the pressure rise (Δp) during a crank angle interval ($\Delta \theta$) is assumed to be made of two parts: a pressure rise due to combustion (Δp_c) and a pressure change due to the volume change (Δp_v):

$$\Delta p = \Delta p_c + \Delta p_v \tag{13.48}$$

As the crank angle (θ_i) increments to its next value (θ_{i+1}) the volume changes from V_i to V_{i+1}, and the pressure changes from p_i to p_{i+1}. It is assumed that the pressure change due to the change in volume can be modelled by a polytropic process with an exponent k. Substituting for Δp_v, equation (13.48) becomes

$$p_{i+1} - p_i = \Delta p_c + p_i \left[\left(\frac{V_i}{V_{i+1}} \right)^k - 1 \right]$$

from which Δp_c can be evaluated:

$$\Delta p_c = p_{i+1} - p_i (V_i / V_{i+1})^k \tag{13.49}$$

The pressure rise due to combustion is not directly proportional to the mass of fuel burned, as the combustion process is not occurring at constant volume. The pressure rise due to combustion has to be referenced to a datum volume, for example the clearance volume at tdc, V_c:

$$\Delta p_c^* = \Delta p_c V_i / V_c \qquad (13.50)$$

The end of combustion occurs after N increments, and is defined by the pressure rise due to combustion becoming zero. If it is assumed that the referenced pressure rise due to combustion is proportional to the mass fraction burned (mfb) then

$$\text{mfb} = \sum_0^i \Delta p_c^* / \sum_0^N \Delta p_c^* \qquad (13.51)$$

The summation of the referenced pressure rise due to combustion is illustrated by figure 13.33, along with the normalisation of the data to give the mass fraction burned.

Since the volume change is small when the piston is in the region of tdc, the computed mass fraction burned is insensitive to slight errors in the positioning of tdc. However, the method does depend on using an appropriate value of the polytropic index, k. Rassweiler and Withrow (1938) evaluated the polytropic index for before and after combustion and used an appropriately averaged value during combustion. For the results shown here in figure 13.33, the polytropic index was only evaluated during compression. This leads to the fall in the referenced pressure due to combustion and the mass fraction burned in figure 13.33 after the end of combustion, as the polytropic index is lower during the expansion process than during compression. This is a consequence of the heat transfer and the presence of combustion products.

During compression the polytropic index is usually within the range of 1.2–1.3 for a spark ignition engine, and a suitable value can be chosen by the user. Alternatively, the polytropic index can be evaluated from the compression process prior to ignition. By evaluating the logarithmic values of pressure and volume, a least squares straight line fit can be used to determine the polytropic index. However, care is needed because of two reasons:

(1) there might be errors in the pressure datum
(2) during the initial part of compression the pressure rise is small, and discretisation errors from the ADC are more significant.

Both effects are minimised if the initial part of the compression is ignored; for example, the polytropic index could be evaluated up to ignition, from half way between the inlet valve closure and ignition.

The Rassweiler and Withrow method contains several assumptions. It is assumed that the referenced pressure rise due to combustion is proportional to the mass fraction burned in each increment. There is no explicit allowance for: heat transfer, dissociation, or change in composition of the gases; though to some extent an allowance is made, as the polytropic

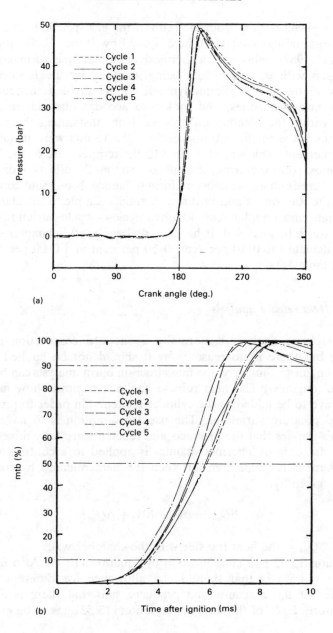

Figure 13.33   The referenced pressure rise attributable to combustion (a), and
the mass fraction burnt (mfb) as a function of time (b), for the
five successive cycles of figure 4.15

index is allowed to vary from the ratio of the gas specific heat capacities. These shortcomings have been investigated by Stone, C. R. and Green-Armytage (1987), who used a thermodynamic analysis to make a direct comparison with the same data being analysed by the Rassweiler and Withrow method. The thermodynamic model divided the combustion chamber into two zones, and took into account dissociation and heat transfer within the combustion process. Notwithstanding the substantial differences between the two approaches, the results were in surprisingly close agreement. This was attributed to the temperature of the burnt gas being almost constant during combustion, so that the effects of dissociation and heat transfer had an almost uniform influence throughout combustion.

Since the Rassweiler and Withrow method is simple to calculate, it is an appropriate and popular method when cycle-by-cycle variations in combustion are to be analysed. It has already been used in chapter 4, section 4.4 to calculate the 0–10 per cent, 0–50 per cent and 0–90 per cent burn times in table 4.3.

13.5.3   Heat release analysis

A heat release analysis is normally applied to combustion in Diesel engines, but there is no reason why it should not be applied to spark ignition engines. (Similarly, the mass fraction burnt analysis can be applied to Diesel engines.) The heat release analyses compute how much heat would have to be added to the cylinder contents, in order to produce the observed pressure variations. The usual assumption is to have a single zone; this implies that the products and reactants are fully mixed.

If the 1st Law of Thermodynamics is applied to a control volume in which there is no mass transfer, then the heat released by combustion (δQ_{hr}) is given by:

$$\delta Q_{hr} = dU + \delta W + \delta Q_{ht} \qquad (13.52)$$

where δQ_{ht} = the heat transfer with the chamber walls.

Equation 13.52 is a simplification of equation (10.1). Also implicit in equation (13.52) is that there is no allowance for differences in the properties of the reactants and products, and that there is a uniform temperature. Each of the terms in equation (13.52) has to be evaluated.

$$dU = mc_v dT \qquad (13.53)$$

From the equation of state ($pV = mRT$)

$$mdT = \frac{1}{R} (pdV + Vdp) \tag{13.54}$$

Substitution of equation (13.54) into equation (13.53) gives

$$dU = \frac{c_v}{R} (pdV + Vdp) \tag{13.55}$$

Substituting equation (13.55) into equation (13.52) and noting that $\delta W = pdV$ gives on an incremental angle basis:

$$\frac{dQ_{hr}}{d\theta} = \frac{c_v}{R} \left(p \frac{dV}{d\theta} + V \frac{dp}{d\theta} \right) + p \frac{dV}{d\theta} + \frac{dQ_{ht}}{d\theta} \tag{13.56}$$

The $dV/d\theta$ term is defined from the geometry of the engine (equation 10.6a), and the $dp/d\theta$ term has been recorded from the engine. However, c_v and R are functions of temperature, and the heat transfer cannot be readily evaluated. But, if semi-perfect gas behaviour is assumed (such that $c_p/c_v = \gamma$ and $R = c_p - c_v$), then equation (13.56) can be written as

$$\frac{dQ_{hr}}{d\theta} - \frac{dQ_{ht}}{d\theta} = \frac{1}{\gamma - 1} \left(p \frac{dV}{d\theta} + V \frac{dp}{d\theta} \right) + p \frac{dV}{d\theta}$$

$$\frac{dQ_n}{d\theta} = \frac{\gamma}{\gamma - 1} p \frac{dV}{d\theta} + \frac{1}{\gamma - 1} V \frac{dp}{d\theta} \tag{13.57}$$

where $dQ_n/d\theta$ = the net heat release.

The gas temperature can be found from the equation of state ($pV = mRT$), since the pressure and volume are known, and it has been assumed that the mass is constant. The gas properties vary with temperature, but as the variation is modest, it is acceptable in most cases to evaluate the properties at the gas temperature computed in the previous increment. Equations such as (10.10) and (10.11) can be used to evaluate u and R, from which γ can then be evaluated. Once the gas temperature has been evaluated, then it is possible to estimate the heat transfer, by assuming a wall temperature and employing a heat transfer correlation (see chapter 10, section 10.2.4 In-cylinder heat transfer). Sometimes the heat release analysed in this way is known as the gross heat release, to emphasise the difference from the net heat release.

Examples of heat release are provided here by the BICCAS (*BICERI*

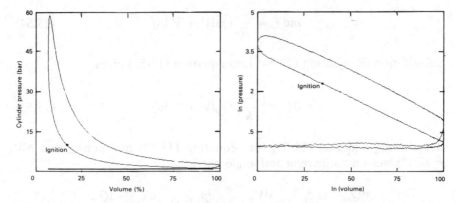

Figure 13.34   The indicator diagrams on a linear and logarithmic basis, for a
gas engine operating at 1500 rpm with wide open throttle and an
equivalence ratio of 0.6 with ignition 28° btdc

Combustion *A*nalysis *S*oftware), which can be used in conjunction with the
Computerscope data acquisition system (Whiteman (1989)).

The data used here are from a high compression ratio lean-burn gas
engine, which has been described by Ladommatos and Stone (1991).
Figure 13.34 shows the indicator diagram for operation at 1500 rpm with
wide open throttle and an equivalence ratio of 0.6, and ignition 18° btdc.
Plotting the pressure/volume diagram on a logarithmic basis provides
several insights:

(a) The pumping loop can be seen more clearly (and this is an alternative
to the linear magnification, for example, in figure 13.15).
(b) As the compression process is a straight line, this demonstrates that
assuming a polytropic process is a good model.
(c) The departure from a straight line just before tdc indicates the start of
combustion.
(d) The combustion is fairly symmetric about tdc, but the end of combustion is in fact much later and less clearly defined.
(e) The shallow curve of the expansion stroke is initially a consequence of
the final stage of combustion, and later on, a consequence of heat
transfer.

Although the engine being discussed here is a spark ignition engine,
there is no reason why a heat release analysis cannot be applied. Figure
13.35 shows both the net heat release rate, and the cumulative net heat
release. The heat release analysis evaluates the data on a differential basis
(equation 13.57), and this leads to noise in the computed result — especially at the lower pressures where the discretisation (ADC steps) are a
larger proportion of the signal. The negative net heat release rate (figure

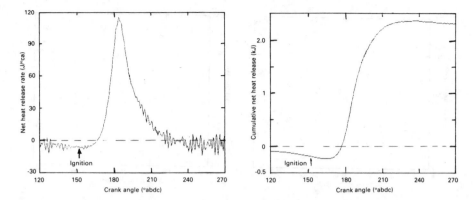

Figure 13.35 The pressure data for figure 13.34(a) have been analysed for:
(a) the net heat release rate, and (b) the cumulative net heat
release

13.35a) implies that there is heat transfer to the cylinder surfaces, and
ignition (28° btdc) should be close to the minimum net heat release rate.
This is because the heat transfer rate increases as the temperature and
density of the gas rise during combustion, but once combustion commences, then the heat release rate will add to the (negative) heat transfer.
The minimum net heat release rate is ill-defined in a spark ignition engine,
because the initial rate of combustion is very low. In a Diesel engine, the
fuel vaporisation also contributes to the negative heat release rate prior to
combustion. However, the initial combustion rate in a Diesel engine is very
rapid (because of the combustion of the pre-mixed reactants formed during
the ignition delay period). The initial combustion is so rapid, in a Diesel
engine, that the start of combustion is often defined as when the net heat
release rate becomes positive (see figure 10.2). Figure 13.35a illustrates
that this would be an unsatisfactory way of defining the start of combustion
in a spark ignition engine.

The cumulative net heat release in figure 13.35b is an integration of the
results in figure 13.35a; the process of integration smooths the noise
present in figure 13.35a. The minimum in the cumulative heat release
corresponds to the zero-crossing of the net heat release rate. The zero-
crossing of the cumulative net heat release occurs somewhat later (about
10°); this has no physical significance and should not be used to define the
start of combustion — even with a Diesel engine.

Near the end of combustion, the heat transfer to the combustion chamber becomes greater than the heat release from the combustion process.
Thus, the net heat release rate becomes negative (in other words the
cumulative heat release starts to fall) and this can be used to estimate the
end of combustion.

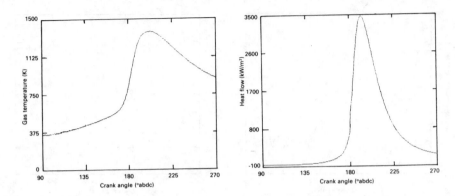

Figure 13.36   The calculated mean gas temperature and the heat flux calculated
by the Woschni (1967) correlation

Figure 13.36 shows the gas temperature (calculated by applying the equation of state: $pV = mRT$), and an estimate of the heat transfer from the Woschni correlation (equations 10.30–10.32). The gas temperature in figure 13.36a assumes a single zone for the combustion, and thus represents an estimate of the mean temperature. The heat flux calculation (figure 13.36b) requires an estimate of the combustion chamber surface temperature; this, and the choice of heat transfer correlation, influence the calculated heat flux. Because of these sources of uncertainty, the (gross) heat release and its rate have not been calculated here.

The final graph in this sequence (figure 13.37) is the mass fraction burnt calculated by the Rassweiler and Withrow method (equations 13.48–13.51). When figure 13.37 is compared with the cumulative net heat release (figure 13.36b) it can be seen that they have a very similar form.

13.6 Advanced test systems

Engine test cells are becoming increasingly complex for several reasons. Additional instrumentation such as exhaust gas analysis has become necessary and, in the search for the smaller gains in fuel economy, greater accuracy is also necessary. Consequently the cost of engine test cells has escalated, but fortunately the cost of computing equipment has fallen.

Computers can be used for the control of a test and data acquisition, thus improving the efficiency of engine testing. The computer can also process all the data, carry out statistical analyses, and plot all the results. The design of the test facility and computer system will depend on its use —

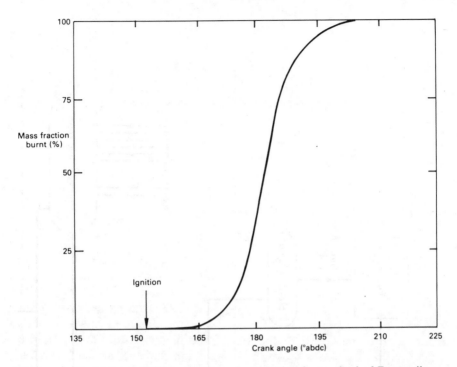

Figure 13.37   The mass fraction burnt calculated by the method of Rassweiler and Withrow (1938) for the pressure data in figure 13.35(a)

whether it is for research, development, endurance running or production testing. There are, of course, over-laps in these areas, but the facility described here will be for development work.

Descriptions of computer-controlled test facilities are given by Watson *et al.* (1981) and Donnelly *et al.* (1981). Complete facilities are marketed by firms such as Froude Consine and Schenck; a Schenck system is described here. A block diagram of a computerised test facility is given in figure 13.38. Separate microprocessors (micros) are used for data acquisition and test point control. A single microprocessor could be used but that would reduce the data-acquisition rate, and reduce the storage space for data and test cycles. The printer and VDU are linked to the micros so that the test pattern can be chosen and then monitored. The host computer can be linked to many such microsystems, and it provides a more powerful data-processing system with greater storage. Data from many tests can then be archived on magnetic disks or tape. The host computer can also provide sophisticated graphics facilities and plotters that would be underused if dedicated to a single test cell.

A typical requirement is to produce an engine fuel-consumption map. This requires running the engine over a wide range of discrete test points.

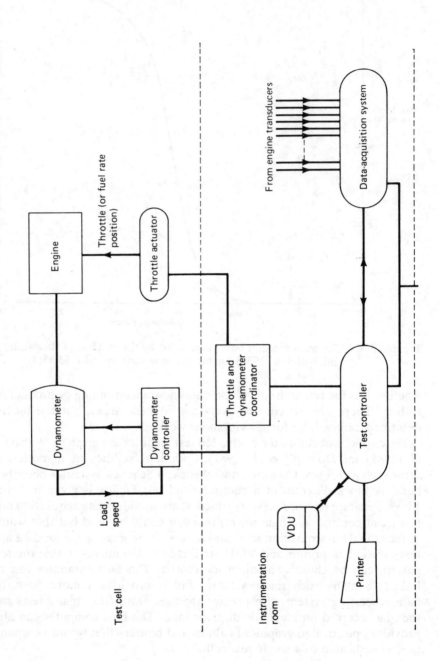

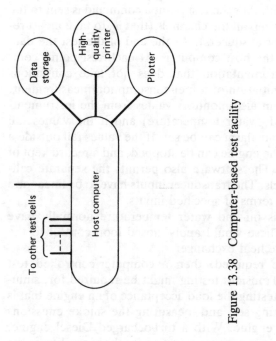

Figure 13.38 Computer-based test facility

Each test point is specified in terms of speed and load (or throttle position), and the sequence and duration of each test point is stored in the micro. The software for setting up a test programme is stored in an EPROM (erasable programmable read only memory) and is designed so that new or existing test programmes can be readily used. At each test point the dynamometer controller and throttle controller work in conjunction to obtain the desired test condition. Additional parameters such as ignition timing and air/fuel mixture can also be controlled. At each test point a command is sent to the data-acquisition system to sweep all the channels (that is, to take measurements from all the specified transducers). At the end of the test the data can be sent for analysis on the host computer. At each test point, commands can be sent to instrumentation that does not take continuous readings; for instance, the initiation of a fuel-consumption measurement.

The data-acquisition system also monitors values from the instrumentation (such as engine speed, water temperature) and if the values fall outside a certain range then an alarm can be set. If the values fall outside a specified wider range, then the engine can be stopped, and a record kept of what caused the shutdown. The software also permits the separate calibration of transducer channels. The transducer inputs have to be linear and the calibration is defined in terms of specified inputs.

Other parameters such as oil and water temperature normally have separate control systems. These would apply closed loop control to the flow through the appropriate heat exchanger.

When transient tests are required, then a computer-controlled test facility is almost essential. Transient testing might be required for: simulating an urban drive cycle, testing the load acceptance of an engine that is intended to drive a generating set, and measuring the smoke emissions from a turbocharged truck engine. With a turbocharged Diesel engine, transient testing is particularly important, since under these conditions the emissions of smoke and noise are higher.

Most instrumentation in a test cell will have a suitable response for measuring transients, but the most notable exception is likely to be the torque measurement. If the torque measurement is from the torque reaction on the dynamometer case (T_d), then this does not include the torque required to accelerate or decelerate the dynamometer:

$$T_e = T_d + I_d \dot{\omega} \qquad (13.58)$$

where　T_e = engine torque output
　　　　I_d = inertia of the dynamometer and the coupling
　　　　$\dot{\omega}$ = angular acceleration.

The two options for measuring the torque output are:

(i) To measure the torque in the coupling.

(ii) If the inertia of the dynamometer is known, then differentiate the speed, and add to the torque reaction on the dynamometer.

Unfortunately, both options have problems associated with them: there are torque fluctuations occurring in the coupling, and there is noise associated with differentiating a speed signal.

In test cells for production testing it is usual to have a pallet-mounting system for the engine, in order to minimise the cell downtime. Such systems may also be justified in development testing where many different engines are being tested.

13.7 Conclusions

Engine testing is an important aspect of internal combustion engines, as it leads to a better understanding of engine operation. This is true whether the engine is a teaching experiment or part of an engine-development programme. As in any experiment, it is important to consider the accuracy of the results. The first decision is the accuracy level that is required; too high a level is expensive in both time and equipment. The second decision is to assess the accuracy of a given test system; this is of ever-increasing difficulty owing to the rising sophistication of the test equipment.

The nature of development testing is also changing. The use of computer control and data acquisition have been complemented by increasing levels of engine instrumentation. This leads to large quantities of data, and a need for effective post-processing. Needless to say, such systems are expensive, but the need for their use is held back by two factors:

(1) The number of different engines produced is reduced by manufacturers standardising on engine ranges, and by collaboration between companies.

(2) The decreasing cost of computing time, and the use of increasingly powerful computer models, leads to greater optimisation prior to the start of development testing.

However, these factors are balanced by increasing restrictions on engine emissions, and the ever-rising difficulty of improving engine fuel economy.

14   Case Studies

14.1   Introduction

The three engines that have been chosen as case studies are the Jaguar V12 HE spark ignition engine, the Chrysler 2.2 litre spark ignition engine, and the Ford (high-speed) 2.5 litre DI Diesel engine. Each engine has been chosen because of its topicality, and characteristics that are likely to be seen also in subsequent engines. The Jaguar engine uses a May combustion chamber; this permits the use of a high compression ratio and the combustion of lean mixtures, both of which lead to economical operation. The Chrysler 2.2 litre spark ignition engine is typical of current practice in the USA and thus has low emissions of carbon monoxide, unburnt hydrocarbons and oxides of nitrogen.

The Ford compression ignition engine achieves short combustion times and thus high engine speeds by meticulous matching of the air motion and fuel injection. By utilising direct injection, as opposed to indirect injection into a pre-chamber, the pressure drop and heat transfer in the throat to the pre-chamber are eliminated. This immediately leads to a 10–15 per cent improvement in economy and better cold starting performance.

14.2   Jaguar V12 HE engine

14.2.1   Background

The design and development of the original version of the Jaguar V12 engine are described by Mundy (1972), in a paper published at the time of the engine's introduction. In 1981 the HE (high-efficiency) version was introduced with the May combustion chamber. The different compression ratios that have been used are shown in table 14.1.

522

Table 14.1   Compression ratios for different engine builds

Market	Fuel	Compression ratios	
		V12 (1972)	V12 HE (1981)
Europe	97 RON	9:1	12.5:1
USA	91 RON (lead free)	7.8:1	11.5:1

The V12 engine was designed as an alternative to a six-cylinder in-line engine that had originated some 25 years earlier. An engine capacity of about 5 litres was needed, and the trial engine had to fit into the same space as the six-cylinder engine. While a V8 engine would have been feasible, a V12 engine has complete freedom from all primary and secondary forces and moments, as well as closer firing intervals. Figure 14.1 shows the relative smoothness of six-cylinder, eight-cylinder and twelve-cylinder engines. Marketing considerations also favoured a V12 engine: Jaguar would be the only volume producer of a V12 engine, while V8 engines are quite common in the USA. Another consideration was that, for competition use, a 5 litre V12 engine with a short stroke (70 mm) would permit ultimate power output since the engine would be able to run safely at 8000–8500 rpm.

14.2.2   Initial engine development

Initial development work was with the classic arrangement of twin overhead camshafts and hemispherical combustion chambers, as had been used

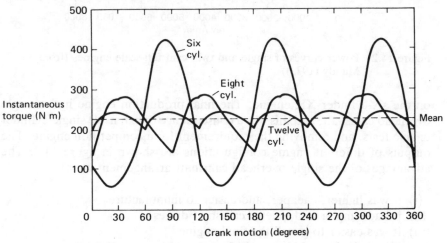

Figure 14.1   Torque characteristics of 6-cylinder, 8-cylinder and 12-cylinder engines (from Campbell (1978))

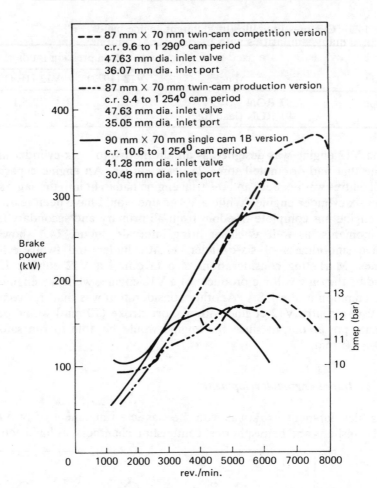

Figure 14.2   Power curves of single and twin cam full-scale engines (from Mundy (1972))

on the six-cylinder XK engine. The final production engine had a single overhead camshaft, and this was a consequence of extensive single-cylinder engine tests, and removing the requirement for a competition engine. The outputs of different engine configurations are shown in figure 14.2. The advantages of the single overhead camshaft arrangement were:

 (i)   it was lighter, cheaper, and easier to manufacture
 (ii) it had simpler and quieter camshaft drives
(iii) it was easier to install the final engine
(iv)   it had significantly better fuel consumption
 (v)   it gave better performance below about 5000 rpm.

The single overhead camshaft cylinder head is shown in figure 14.3, and there are many similarities with figure 6.2. To minimise the weight of the engine, alloy castings are used for the cylinder head, block and sump. The alloy cylinder head necessitates the use of valve guides and sintered iron valve seat inserts. The camshaft is mounted directly over the valve stems and the cam follower is an inverted bucket-type tappet. The valve clearances are adjusted by interchangeable hardened steel shims placed between the valve stem and the tappet. This so-called Ballot type of valve train provides a very stiff valve gear that is highly suited to high-speed operation. The main disadvantage is that the camshaft has to be removed in order to adjust the valve clearances; however, adjustment is needed much less frequently than for many other arrangements.

The design of the induction and exhaust passages, and even the valve seat inserts, were the subjects of much development work. The cylinder block is also alloy, and the cuff-type liners, which are like wet liners at the

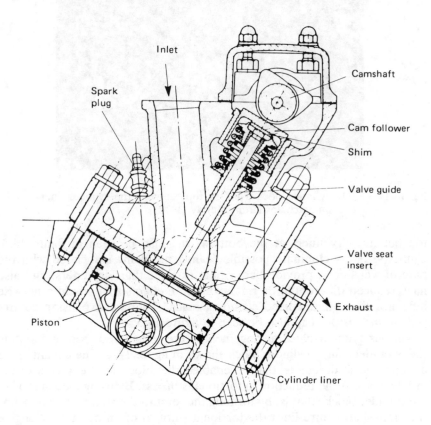

Figure 14.3   Jaguar V12 single overhead camshaft cylinder head (from Mundy (1972))

Figure 14.4   V12 cylinder block from the top, showing the cylinder liners and
the timing chains (from Mundy (1972))

top but like dry liners at the bottom, can also be seen in figure 14.3.
Cuff-type liners eliminate possible sealing problems associated with the
base of wet liners, provide better heat transfer than dry liners, and also
have reduced differential thermal expansion effects that would otherwise
influence the cylinder head clamping loads. Thermal expansion control
slots in the pistons can also be seen in figure 14.3.

Various cylinder block arrangements were considered, before adopting
the open-deck alloy cylinder block shown in figure 14.4. The advantage of
an open-deck design is that it enables the block to be die-cast; the
disadvantage is the reduction in torsional stiffness. By using a deep skirt to
the cylinder block (that is, by forming the joint to the sump 100 mm below
the crankshaft centre-line) the torsional stiffness of an open-deck engine
was greater than that of an alternative design with a closed deck. A cast
iron cylinder block was also tested; it weighed 50 kg more than an alloy

cylinder block and did not produce a detectably quieter engine. The crankshaft is made from a Tufftrided manganese–molybdenum steel.

The crescent-type oil pump is mounted directly on the crankshaft, in the space created at the front of the crankcase by the offset between the two banks of cylinders. With any fixed displacement pump driven directly by the engine, the pump delivery is too great at high speeds if the pump displacement is sufficient to provide adequate supply at low speeds. This characteristic is turned to advantage by using the flow from the relief valve to divert through an oil cooler. Thus at high engine speeds when oil cooling is most needed, the flow will be highest through the oil cooler. The oil-cooler circuit is shown in figure 14.5. The oil flow, directed by the relief valve, passes through the oil cooler and back to the pump inlet. This system minimises the flow from the oil pick-up, and also minimises the oil temperature in the pump. The oil is cooled by water that comes direct from the radiator, before entering the water pump. Figure 14.6 shows the oil flow as a function of engine speed. The pump delivery increases almost linearly with engine speed (thus showing almost no dependence on pressure), and the flow is unaffected by the oil viscosity. The oil flow to the engine also increases with engine speed, but with high oil temperatures the reduced viscosity increases the oil demand. The difference between these two flows is the relief flow that passes through the oil cooler.

14.2.3 Jaguar V12 ignition and mixture preparation development

During the initial development, twin six-cylinder ignition distributors were used. These were necessary to ensure adequate spark energy with the very high spark rates from a mechanically operated contact breaker. One distributor incorporated two sets of contact breakers, with the usual vacuum advance and centrifugal advance mechanisms. Even so, it was difficult to match the settings of the contact breakers, and this led to unacceptable variations in ignition timing, especially at higher speeds.

Fortunately, by the time the engine was introduced the Lucas Opus (oscillating pick-up) ignition system was available with a suitable twelve-cylinder distributor. In the Opus ignition system the distributor cam is replaced by a timing rotor with ferrite rod inserts, and the contact breaker is replaced by a pick-up module. An oscillating signal is sent to the primary winding in the pick-up, and a small voltage is induced in the secondary winding. When a ferrite rod passes the pick-up, the coupling between the two coils increases, and a much larger voltage is induced in the pick-up secondary winding. The signal from the secondary winding is used to switch transistors that control the current to the primary winding of the coil.

When the May combustion chamber was introduced, a higher-energy ignition system was needed, and the Lucas constant-energy ignition system

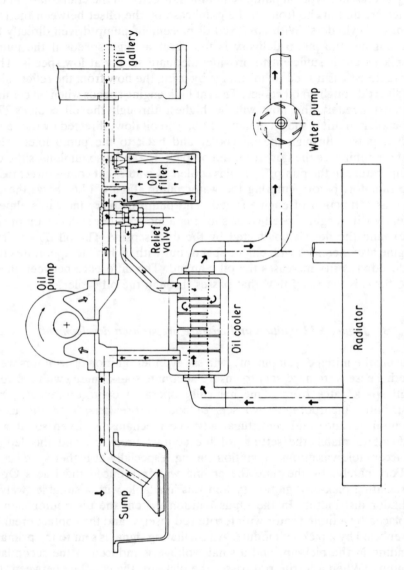

Figure 14.5 Diagrammatic layout of the oil cooler circuit (from Mundy (1972))

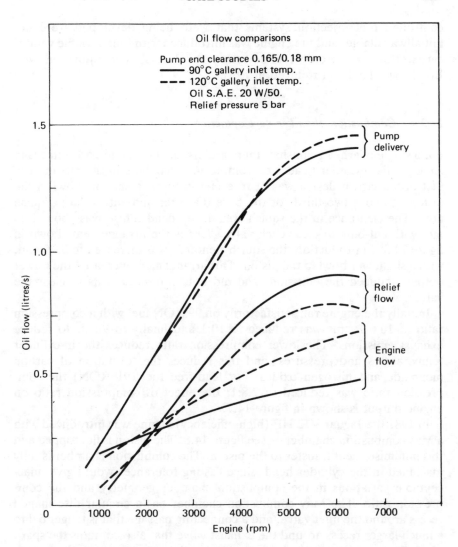

Figure 14.6   Oil pump flow characteristics with oil temperature changes (from Mundy (1972))

was adopted. This system has a variable-reluctance pick-up; a coil is wound around a magnet, and when the 'reluctor' (a toothed cam) changes the magnetic circuit a voltage is induced in the pick-up coil. The primary coil current of 8 amps is then controlled by transistor switching circuits.

Development work proceeded with both carburettor and fuel injection systems. Mechanically controlled fuel injection systems gave good maximum power and torque, but poor engine emissions. Electronically controlled fuel injection systems that were being developed were not initially available, and the engine was introduced with four variable venturi carburettors. Electronic fuel injection (by Lucas, in conjunction with Bosch) was finally introduced in 1975.

14.2.4 Combustion chamber development

For a single overhead camshaft there are distinct production advantages in having a flat cylinder head, and combustion chambers in the piston. The first piston crown design was of true Heron form: a shallow bowl in the piston of about two-thirds of the bore diameter and with a large squish area. The clearance in the squish area at top dead centre was about 1.25 mm with cut-outs to clear each valve. This is the arrangement shown in figure 14.3. In production, the squish clearance was increased to 3.75 mm, with a shallower bowl in the piston. This arrangement increased the power output, reduced the emissions and did not require cut-outs to clear the valves.

Initially the engine ran satisfactorily on 99 RON fuel with a compression ratio of 10.6:1; this was reduced to 10:1 and finally to 9.0:1, to reduce exhaust emissions. The lower compression ratio reduces the in-cylinder temperatures and pressures, and this reduces the formation of carbon monoxide and nitrogen oxides. For lead-free fuel (91 RON) the compression ratio was reduced to 7.8:1; the effect of compression ratio on engine output is shown in figure 14.7.

In 1981 the Jaguar V12 HE (high-efficiency) engine was introduced with a May combustion chamber — see figure 14.8. The piston is flat topped and this minimises heat transfer to the piston. The combustion chamber is fully machined in the cylinder head, since casting tolerances would give unacceptable variations in the combustion chamber geometry and the compression ratio. The May combustion chamber has a circular disc-shaped recess around the inlet valve, and a connecting passage that is tangential to a much larger recess around the exhaust valve that also contains the spark plug. The swirl generated by the tangential passage, and the squish generated by the flat-topped piston, ensure rapid and controlled combustion. Prior to ignition, the charge helps to cool the exhaust valve, and after ignition the very compact combustion chamber around the exhaust valve produces rapid combustion to reduce the risk of knock. Furthermore, the compact combustion chamber concentrates the charge so that combustion of a lean mixture can be self-sustaining. The result is a combustion chamber that permits the engine to use a 12.5:1 compression ratio with 97 RON fuel (11.5:1 with 91 RON fuel in the USA), and to burn mixtures with an air/fuel ratio as lean as 23:1. The cooler-running piston allows the top-land to be reduced, and this reduces the quench areas and unburnt

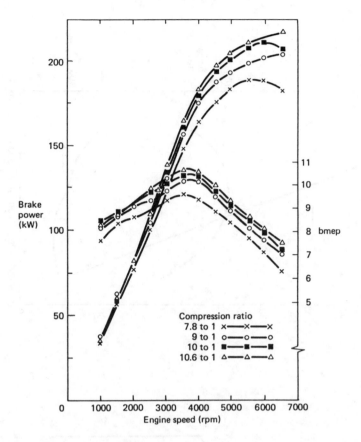

Figure 14.7   Effect of compression ratio on Jaguar V12 engine performance
(from Mundy (1972))

hydrocarbon emissions. Low CO and NO_x are also inherent features of the
V12 HE engine (Crisp (1984)).

The original objective was to produce an engine with comparable power
output; the success in achieving this target is shown by figure 14.9.
Furthermore, the maximum torque occurred at a lower speed; this permitted a change in the vehicle final drive ratio. The combination of more
efficient engine and higher gear ratio gave an overall improvement of over
20 per cent in the vehicle fuel economy.

14.2.5   V12 engine development

The most widely known development of the Jaguar V12 engine has been
undertaken by Tom Walkinshaw Racing, in connection with the Jaguar

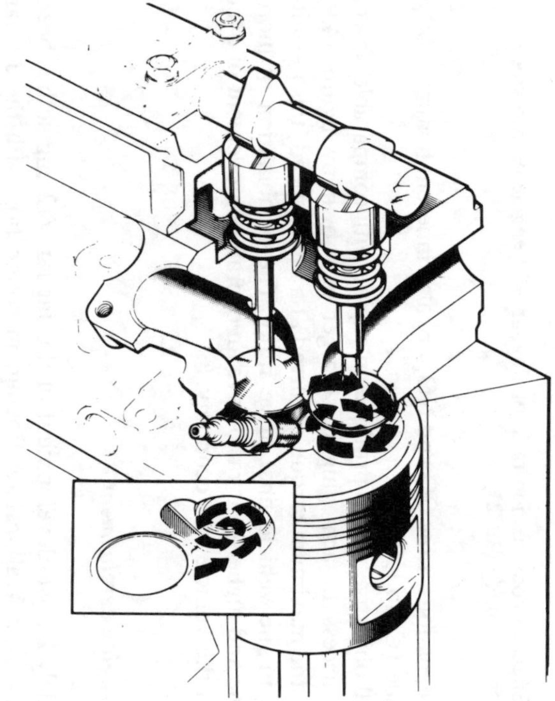

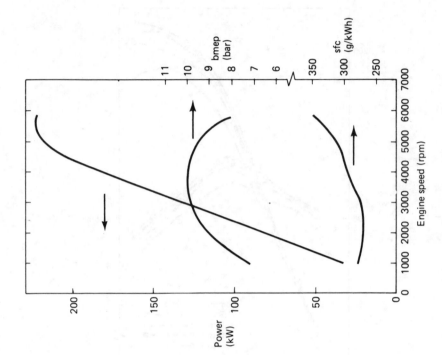

Figure 14.9 Performance of Jaguar V12 HE engine

Figure 14.8 Jaguar V12 HE combustion chamber. Insert: underside view of the swirl pattern (with acknowledgement to Newton et al. (1983))

wins at Le Mans (firstly in 1988). Racing engines have been developed according to both Group A regulations and Group C regulations. The pertinent regulations are:

Group A Production inlet manifolds, exhaust manifolds and valve diameters had to be preserved. Pistons and the compression ratio could be changed, as could the valve timing (but with the standard valve lift).

Group C Swept volume could be increased, no restrictions on the valve gear. A limit of 510 litres of fuel for a race distance of 1000 km.

The increase in the engine performance is summarised in table 14.2.

Table 14.2 Development of the Jaguar V12 racing engine, from Scott (1991)

Year	Group	Swept volume (litres)	Power (kW)	Speed (rpm)	bmep (bar)	bore (mm)	stroke (mm)
1982	Standard	5.3	205	6500	7.1	90	70
1982	A	5.3	291	—	—	90	70
1984	A	5.3	343	7500	10.4	90	70
1986	C	6.5	522	7500	12.8	94	78
1987	C	7.0	616	7500	14.1	—	—

The development of the Group A engine was achieved mostly by camshaft design and a more versatile engine management system which could be easily reprogrammed.

The Group C engine used a Zytek engine management system which had sequential fuel injection control, and a facility for reprogramming the management strategy from a PC. The air/fuel ratio could also be controlled by the driver during the race, and the engine could be operated at full load with an air/fuel ratio of 16:1 ($\phi = 0.9$). Normally such an engine would be operated with a rich mixture to reduce the thermal loading on the engine (see chapter 12, section 12.2.2 and figure 12.8), but the fuel limitations in Group C made it useful to be able to run with a weak mixture, to reduce the fuel consumption. At 7000 rpm the engine was capable of a specific fuel consumption of 274 g/kWh. The winning car in the 1988 Le Mans had just 15 litres left from the allocation of 2550 litres.

14.3 Chrysler 2.2 litre spark ignition engine

14.3.1 Background

The increasing emphasis on fuel economy and reduction of polluting emissions, and a trend towards compact, front wheel drive (fwd) cars all favour the adoption of a four-cylinder in-line spark ignition engine. This particular engine was specifically designed for fwd applications, and attention was paid to performance, fuel economy, durability, cost, emissions, engine weight and size, serviceability and manufacturing. The philosophy, development and manufacture of this engine are described by Weertman and Dean (1981) in SAE 810007.

For fwd applications with the engine mounted transversely, a four-cylinder configuration provides a compact package. The overall length of the engine can be reduced by siamesing the bores and having an under-square cylinder (the stroke being greater than the bore). An isometric cut-away view of the engine is shown in figure 14.10; the bore is 87.5 mm and the stroke is 92.0 mm, giving a swept volume of 2.213 litres.

The single overhead camshaft is driven by a toothed belt, and operates the in-line valves via rocker arms. The cylinder head is alloy, and the cylinder block is cast iron; the induction and exhaust manifolds are on the same side of the engine to provide a large clear space for accessories. The timing belt also drives an accessory shaft, which drives the distributor and oil pump. Since these have been removed from figure 14.10, they are shown separately in figure 14.11. The engine is used in a very wide range of Chrysler vehicles in the following forms: carburetted with normal and high compression ratios, electronic fuel injection with normal compression ratio, and as a turbocharged engine with multi-point electronic fuel injection.

14.3.2 The cylinder head

The cylinder head is cast from aluminium alloy, and the camshaft runs directly in the cylinder head without shell bearings. A detailed view of the cylinder head and combustion chamber is shown in figure 14.12.

The camshaft is made of a hardenable cast iron; the cam lobes are induction hardened and phosphate treated. The five camshaft journals are pressure lubricated, and 0.8 mm drilled holes also direct an oil jet on to the cam-lobe/rocker-arm contact area. The rocker arm is cast iron, and pivots on a hydraulic lash adjuster (hydraulic tappet). The tappet adjustment range is 4.5 mm, and the valve geometry is such that the valves can never hit the piston.

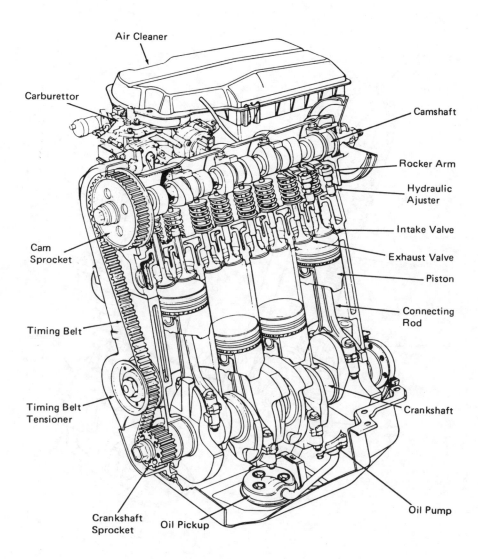

Air Cleaner

Carburettor

Camshaft

Rocker Arm

Hydraulic
Ajuster

Intake Valve

Cam
Sprocket

Exhaust Valve

Piston

Connecting
Rod

Timing Belt

Timing Belt
Tensioner

Crankshaft

Crankshaft
Sprocket

Oil Pickup

Oil Pump

Figure 14.10   Chrysler 2.2 litre four-cylinder engine – longitudinal section
(reprinted with permission, © 1981 Society of Automotive
Engineers, Inc.)

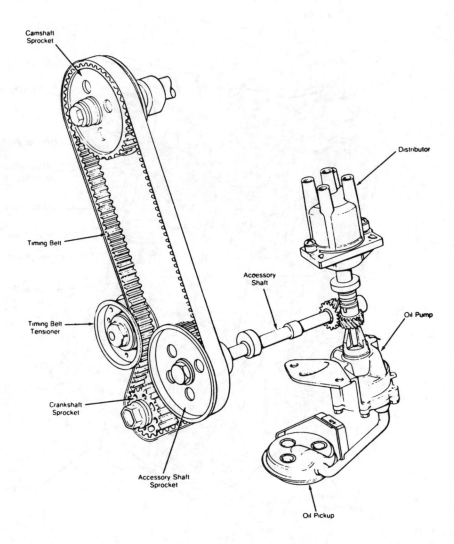

Figure 14.11   Chrysler 2.2 litre four-cylinder engine: timing belt drive –
accessory shaft – distributor and oil pump (reprinted with
permission, © 1981 Society of Automotive Engineers, Inc.)

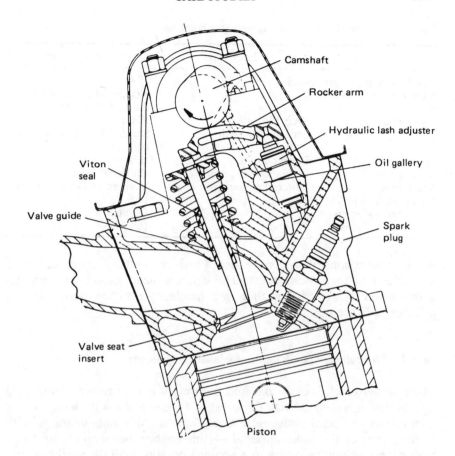

Figure 14.12 Chrysler 2.2 litre engine cylinder head (reprinted with permission, © 1981 Society of Automotive Engineers, Inc.)

The valves and associated components have also been designed for long life. The valve spring rests on a steel washer to prevent the spring embedding in the cylinder head. The Viton (a heat-resistant synthetic rubber) seals on the valve spindles prevent oil entering the combustion chamber through the valve guides; this could be a problem with the copious supply of lubricant to the overhead camshaft. Table 14.3 highlights some of the design and materials aspects of the valve gear.

The valve guides and seat inserts are all inserted after pre-cooling in liquid nitrogen.

The compact combustion chamber is as cast, and the piston is flat topped; there are two separate squish or quench areas. The spark plug is located very close to the centre of the combustion chamber.

To ensure predictable and reliable clamping of the cylinder head to the

Table 14.3   Chrysler 2.2 litre engine valve components

	Inlet valve	Exhaust valve
Head diameter (mm)	40.6	35.4
Stem diameter (mm)	8	8 (tapered to allow for thermal expansion)
Valve material	SAE 1541 steel with hardened tip	Head: 21–2N steel Stem: SAE 4140 steel with hardened tip
Stem finish	Flash chrome	Heavy chrome
Valve guide	Low alloy cast iron	Hardenable iron
Valve seat	Sintered copper-iron alloy	Sintered cobalt-iron alloy
Valve lift (mm)	10.9	10.9
Valve timing: open	12° btdc	48° bbdc
close	52° abdc	16° atdc

block, a 'joint control' system is used, in which carefully toleranced bolts are loaded to their yield point. The cylinder head gasket has minimal compression relaxation, and there are stainless steel bore flanges for the cylinder bore sealing.

14.3.3   The cylinder block and associated components

The cylinder block is cast iron, and the weight is kept to a minimum (35 kg), by having a nominal wall thickness of 4.5 mm and a skirt depth of only 3 mm below the crankshaft centre-line. There are five main bearings, with bearing caps of the same material — this enables bore circularity to be maintained without resorting to a honing operation. All the shell bearings are either a lead-aluminium alloy or tri-metal on a steel backing strip.

The crankshaft is of nodular cast iron, with main bearing journal diameters of 60 mm, and crankpin or big-end journal diameters of 50 mm. All the crankshaft journals have under-cut radiused fillets; these are deep rolled to give a fatigue strength improvement of 35 per cent.

The connecting-rod and cap are made from forged steel, with a centre distance of 151 mm. A squirt hole in the connecting-rod provides cylinder lubrication from the big-end.

The piston design is the result of finite element modelling and high-speed engine tests. The piston is made from aluminium alloy, and has cast-in-steel struts at the gudgeon pin bosses to control the thermal expansion. The top ring is made from nodular cast iron with a molybdenum-filled, radiused face for wear and scuff resistance. In contrast the second ring has a phosphate-coated, tapered face; both rings are 1.5 mm thick. The oil control ring is of a three-piece construction with a stainless steel expander, and chromium-faced side rails.

14.3.4 Combustion control

Combined electronic control is used for the ignition timing and the air/fuel mixture preparation. Signals from seven transducers are processed electronically:

 (i) ambient air temperature
 (ii) engine load
(iii) carburettor throttle plate, open or closed
 (iv) engine speed
 (v) engine coolant temperature
 (vi) exhaust gas oxygen level
(vii) engine starting.

The air–fuel mixture is prepared in a twin choke carburettor with progressive choke operation. The carburettor has an electric choke, and an electronic feedback system for mixture strength control. The electronic engine-management system ensures optimum fuel-economy performance, and driveability, while still meeting stringent exhaust emission regulations.

The induction and exhaust passages were optimised by air flow testing with plastic models. The compact combustion chamber, in an aluminium alloy cylinder head, with significant squish or quench areas permits a compression ratio of 8.9:1 to be used with 91 RON fuel; the final performance is shown in figure 14.13. The engines in current production have a

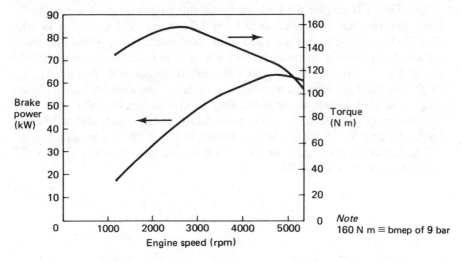

Figure 14.13   Chrysler 2.2 litre engine output (reprinted with permission, © 1981 Society of Automotive Engineers, Inc.)

compression ratio of 9.6:1, or 9:1 if naturally aspirated, while the turbo-charged engine has the compression ratio reduced to 8.1:1.

In all versions of the engine, emission control is by a combination of exhaust gas recirculation (EGR), air injection (exhaust manifold outlet cold; catalytic converter hot), and a catalytic converter below the exhaust manifold. The catalytic converter uses a three-way catalyst on a monolithic substrate.

14.3.5 Engine development

The output of the Chrysler 2.2 litre engine has been increased by turbo-charging. The performances of the three different turbocharged derivatives are compared in table 14.4 with the baseline engine performance.

Table 14.4 Performance of the Chrysler 2.2 litre engine and its derivatives

Engine description	Max. torque: speed		Max. power: speed	
	(N m)	(rpm)	(kW)	(rpm)
Standard, EDE	161	3200	72	5200
Turbo I	217	3200	106	5600
Turbo II	256	3200	130	4800
Lotus/Shelby	305	3200	169	6000

The Turbo II engine uses an air to air inter-cooler, and the maximum boost pressure has been increased from 0.7 bar to 0.83 bar (Anon 1986)). The Turbo II engine uses a forged steel crankshaft, and a tuned induction system that raises the torque curve by about 10 per cent (this is particularly useful for the low speeds at which the turbocharger performance is limited). The primary runners from each cylinder to the manifold are 370 mm long, and give a torque improvement in the range of 4800 rpm. The secondary runner (from the inlet manifold to the throttle body) is 300 mm long, and provides a torque improvement in the 2000 rpm range.

The Lotus/Shelby engine employs 4 valves per cylinder and a Garrett T3 turbocharger (Birch (1988)).

14.4 Ford 2.5 litre DI Diesel engine

14.4.1 Background

Compression ignition engines with direct fuel injection (DI) have typically a 10–15 per cent fuel economy benefit over indirect engines and are easier to start. However, indirect engines are normally used in light automotive applications, because the restricted speed range of direct injection engines (say up to 3000 rpm) leads to a poor power-to-weight ratio, and necessitates a multi-ratio gearbox. Indirect injection engines achieve their greater speed range by injecting fuel into a pre-chamber where there is rapid air motion. The air motion is generated in a throat that connects the pre-chamber and the main chamber; but this also leads to pressure drops and the high heat transfer coefficients that account for the reduced efficiency.

The Ford high-speed direct injection engine achieves a high engine speed (4000 rpm) by meticulous attention to the fuel injection and air motion. This four-cylinder engine is naturally aspirated, slightly over-square, with a bore of 93.7 mm and a stroke of 90.5 mm. Since there is no need to pump air into and out of the pre-chamber the compression ratio can be lower, while still maintaining good starting performance. Consequently, the compression ratio is 19:1 while the compression ratio for an indirect injection engine is typically 22:1.

14.4.2 Description

An isometric sectioned drawing of the engine is shown in figure 14.14. A single toothed belt drives the camshaft and fuel injection pump, while poly-vee belt drives are used for the other engine auxiliaries. The inlet and exhaust valves are all in-line, but are offset slightly from the cylinder bore axes; the induction and exhaust manifolds are on opposite sides of the engine to provide a crossflow system. The engine is inclined at an angle of $22\frac{1}{2}°$ to the vertical, and this is shown more clearly in figure 14.15, a sectioned view of the engine.

The camshaft is mounted at the side of the cylinder block and the cam followers operate the valves through short push rods and rocker arms. The valve clearance is adjusted by spherically ended screws in the rocker arms, which engage with the push rods. To ensure valve train rigidity, the rocker shaft is carried by five pedestal bearings. The single valve spring rests on a hardened steel washer (to reduce wear), and the spring retainer is connected to the valve stem by a multi-groove collet. The valves operate in valve guides inserted into the cast iron cylinder head; the valve seats are hardened by induction heating after machining. The valve stem seals

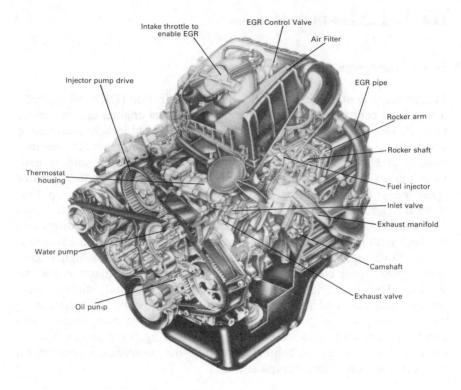

Figure 14.14   Ford 2.5 litre HSDI low emission engine (courtesy of Ford)

restrict the flow of lubricating oil down the valve stem, in order to reduce the build-up of carbon deposits.

The cylinder head for a direct injection engine is simpler than a comparable indirect injection design since there is no pre-chamber or heater plug. The cast iron cylinder head contains a dual-acting thermostat, which restricts water circulation to within the engine and heater until a temperature of about 82°C is attained. The water flow to the cylinder head from the cylinder block is controlled by graduated holes in the cylinder head gasket. To ensure optimum clamping of the cylinder head to the block, a torque-to-yield system is used on the cylinder head bolts.

The cylinder bores are an integral part of the cast iron cylinder block. A finite element vibration analysis was used in the design of the cylinder block to help minimise the engine noise levels. The main and big-end bearing shells use an aluminium-tin alloy. The crankshaft is made from nodular cast iron, with induction hardened journals to reduce wear. All the crankshaft journals are under-cut and fillet-rolled to improve the fatigue

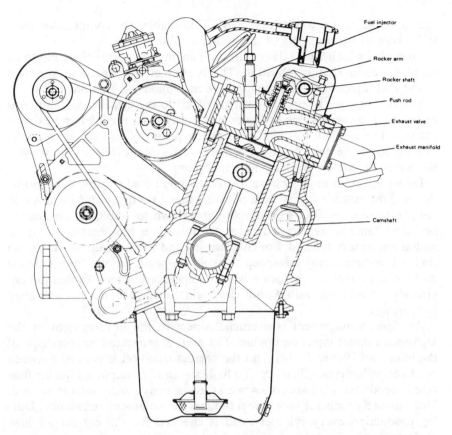

Figure 14.15   Cross-section of the Ford 2.5 litre DI Diesel (courtesy of Ford)

life. The connecting-rods are made from an air hardening steel that is relatively easy to machine, despite strengths of up to 930 MN/m^2.

The pistons are made from an aluminium alloy and are expansion-controlled; this enables the bore clearance to be reduced to a minimum of 8 μm, thereby minimising piston slap and its ensuing noise. This close tolerance is obtained by selective assembly during manufacture with four grades of piston skirt diameter. The piston carries an oil control ring and two compression rings; the top ring is located in a cast iron insert to achieve long life.

The piston clearance at top dead centre is carefully controlled to ensure the correct power output and smoke conformity at the higher speeds. This clearance is controlled by selective assembly of the pistons and connecting-rods to match a given crankshaft and block. Four grades of connecting-rod (by length) are matched with five grades of piston height (gudgeon pin centre to piston crown). This enables the compression ratio to be kept to within ± 0.7 of the nominal value.

To enable a high-speed direct injection engine to operate there has to be rapid, yet controlled, combustion. This is attained by meticulous attention to the fuel injection equipment and the in-cylinder air motion.

The fuel injector is inclined at angle of 23° to the cylinder axis, and it is offset from the centre-line of the engine (away from the inlet and exhaust valves). The shallow toroidal bowl in the piston is also offset from the cylinder axis. The peak fuel injection pressure is about 700 bar; this is twice the normal figure for a rotary type fuel pump.

Direct injection engines have inherently good cold starting characteristics, and the automatic provision of excess fuel gives quick starting down to temperatures of −10°C. For temperatures down to −20°C an electrically operated flame heater is used. Also, for this lower temperature range an additional battery is fitted. For cold starting and warming-up, the injection timing is automatically advanced to avoid the white smoke from unburnt fuel. Independently, a separate water-temperature sensor causes the engine idle speed to be raised until the engine reaches its normal operating temperature.

Air flow management is a crucial aspect in the development of the high-speed direct injection engine. The swirl is generated by the shape of the inlet port (figure 14.16), and the correct trade-off is needed between swirl and volumetric efficiency. To limit the smoke output a high air flow rate is needed at high speeds, while at low speeds a high swirl is needed. Unfortunately inducing more swirl reduces the volumetric efficiency. During production every cylinder head is checked for the correct air flow performance.

The aluminium induction manifold is the result of a CAD study, and the manufacture is tightly controlled to ensure accurate alignment with the ports in the cylinder head. Any misalignment between the passages in the inlet manifold and the cylinder head has a disproportionately serious effect on the flow efficiency. The consequences would be particularly serious on a direct injection engine. To improve the air flow at the valve throat, the valve seat is angled at 30°, and the valve throat is angled at 45°. During development a decrease in valve diameter was found to improve the air flow — this is explained by the reduced interference between the inlet air flow and the cylinder wall.

A turbocharged version of the engine has been produced using a Garrett T2 turbocharger. Table 14.5 compares the performance of the naturally aspirated and turbocharged engines. The higher torque at a lower speed improves the driveability.

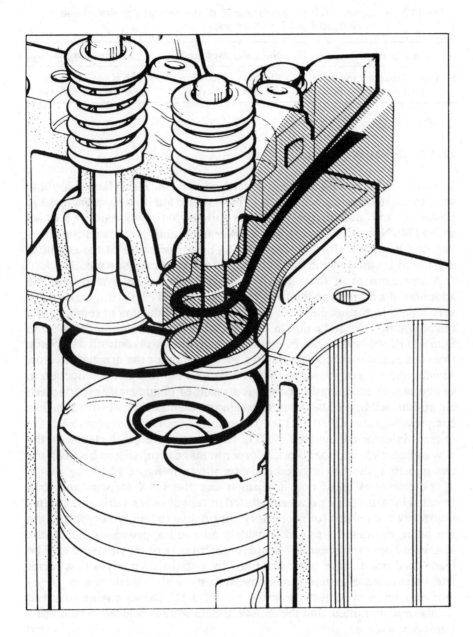

Figure 14.16   Swirl generation in the Ford 2.5 litre DI Diesel (courtesy of Ford)

Table 14.5   Comparison of the performance of the naturally aspirated and
turbocharged Ford 2.5 DI engines

Engine	Naturally aspirated	Turbocharged
Maximum torque; speed	146 N m; 2500 rpm	173 N m; 2300 rpm
Maximum power; speed	52 kW; 4000 rpm	60 kW; 3900 rpm

14.4.3   Combustion system

The combustion system has been developed since the initial launch in 1984, so as to meet emissions legislation. This has also led to: a slight increase in power (52 kW at 4000 rpm), a higher torque that now occurs at a lower speed (146 N m at 2500 rpm), and a lower minimum specific fuel consumption of 220 g/kWh (Bird *et al.* (1989)). The performance of the engine is illustrated by the specific fuel consumption map shown in figure 14.17.

A key element in the refinement of the combustion system was the adoption of a Stanadyne Slim Tip Pencil Nozzle (STPN). This nozzle has a very long injector needle, with the spring and its adjustment remote from the nozzle tip. The reduction in the injector body diameter means that the Slim Tip Pencil Nozzle can be mounted so that its axis is only offset 5.5 mm from the cylinder axis; this compares with 9.5 mm for the original injector. Moving the injector closer to the cylinder axis led to a direct reduction in smoke at a given operating point, as a result of better fuel/air mixing and air utilisation. In parallel, a new design of re-entrant bowl with a higher compression ratio (20.7) had led to performance improvements with the original injector (designated 17/21). Thus tests were conducted with the new combustion chamber bowl, in order to make comparisons between the two injectors; the two engine builds are shown in figure 14.18.

The emissions legislation in Europe and the USA for cars and light trucks is in terms of engines being tested in vehicles over a drive cycle. This is not very convenient (or necessarily informative) when prototype engines are being evaluated. Instead, a widely adopted approach is to identify load/speed operating points that are representative of the drive cycles. The load/speed points have to be identified by a statistical analysis that correlates steady-state engine/dynamometer tests with vehicle/chassis dynamometer drive cycle tests. For the Ford 2.5 DI Diesel engine four test points were identified, and the emissions data were expressed in terms of a four-point average.

The emissions results for the two engine builds (figure 14.18) are shown in figure 14.19. Each build was tested with the standard injection timing, and with 2° advance and 2° retard. The slim tip injector gave lower gaseous

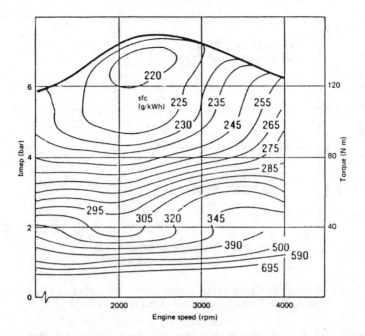

Figure 14.17 Specific fuel consumption contours (g/kWh) for the Ford 2.5 litre DI Diesel engine (adapted from Brandstetter and Howard (1989))

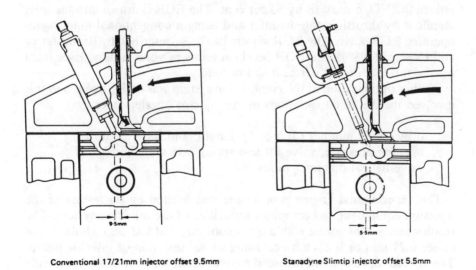

Conventional 17/21mm injector offset 9.5mm   Stanadyne Slimtip injector offset 5.5mm

Figure 14.18 Combustion system development for the Ford 2.5 litre DI Diesel engine (Bird *et al.* (1989))

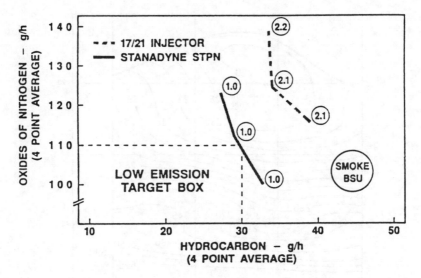

Figure 14.19   Comparison of gaseous emissions of conventional 17/21 mm and
Stanadyne Slimkip injectors (Bird *et al.* (1989))

emissions, and a smoke level that was always more than 1 Bosch Smoke
Unit lower. The low emission target box in figure 14.19 that represents the
emissions legislation requirements, assumed that EGR could be applied to
reduce the NO_x emissions by 50 per cent. The EGR is drawn into the inlet
manifold by throttling the air inlet and using a conventional diaphragm-
operated EGR valve. The EGR system has been described by Brandstetter
and Howard (1989). The EGR level (in terms of volume flow) rises from
zero at full load, to 50 per cent at low load.

Further optimisation of the combustion system was undertaken, and this
involved the use of 60 variations in the injector nozzle, covering:

> orifice size, number (4, 5 or 6), length, and spray pattern
> opening pressure, valve lift and spring rate sac volumes
> valve to seat diameter ratios

The experimental engine programme was backed by rig testing of the
injection equipment and computer modelling of the injection system. The
result was a 5-hole nozzle with a spray cone angle of 150° and a hole size of
either 0.21 mm or 0.22 mm according to the selection of injector pump.
The injector development included a novel extension to the injector needle
that further reduced the sac volume in the injector tip. This reduction in
sac volume reduced the unburnt hydrocarbon emissions.

Other refinements concerned the piston and ring pack. An AEcono-
guide low-friction piston was adopted, and the piston bowl was moved

towards the cylinder axis — this led to a higher combustion efficiency and a potential for reduced noise with retarded injection. The oil control ring has been improved and the ensuing lower lubricant consumption also leads to lower emissions of particulates and unburned hydrocarbons. The piston ring pack has also been moved up, so the top ring is now 7 mm away from the piston crown. As with spark ignition engines, reductions in crevice volumes led to lower hydrocarbon emissions. The response of this engine to changes in the fuel cetane rating and changes in fuel volatility have been reported by Cooke *et al.* (1990). They conclude that there is limited potential for reducing exhaust emissions with conventional diesel fuel blending components.

Appendix A: The Use of SI Units

SI (Système International) Units are widely used, and adopt prefixes in multiple powers of one-thousand to establish the size ranges. Using the watt (W) as an example of a base unit:

picowatt	(pW)	10^{-12}	W
nanowatt	(nW)	10^{-9}	W
microwatt	(μW)	10^{-6}	W
milliwatt	(mW)	10^{-3}	W
watt	(W)	1	W
kilowatt	(kW)	10^{3}	W
megawatt	(MW)	10^{6}	W
gigawatt	(GW)	10^{9}	W
terawatt	(TW)	10^{12}	W

It is unusual for any single unit to have such a size range, nor are the prefixes nano (10^{-9}) and giga (10^{9}) very commonly used.

An exception to the prefix rule is the base unit for mass — the kilogram. Quantities of 1000 kg and over commonly use the tonne (t) as the base unit (1 tonne (t) = 1000 kg).

Sometimes a size range using the preferred prefixes is inconvenient. A notable example is volume; here there is a difference of 10^{9} between mm^{3} and m^{3}. Consequently it is very convenient to make use of additional metric units:

$$1 \text{ cm} = 10^{-2} \text{ m}$$

thus

$$1 \text{ cm}^{3} = 10^{3} \text{ mm}^{3} = 10^{-6} \text{ m}^{3}$$

$$1 \text{ litre (1)} = 1000 \text{ cm}^{3} = 10^{6} \text{ mm}^{3} = 10^{-3} \text{ m}^{3}$$

Pressure in SI units is the unit of force per unit area (N/m^{2}), and this is sometimes denoted by the Pascal (Pa). A widely used unit is the bar (1 bar = 10^{5} N/m^{2}), since this is nearly equal to the standard atmosphere:

$$1 \text{ standard atmosphere (atm)} = 1.01325 \text{ bar}$$

A unit commonly used for low pressures is the torr:

$$1 \text{ torr} = \frac{1}{760} \text{ atm}$$

In an earlier metric system (cgs), 1 torr = 1 mm Hg.

The unit for thermodynamic temperature (T) is the kelvin with the symbol K (*not* °K). Through long established habit a truncated thermodynamic temperature is used, called the Celsius temperature (t). This is defined by

$$t = (T - 273.15) \text{ °C}$$

Note that (strictly) temperature differences should always be expressed in terms of kelvins.

Some additional metric (non-SI) units include:

Length 1 micron = 10^{-6} m
1 angstrom (Å) = 10^{-10} m
Force 1 dyne (dyn) = 10^{-5} N
Energy 1 erg = 10^{-7} N m = 10^{-7} J
1 calorie (cal) = 4.1868 J
Dynamic viscosity 1 poise (P) = 1 g/cm s = 0.1 N s/m^2
Kinematic viscosity 1 stokes (St) = 1 cm^2/s = 10^{-4} m^2/s

A very thorough and complete set of definitions for SI Units, with conversions to other unit systems, is given by Haywood (1972).

Conversion factors for non-SI units

Exact definitions of some basic units:

Length   1 yard (yd) = 0.9144 m
Mass   1 pound (lb) = 0.453 592 37 kg

$$\text{Force} \quad 1 \text{ pound force (lbf)} = \frac{9.806\ 65}{0.3048} \text{ pdl}$$

(1 poundal (pdl) = 1 lb ft/s^2)

Most of the following conversions are approximations:

Length   1 inch (in) = 25.4 mm
1 foot (ft) = 0.3048 m
1 mile (mile) ≈ 1.61 km
Area   1 square inch (sq. in) = 645.16 mm^2
1 square foot (sq. ft) ≈ 0.0929 m^2
Volume   1 cubic inch (cu. in) ≈ 16.39 cm^3
1 gallon (gal) ≈ 4.546 1
1 US gallon ≈ 3.785 1
Mass   1 ounce (oz) ≈ 28.35 g
1 pound (lb) ≈ 0.4536 kg
1 ton (ton) ≈ 1016 kg
1 US short ton ≈ 907 kg

Density   1 lb/ft^3 $\approx$ 16.02 kg/m^3

Force   1 pound force (lbf) $\approx$ 4.45 N

Pressure   1 lbf/in^2 $\approx$ 6.895 kN/m^2

1 in Hg $\approx$ 3.39 kN/m^2

1 in H$_2$O $\approx$ 0.249 kN/m^2

Dynamic viscosity   1 lb/ft s $\approx$ 1.488 kg/m s

N s/m^2

Kinematic viscosity   1 ft^2/s $\approx$ 0.0929 m^2/s

Energy   1 ft lbf $\approx$ 1.356 J

Power   1 horse power (hp) $\approx$ 745.7 W

Specific fuel consumption   1 lb/hp h $\approx$ 0.608 kg/kW

$\approx$ 0.169 kg/MJ

Torque   1 ft lbf $\approx$ 1.356 N m

Energy   1 therm (= 10^5 Btu) $\approx$ 105.5 MJ

Temperature   1 rankine (R) $= \dfrac{1}{1.8}$ K

$$\begin{cases} t_F = (T_R - 459.67)°F \\ \text{thus } t_F + 40 = 1.8\,(t_C + 40) \end{cases}$$

Specific heat capacity ⎫
Specific entropy ⎭   1 Btu/lb R = 4.1868 kJ/kg K

Specific energy   1 Btu/lb = 2.326 kJ/kg

Appendix B: Answers to Numerical Problems

2.1 0.678, 0.623, 0.543, 0.405
2.3 6.0 bar, 33.5 per cent, 22.3:1
2.4 6.9 bar, 20.7 per cent, 73 per cent
2.5 (a) 74.1 kW, 173.6 N m: (b) 9.89 bar, 7.27 bar, 10.99 bar, 8.08 bar; (c) 25.25 per cent, 58.3 per cent; (d) 90.7 per cent, 12.0:1

3.2 C:0.855, H:0.145; 13.50:1
3.3 C:0.848, H:0.152; 0.833; 15:1
3.4 Insufficient data are given to answer the question accurately, so clearly stated assumptions (and their significance) are more important than the numerical values.
702 K. 21.3 bar; 4500 K, 137 bar
3.5 (i) CO:2.15, O_2:1.08, CO_2:6.77 bar; (ii) CO:4.12, O_2:2.06, CO_2:3.81 bar
3.6 51.1
3.9 (i) 8.93 per cent CO_2, 13.85 per cent H_2O, 3.68 per cent O_2, 73.53 per cent N_2; (ii) 29.2 per cent; (iii) 1.23 kg CH_4:1 kg $C_{16}H_{34}$
3.10 (i) 2.63 per cent H_2; (ii) CO — 16.9 per cent, H_2 — 9.5 per cent
3.11 Assume the fuel is a hydrocarbon with only CO_2, O_2, H_2O and N_2 in the exhaust. (i) 21.75; (ii) 1.805 H:1 C, 1 kg H to 6.65 kg C; (iii) 14.51; (iv) 0.667
3.12 (i) 5.9 MJ/kg fuel, 31.7 per cent; (ii) 38.9 kW
3.13 (i) 13.12; (ii) 85.2 per cent C, 14.8 per cent H; (iii) 14.83:1
3.14 $k = 2166$

4.7 (i) 4.51 bar; (ii) 28.1 per cent (iii) 8.62 litres/100 km; (iv) 61.3 per cent
4.8 (i) 45 K; (ii) 13.9 K; (iii) CO_2 — 11.94 per cent O_2 — 4.34 per cent, N_2 — 83.7 per cent; 0.805
4.9 (i) 0.594; (ii) 0.32; (iii) 11.2 bar

5.6 (i) $\alpha = 1 + 50/F$; (ii) 0.543, 0.473
5.7 (i) bsp $= \bar{p}_b \times \bar{v}_p/4$; (ii) AFR $= (\rho \times \eta_{vol})/(\bar{p}_b \times$ bsfc); (iii) bsac $=$ bsfc AFR
5.8 (i) 56.5 kW, 19.3; (ii) 6.78, 7.3 bar; (iii) 0.341

6.1 29.4 per cent

9.1 54°C, 671°C, 92 per cent

9.4
$$\left[\left(\frac{p_2}{p_1}\right)^{(\gamma_a-1)/\gamma_a} - 1\right] = \left[1 - \left(\frac{p_4}{p_3}\right)^{(\gamma_e-1)/\gamma_e}\right] \eta_{mech}\,\eta_c\,\eta_t$$

$$\left(1 + \frac{1}{AFR}\right)\left(\frac{c_{p_e}}{c_{p_a}}\right)\left(\frac{T_3}{T_1}\right)$$

9.8 (a) 0.932, 8.63 kW; (b) 0.51 kW
9.9 2.2, 80 000 rpm, 414 K, 6.1 kW
9.10 (a) 628 kW; (b) 0.404; (c) 0.57; (d) 0.78

Bibliography

The most prolific source of published material on internal combustion engines is the Society of Automotive Engineers (SAE) of America. Some of the individual papers are selected for inclusion in the annual *SAE Transactions*. Other SAE publications include the *Progress in Technology* (PT) and *Specialist Publications* (SP), in which appropriate papers are grouped together. Examples are

SP-532 Aspects of Internal Combustion Engine Design
PT-24 Passenger Car Diesels

The SAE also organise a wide range of meetings and conferences, and publish the magazine *Automotive Engineering*.

In the United Kingdom the Institution of Mechanical Engineers (I. Mech. E.) publish *Proceedings* and hold conferences, some of which relate to internal combustion engines. The Automobile Division also publishes the bi-monthly *Automotive Engineer*.

The other main organisers of European conferences include:

CIMAC Conseil International des Machines à Combustion
FISITA Fédération International des Sociétés d'Ingénieur et de Techniciens de l'Automobile
IAVD International Association for Vehicle Design
ISATA International Symposium on Automotive Technology and Automation

Many books are published on internal combustion engines, and this can be seen in the list of references. However, since books can become dated, care and discretion are necessary in the use of old material.

The following books are useful texts:

Rowland S. Benson, *The Thermodynamics and Gas Dynamics of Internal-Combustion Engines*, Vol. I (eds J. H. Horlock and D. E. Winterbone), Clarendon, Oxford, 1982.
Rowland S. Benson, *The Thermodynamics and Gas Dynamics of Internal-Combustion Engines*, Vol. II (eds J. H. Horlock and D. E. Winterbone), Clarendon, Oxford, 1986.
Rowland S. Benson and N. D. Whitehouse, *Internal Combustion Engines*, Vol. I, Pergamon, Oxford, 1979.

Rowland S. Benson and N. D. Whitehouse, *Internal Combustion Engines*, Vol. II, Pergamon, Oxford, 1979.

G. P. Blair, *The Basic Design of Two-Stroke Engines*, SAE, Warrendale, Pennsylvania, USA, 1989.

Colin R. Ferguson, *Internal Combustion Engines — Applied Thermosciences*, Wiley, Chichester, 1986.

E. M. Goodger, *Hydrocarbon Fuels — Production, Properties and Performance of Liquids and Gases*, Macmillan, London, 1975.

John B. Heywood, *Internal Combustion Engine Fundamentals*, McGraw-Hill, Maidenhead, 1988.

L. R. C. Lilly (ed.), *Diesel Engine Reference Book*, Butterworth, London, 1984.

Charles Fayette Taylor, *The Internal-Combustion Engine in Theory and Practice, Vol. I: Thermodynamics, Fluid Flow, Performance*, 2nd edn (revised), MIT Press, Cambridge, Massachusetts, 1985.

Charles Fayette Taylor, *The Internal-Combustion Engine in Theory and Practice, Vol. II: Combustion, Fuels, Materials, Design*, Revised Edition, MIT Press, Cambridge, Massachusetts, 1985.

N. Watson and M. S. Janota, *Turbocharging the Internal Combustion Engine*, Macmillan, London, 1982.

J. H. Weaving (ed.), *Internal Combustion Engineering*, Elsevier Applied Science, London and New York, 1990.

Finally, *Engines — the search for power* by John Day (published by Hamlyn, London, 1980) is a copiously illustrated book describing the development of all types of engine.

References

Adams L. F. (1975). *Engineering Measurements and Instrumentation*, EUP, London

Ahmad T. and Theobald M. A. (1989). 'A survey of variable valve actuation technology', *SAE Paper* 891674

Alcock J. F., Robson F. V. B. and Mash C. (1958). 'Distribution of heat flow in high duty internal combustion engines', *CIMAC*, pp. 723–49

Allard A. (1982). *Turbocharging and Supercharging*, Patrick Stephens, Cambridge

Annand W. J. D. (1963). 'Heat transfer in the cylinders of reciprocating internal combustion engines', *Proc. I. Mech. E.*, Vol. 177, No. 36, pp. 973–90

Annand W. J. D. (1986). 'Heat transfer in the cylinder and porting', in Horlock J. H. and Winterbone D. E. (eds), *The Thermodynamics and Gas Dynamics of Internal Combustion Engines*, Vol. II, Oxford University Press (*see also* Benson (1982))

Annand W. J. D. and Ma T. (1971). 'Instantaneous heat transfer rates to the cylinder head surface of a small compression ignition engine', *Proc. I. Mech. E.*, Vol. 185, No. 72, pp. 976–87

Annand W. J. D. and Roe G. E. (1974). *Gas Flow in the Internal Combustion Engine*, Foulis, Yeovil

Annual Book of ASTM Standards, Part 47 — Test Methods for Rating Motor, Diesel and Aviation Fuels, ASTM, Philadelphia, Pennsylvania

Anon (1979). 'Programs for digital signal processing', IEEE/Wiley, Chichester

Anon (1984a). 'Catalytic exhaust-purification for Europe?', *Automotive Engineer*, Vol. 9, No. 1

Anon (1984b). 'Variable inlet valve timing aids fuel economy', *CME*, Vol. 31, No. 4, p. 19

Anon (1986). 'Chrysler, Turbo II', *Automotive Engineering*, Vol. 94, No. 10, pp. 56–7

Anon (1990). 'Isuzu reveals 'bolt-on' compact engine, *Vehicle Engineering and Design*, p. 3, Design Council, London

Arai M. and Miyashita S. (1990). 'Particulate regeneration improvement on actual vehicle under various conditions', Paper C394/012, *Automotive Power Systems — Environment and Conservation*, I. Mech. E. Conf. Proc., MEP, London

Asmus T. W. (1984). 'Effects of valve events on engine operation', in Hilliard J. C. and Springer G. S. (eds), *Fuel Economy in Road Vehicles Powered by Spark Ignition Engines*, Plenum, New York

Assanis D. N. and Heywood J. B. (1986). 'Development and use of computer simulation of the turbocompounded diesel system for engine performance and components heat transfer studies', *SAE Paper* 860329

Baker A. (1979). *The Component Contribution*, Hutchinson Benham, London

557

Barnes-Moss H. W. (1975). 'A designer's viewpoint', in *Passenger Car Engines*, I. Mech. E. Conf. Proc., pp. 133–47, MEP, London

Baruah P. C. (1986). 'Combustion and cycle calculations in spark ignition engines', in Horlock J. H. and Winterbone D. E. (eds), *The Thermodynamics and Gas Dynamics of Internal Combustion Engines*, Vol. II, Oxford University Press (*see also* Benson (1982))

Beard C. A. (1958). 'Some aspects of valve gear design', *Proc. I. Mech. E.*, Vol. 2, pp. 49–62

Beard C. A. (1984). 'Inlet and exhaust systems' in Lilly L. R. C. (ed.), *The Diesel Engine Reference Book*, Butterworth, London

Beckwith P., Denham M. J., Lang G. J. and Palmer F. H. (1986). 'The effects of hydrocarbon in methanol automotive fuels', *7th International Symposium on Alcohol Fuels, Paris*

Belardini P., Bertoli C., Corcoine F. and Police G. (1983). 'Ignition delay measurement in a direct injection diesel engine', Paper C86/83, *Int. Conf. on Combustion in Engineering*, Vol. II, I. Mech. E. Conf. Proc., pp. 1–8, MEP, London

Benjamin S. F. (1988). 'The development of the GTL 'barrel swirl' combustion system with application to four-valve spark ignition engines', Paper C54/88, *Int. Conf. Combustion in Engines — Technology and Applications*, I. Mech. E. Conf. Proc., MEP, London

Benson R. S. (1960). 'Experiments on a Piston controlled port', *The Engineer*, Vol. 210, pp. 875–80

Benson R. S. (1977). 'A new dynamic model for the gas exchange process in two-stroke loop and cross scavenged engines', *Int. J. Mech. Sci.*, Vol. 19, pp. 693–711

Benson R. S. (1982). *The Thermodynamics and Gas Dynamics of Internal Combustion Engines*, Vol. I (eds Horlock J. H. and Winterbone D. E.), OUP, p. 9 (*see also* Horlock and Winterbone (1986))

Benson R. S. and Brandham P. T. (1969). 'A method for obtaining a quantitative assessment of the influence of charging efficiency on two-stroke engine performance', *Int. J. Mech. Sci.*, Vol. 11, pp. 303–12

Benson R. S. and Whitehouse N. D. (1979). *Internal Combustion Engines*, vols 1 and 2, Pergamon, Oxford

Benson R. S., Garg R. D. and Woollatt D. (1964). 'A numerical solution of unsteady flow problems', *Int. J. Mech. Sci.*, Vol. 6, pp. 117–44

Beretta G. P., Rashidi M. and Keck J. C. (1983). 'Turbulent flame propagation and combustion in spark ignition engines', *Combustion and Flame*, Vol. 52, pp. 217–45

Biddulph T. W. and Lyn W. T. (1966). 'Unaided starting of diesel engines', *Proc. I. Mech. E.*, Vol. 181, Part 2A.

Birch S. (1988). 'Lotus/Shelby', *Automotive Engineering*, Vol. 96, No. 2, p. 172

Bird G. L., Duffy K. A. and Tolan L. E. (1989). 'Development and application of the Stanadyne new slip tip pencil injector', *Diesel Injection Systems*, I. Mech. E. Seminar, MEP, London

Blackmore D. R. and Thomas A. (1977). *Fuel Economy of the Gasoline Engine*, Macmillan, London

Blair G. P. (1990). *The Basic Design of Two-Stroke Engines*, SAE, Warrendale, Pennsylvania, USA

Borgnakke C. (1984). 'Flame propagation and heat-transfer effects in spark-ignition engines', in Hilliard J. C. and Springer G. S. (eds), *Fuel Economy of Road Vehicles Powered by Spark Ignition Engines*, Plenum, New York

Boucher R. F. and Kitsios E. E. (1986). 'Simulation of fluid network dynamics by transmission line modelling', *Proc. I. Mech. E.*, Vol. 200, No. C1, pp. 21–9

Brandstetter W. and Howard J. (1989). 'The second generation of the Ford 2.5 litre direct injection diesel engine', *ATZ*, Vol. 91, No. 6, pp. 327–9

Brown A. G. (1991). PhD Thesis, Brunel University, London

Brown C. N. (1987). 'An investigation into the performance of a Waukesha VRG220 SI engine, fuelled by CH_4–CO_2 mixtures', *Final Year Project Report*, Brunel University, London

Brown C. N. and Ladommatos N. (1991). 'The effects of mixture preparation and trapped residuals on the performance of a port-injected spark-ignition engine at low load and low speed', *Proc. I. Mech. E.*, Vol. 205

BS2637: 1978 Motor and aviation-type fuels — Determination of knock characteristics — Motor method, BSI, London

BS2638: 1978 Motor fuels — Determination of knock characteristics — Research method, BSI, London

BS2869: 1988 Petroleum fuels for oil engines and burners, BSI, London

BS4040: 1978 Petrol (gasoline) for motor vehicles, BSI, London

BS7070: 1988 Unleaded petrol (gasoline) for motor vehicles, BSI, London

Campbell C. (1978). *The Sports Car*, 4th edn, Chapman and Hall, London

Caris D. F. and Nelson E. E. (1958). 'A new look at high compression engines', *SAE Paper* 61A

Caton J. A. and Heywood J. B. (1981). 'An experimental and analytical study of heat transfer in an engine exhaust port', *Int. J. Heat Mass Transfer*, Vol. 24, No. 4, pp. 581–95

Charlton S. J. (1986). *SPICE User Manual*, School of Engineering, University of Bath

Charlton S. J., Keane A. J., Leonard H. J. and Stone C. R. (1990). 'Application of variable valve timing to a highly turbocharged diesel engine', *Turbochargers and Turbocharging*, I. Mech. E. Conf. Proc., MEP, London

Chen J. C. (1966). 'A correlation for boiling heat transfer to saturated fluids in convective flow', *Industrial and Engineering Chemistry — Process Design and Development*, Vol. 5, No. 3, pp. 322–9

Chen S. K. and Flynn P. (1965). 'Development of a compression ignition research engine', *SAE Paper* 650733

Cheng W. K., Galliott F., Chen T., Sztenderowiicz M. and Collings N. (1990). 'In cylinder measurements of residual gas concentration in a spark ignition engine', *SAE Paper* 900485

Cockshott C. P., Vernon J. P. and Chambers P. (1983). 'An air mass flowmeter for test cell instrumentation', *4th Int. Conf. on Automotive Electronics*, pp. 20–6, I. Mech. E. Conf. Proc., MEP, London

Cohen, H., Rogers G. F. C. and Saravanamuttoo H. I. H. (1972). *Gas Turbine Theory*, 2nd edn, Longman, London

Collings N. and Willey J. (1987). 'Cyclically resolved emissions from a spark ignition engine', *SAE Paper* 871691

Collins D. and Stokes J. (1983). 'Gasoline combustion chambers — compact or open?' *SAE Paper* 830866

Collis D. C. and Williams M. J. (1959). 'Two-dimensional convection from heated wires at low Reynolds numbers', *Journal of Fluid Mechanics*, Vol. 6, pp. 357–84

Cooke J. A., Roberts D. D., Horrocks R. and Ketcher D. A. (1990). 'Automotive diesel emissions: a fuel appetite study on two light duty direct injection diesel engines', Paper C394/058, *Automotive Power Systems — Environment and Conservation*, I. Mech. E. Conf. Proc., MEP, London

Crisp T. (1984). Jaguar V12 HE engine (private communication)

Cuttler D. H., Girgis N. S. and Wright C. C. (1987). 'Reduction and analysis of combustion data using linear and non-linear regression techniques', Paper C17187, *I. Mech. E. Conf. Proc.*, MEP, London

Dale A., Cole A. C. and Watson N. (1988). 'The development of a turbocharger turbine test facility', in *Experimental Methods in Engine Research and Development*, I. Mech. E. Seminar, MEP, London

Daneshyar H. (1976). *One-Dimensional Compressible Flow*, Pergamon, Oxford

Davies G. O. (1983). 'The preparation and combustion characteristics of coal derived transport fuels', Paper C85/83, *Int. Conf. on Combustion in Engineering*, Vol. II, I. Mech. E. Conf. Proc., MEP, London

Davies P. O. A. L. and Fisher M. J. (1964). 'Heat transfer from electrically heated cylinders', *Proc. Royal Soc.*, Vol. A280, pp. 486–527

Desantes J. M., Benajes J. V. and Lapuerta M. (1989). 'Intake pipes evaluation: comparison between paddle-wheel and hot-wire anemometry methods', *Proc. I. Mech. E.*, Vol. 203, No. A2, pp. 105–11.

Donnelly M. J., Junday J. and Tidmarsh D. H. (1981). 'Computerised data acquisition and processing system for engine test beds', *3rd Int. Conf. on Automotive Electronics*, MEP, London

Downs D. and Wheeler R. W. (1951–52). 'Recent developments in knock research', *Proc. I. Mech. E. (AD)*, Pt III, p. 89

Downs D., Griffiths S. T. and Wheeler R. W. (1961). 'The part played by the preparational stage in determining lead anti-knock effectiveness', *J. Inst. Petrol.*, Vol. 47, p. 1

Dudley W. M. (1948). 'New methods in valve cam design', *SAE Quarterly Transactions*, Vol. 2, No. 1, pp. 19–33

Dunn J. (1985). 'Top-hat piston engine set to make a comeback', *The Engineer*, 12 December 1985, p. 38

Eichelberg G. (1939). 'Some new investigations on old combustion-engine problems' (in four parts), *Engineering*, Vol. 149 [original work published in German, *ZDVI*, Vol. 67, in 1923]

Eltinge L. (1968). 'Fuel–air ratio and distribution from exhaust gas composition', *SAE Paper* 680114, *SAE Transactions*, Vol. 77

Enga B. E., Buchman M. F. and Lichtenstein I. E. (1982). 'Catalytic control of diesel particulates', *SAE Paper*, 820184 (*also in SAE* P-107)

Ernest R. P. (1977). 'A unique cooling approach makes aluminium alloy cylinder heads cost effective', *SAE Paper* 770832

Ferguson C. R. (1986). *Internal Combustion Engines — Applied Thermosciences*, Wiley, New York

Finlay I. C., Harris D., Boam D. J. and Parks B. I. (1985). 'Factors influencing combustion chamber wall temperatures in a liquid-cooled automotive, spark-ignition engine', *Proc. I. Mech. E.*, Vol. 199, No. D3, pp. 207–14

Finlay I. C., Boyle R. J., Pirault J-P. and Biddulph T. (1987a). 'Nucleate and film boiling of engine coolants, flowing in a uniformly heated duct of small cross section', *SAE Paper* 870032

Finlay I. C., Tugwell W., Pirault J-P. and Biddulph T. (1987b). 'The influence of coolant temperature on the performance of a 1100cc engine employing a dual circuit cooling system', *XIXth International Symposium, International Centre for Heat and Mass Transfer, Heat and Mass Transfer in Gasoline and Diesel Engines*, Dubrovnik, August 1987

Finlay I. C., Boam D. J., Bingham J. F. and Clark T. (1990). 'Fast response FID measurements of unburned hydrocarbons in the exhaust port of a firing gasoline engine', *SAE Paper* 902165

Fisher E. H. (1989). 'Means of improving the efficiency of automotive cooling systems', Paper C372/001, *Small Engines Conference*, I. Mech. E. Conf. Proc., pp. 71–6, MEP, London

Ford (1982). *Ford Energy Report*, Interscience Enterprises, Channel Islands, UK

Forlani E. and Ferrati E. (1987). 'Microelectronics in electronic ignition — status and evolution', Paper 87002, *16th ISATA Proceedings*

Fortnagel M. (1990). 'The Mercedes–Benz diesel engine range for passenger cars and light duty trucks as viewed from emissions aspects', *Automotive Power Systems — Environment and Conservation*, I. Mech. E. Conf. Publication 1990–9, MEP, London

Fosberry R. A. C. and Gee D. E. (1961). 'Some experiments on the measurement of exhaust smoke emissions from diesel engines', *MIRA Report* 1961/5

Fraidl G. K. (1987). 'Spray quality of mixture preparation systems' *EAEC Int. Conf. on New Developments in Powertrain and Chassis Engineering* Strasbourg, Vol. 1, pp. 232–46

Francis R. J. and Woollacott P. N. (1981). 'Prospects for improved fuel economy and fuel flexibility in road vehicles', *Energy Paper No. 45*, Department of Energy, HMSO, London

Frankl G., Barker B. G. and Timms C. T. (1989). 'Electronic unit injectors for direct injection engines', *Diesel Fuel Injection Systems*, I. Mech. E. Seminar, MEP, London

French C. C. J. (1970). 'Problems arising from the water cooling of engine components', *Proc. I. Mech. E.*, Vol. 184, Pt 1, No. 29, pp. 507–42

French C. C. J. (1974). 'Thermal loading', in Lilly L. R. C. (ed.), *Diesel Engine Reference Book*, Butterworth, London

Freudenschuss O. (1988). 'A new high speed light duty DI diesel engine', *SAE Paper 881209*

Garrett K. (1990). 'Fuel quality, diesel emissions and the City Filter', *Automotive Engineer*, Vol. 15, No. 5, pp. 51, 55

Gaydon A. G. and Wolfhard H. G. (1979). *Flames, their Structure, Radiation and Temperature*, 4th edn, Chapman and Hall, London

Gilchrist J. M. (1947). 'Chart for the investigation of thermodynamic cycles in internal combustion engines and turbines', *Proc. I. Mech. E.*, Vol. 156, pp. 335–48

Glikin P. E. (1985). 'Fuel injection in diesel engines', *Proc. I. Mech. E.*, Vol. 199, No. 78, pp. 1–14

Glikin P. E., Mowbray D. F. and Howes P. (1979). 'Some developments on fuel injection equipment for diesel engine powered cars', *I. Mech. Conf. Publication 1979–13*, MEP, London

Goodger E. M. (1975). *Hydrocarbon Fuels*, Macmillan, London

Goodger E. M. (1979). *Combustion Calculations*, Macmillan, London

Gosman A. D. (1986). 'Flow processes in cylinders', in Horlock J. H. and Winterbone D. E. (eds), *The Thermodynamics and Gas Dynamics of Internal Combustion Engines*, Vol II, Oxford University Press

Greene A. B. and Lucas G. G. (1969). *The Testing of Internal Combustion Engines*, EUP, London

Greenhalgh D. A. (1983). 'Gas phase temperature and concentration diagnostics with lasers', *Int. Conf. on Combustion in Engineering*, Vol. I, I. Mech. E. Conf. Publications 1983–3, MEP, London

Greeves G. and Wang C. H. T. (1981). 'Origins of diesel particulate mass emission', *SAE Paper 810260*

Gruden D. and Kuper P. F. (1987). 'Heat balance of modern passenger car SI engines', *XIXth International Symposium, International Centre for Heat and Mass transfer, Heat and Mass Transfer in Gasoline and Diesel Engines*, Dubrovnik, August 1987

Hahn H. W. (1986). 'Improving the overall efficiency of trucks and buses. *Proc. I. Mech. E.*, Vol. 200, No. D1, pp. 1–13

Hara S., Nakajima Y. and Nagumo S. (1985). 'Effects of intake value closing timing on SI engine combustion', *SAE Paper* 850074

Hall M. J. and Bracco F. V. (1986). 'Cycle resolved velocity and turbulence measurements near the cylinder wall of a firing S.I. engine', *SAE Paper* 861530

de Haller R. (1945). 'The application of a graphic method to some dynamic problems in gases', *Sulzer Technical Review*, Vol. 1, No. 6

Hardenberg H. O. and Hase F. W. (1979). 'An empirical formula for computing the pressure rise delay of a fuel from its cetane number and from the relevant parameters of direct injection diesel engines', *SAE Paper* 790493, *SAE Trans.*, Vol. 88

Harman R. T. C. (1981). *Gas Turbine Engineering*, Macmillan, London

Hartmann V. and Mallog J. (1988). 'In-cylinder flow analysis as a practical aid in the development of internal combustion engines', in *Int. Conf. Combustion in Engines — Technology and Applications*, I. Mech. E. Conf. Proc., MEP, London

Hay N., Watt P. M., Ormerod M. J., Burnett G. P., Beesley P. W. and French B. A. (1986). 'Design study for a low heat loss version of the Dover engine', *Proc. I. Mech. E.*, Vol. 200, No. DI, pp. 53–60

Haywood R. W. (1972). *Thermodynamic Tables in SI (Metric) Units*, 2nd edn, Cambridge University Press

Haywood R. W. (1980). *Analysis of Engineering Cycles*, 3rd edn, Pergamon International Library, Oxford

Heywood J. B. (1988). *Internal Combustion Engine Fundamentals*, McGraw-Hill, New York

Heywood J. B., Higgins J. M., Watts P. A. and Tabaczynski R. J. (1979). 'Development and use of a cycle simulation to predict SI engine efficiency and NO_x emissions', *SAE Paper* 790291

Hoag K. L., Brands M. C. and Bryzik W. (1985). 'Cummins/TACOM adiabatic engine program', *SAE Paper* 850356 (*also in SAE* SP-610)

Hohenberg G. F. (1979). 'Advanced approaches for heat transfer calculations', *SAE Paper* 790825, *SAE Trans.*, Vol. 88

Holmes M., Willcocks D. A. R. and Bridgers B. J. (1988). 'Adaptive ignition and knock control', *SAE Paper* 885065

Horlock J. H. and Winterbone D. E. (eds) (1986). *The Thermodynamics and Gas Dynamics of Internal Combustion Engines*, Vol. II, OUP (*see also* Benson (1982))

Howarth M. H. (1966). *The Design of High Speed Diesel Engines*, Constable, London

Hudson C., Ladommatos N., Schmid F. and Stone R. (1988). 'Digital instrumentation for combustion study and monitoring in internal combustion engines'. *ISATA Paper* 88095.

Ives A. P. and Trenne M. V. (1981). 'Closed loop electronic control of diesel fuel injection', *3rd Int. Conf. on Automotive Electronics*, I. Mech. E. Conf. Publication 1981–10, MEP, London

Iwamoto J. and Deckker B. E. L. (1985). 'Application of random-choice method to the calculation of unsteady flow in pipes', *SAE Paper* 851561

James E. H. (1990). 'Further aspects of combustion modelling in spark ignition engines', *SAE Paper* 900684

Jante A. (1968). 'Scavenging and other problems of two-stroke cycle spark-ignition engines', *SAE Paper* 680468

Jenny E. (1950). 'Unidimensional transient flow with consideration of friction, heat transfer, and change of section', *Brown Boveri Review*, Vol. 37, No. 11, p. 447

Judge A. W. (1967). *High Speed Diesel Engines*, 6th edn, Chapman and Hall, London

Judge A. W. (1970). *Motor Manuals 2: Carburettors and Fuel Injection Systems*, 8th edn, Chapman and Hall, London

Katsoulakos P. S. (1983). 'Effectiveness of the combustion of emulsified fuels in diesel engines', *Int. Conf. on Combustion in Engineering*, Vol. II, I. Mech. E. Conf. Publication 1983–3, MEP, London

Katys G. P. (1964). *Continuous Measurement of Unsteady Flow* (ed. Walker G. E.) (translated by Barrett D. P.), Pergamon, Oxford

Keck J. C. (1982). 'Turbulent flame structure and speed in spark ignition engines', *Proc. 19th Int. Symp. on Combustion*, The Combustion Institute, pp. 1451–66

Keck J. C., Heywood J. B. and Noske G. (1987). 'Easy flame development and burning rates in spark-ignition engines', *SAE Paper 870104*

King L. V. (1914). 'On the convection of heat from small cylinders in a stream of fluid: determination of the convective constants of small platinum wires with application to hot wire anemometry', *Proc. Royal Soc.*, Vol. 214A, No. 14, p. 373

Kittelson D. B. and Collings N. (1987). 'Origin of the response of electrostatic particle probes', *SAE Paper 870476*

Knight B. E. (1960–61). 'Fuel injection system calculations', *Proc. I. Mech. E.*, No. 1

Kobayashi H., Yoshimura K. and Hirayama T. (1984). 'A study on dual circuit cooling for higher compression ratio', *SAE Paper 841294*

Krieger R. B. and Borman G. L. (1986). 'The computation of apparent heat release for internal combustion engines', *ASME Paper 66-WA/DGP-4*

Kubozuka T., Ogawa N., Hirano Y. and Hayashi Y. (1988). 'The development of engine evaporative cooling systems', *SAE Paper 870033*

Kyriakides S. C. and Glover A. R. (1988). 'A study of the correlation between in-cylinder air motion and combustion in gasoline engines', in *Int. Conf. Combustion in Engines — Technology and Applications*, I. Mech. E. Conf. Proc., MEP, London

Ladommatos N. and Stone R. (1991). 'Conversion of a diesel engine for gaseous fuel operation at high compression ratio', *SAE Paper 910849*

Lancaster D. R. (1976). 'Effects of engine variables on turbulence in a spark ignition engine', *SAE Paper 760159*

Leshner M. D. (1983). 'Evaporative engine cooling for fuel economy — 1983', *SAE Paper 831261*

Lewis B. and von Elbe G. (1961). *Combustion Flames and Explosions of Gases*, 2nd edn, Academic Press, New York

Light J. T. (1989), 'Advanced anhydrous engine cooling systems', *SAE Paper 891635*

Liou T-M., Hall M., Santavicca D. A. and Bracco F. V. (1984). 'Laser doppler velocimetry measurements in valved and ported engines', *SAE Paper 840375*

Lyon D. (1986). 'Knock and cyclic dispersion in a spark ignition engine', *Petroleum Based Fuels and Automotive Applications*, I. Mech. E. Conf. Proc., MEP, London

Ma T. (1988). 'Effect of variable valve timing on fuel economy', *SAE Paper 880390*

Maekawa M. (1957). *Text of Course*, JSME No. G36, p. 23

Maly R. R. (1984). 'Spark ignition: its physics and effect on the internal combustion engine'. In Hilliard J. C. and Springer C. S. (eds), *Fuel Economy of Road Vehicles Powered by Spark Ignition Engines*, Plenum, New York

Maly R. R. and Vogel M. (1978). 'Initiation and propagation of flame fronts in lean CH_4–air mixtures by the three modes of the ignition spark', *17th Int. Conf. on Combustion*, pp. 821–31, The Combustion Institute

Mark P. E. and Jetten W. (1986). 'Propylene glycol: a new base fluid for automotive coolants', in Beal R. E. (ed.), *Engine Coolant Testing: Second Symposium*, ASTM STP 87, pp. 61–77

Mattavi J. N. and Amann S. A. (1980). *Combustion Modelling in Reciprocating Engines*, Plenum, New York

May M. (1979). 'The high compression lean burn spark ignited 4-stroke engine', *I. Mech. E. Conf. Publication 1979–9*, MEP, London

McMahon P. J. (1971). *Aircraft Propulsion*, Pitman, London

Mercer A. D. (1980a). 'Experience of the British Standards Institution in the field of engine coolants', in Ailor W. H. (ed.), *Engine Coolant Testing: State of its Art*, ASTM STP 705, pp. 24–41

Mercer A. D. (1980b). 'Laboratory research in the development and testing of inhibited coolant in boiling heat-transfer conditions', in Ailor W. H. (ed.), *Engine Coolant Testing: State of its Art*, ASTM STP 705, pp. 53–80

Metghalchi M. and Keck J. C. (1980) 'Laminar burning velocity of propane–air mixtures at high temperature and pressure', *Combustion and Flame*, Vol. 38, pp. 143–54

Meyer, E. W., Green R. and Cops M. H. (1984). 'Austin–Rover Montego programmed ignition system', Paper C446/84, *VECON '84 Fuel Efficient Power Trains and Vehicles*, I. Mech. E. Conf. Publication 1984–14, MEP, London

Millington B. W. and Hartles E. R. (1968), 'Frictional losses in diesel engines', *SAE Paper* 680590

Monaghan M. L. (1990). 'The diesel passenger car in a green world', *SIA Diesel Congress*, Lyons, France

Moore C. H. and Hoehne J. L. (1986). 'Combustion chamber insulation effect on the performance of a low heat rejection Cummins V-903 engine', *SAE Paper* 860317

Mundy H. (1972). 'Jaguar V12 engine: its design and development history', *Proc. I. Mech. E.*, Vol. 186, paper 34/72, pp. 463–77

Muranaka S., Takagi Y. and Tshida T. (1987). 'Factors limiting the improvement in thermal efficiency of SI engine at higher compression ratio', *SAE Paper* 870548

Nakajima Y., Sugihara K. and Takagi Y. (1979). 'Lean mixture or EGR — which is better for fuel economy and NO_x reduction?', *Proceedings of Conference on Fuel Economy and Emissions of Lean Burn Engines*, I. Mech. E. Conf. Proc., MEP, London

Narasimhan S. L. and Larson J. M. (1985). 'Valve gear wear and materials', *SAE Paper* 851497

Newton K., Steeds W. and Garrett T. K. (1983). *The Motor Vehicle*, 10th edn, Butterworth, London

O'Callaghan T. M. (1974). 'Factors affecting the coolant heat transfer surfaces in an IC engine', *General Engineer*, October 1974, pp. 220–30

Ohata A. and Ishida Y. (1982). 'Dynamic inlet pressure and volumetric efficiency of four-cycle four cylinder engine, *SAE Paper* 820407

Onion G. and Bodo L. B. (1983). 'Oxygenate fuels for diesel engines: a survey of worldwide activity', *Biomass*, Vol. 3, pp. 77–133

Osborne A. G. (1985). 'Low cost FM transducer for diesel injector needle lift', *Developments in Measurements and Instrumentation in Engineering*, I. Mech. E. Conf. Proc., MEP, London

Packer J. P., Wallace F. J., Adler D. and Karimi E. R. (1983). 'Diesel fuel jet mixing under high swirl conditions', *Int. Conf. on Combustion in Engineering*, Vol. II, I. Mech. E. Conf. Publication 1983–3, MEP, London

Palmer F. H. (1986). 'Vehicle performance of gasoline containing oxygenates', Paper C319/86, *Int. Conf. Petroleum Based Fuels and Automotive Applications*, pp. 36–46, I. Mech. E. Conf. Publication 1986–11, MEP, London

Palmer F. H. and Smith A. M. (1985). 'The performance and specification of gasoline' in Hancock E. G. (ed.), *Technology of Gasoline*, Blackwell Scientific Publications, Oxford

Pischinger R. and Cartellieri W. (1972). 'Combustion system parameters and their effect upon diesel engine exhaust emissions', Paper 720756, *SAE Trans.*, Vol. 81

Poloni M., Winterbone D. E. and Nichols J. R. (1987). 'Calculation of pressure and temperature discontinuity in a pipe by the method of characteristics and the two step Lax–Wendroff method', *ASME Conference*, Boston, 14–16 December 1987

Poulos S. G. and Heywood J. B. (1983). 'The effect of chamber geometry on spark-ignition engine combustion', *SAE Paper 830334*, *SAE Trans.*, Vol. 92

Priede T. and Anderton D. (1974). 'Likely advances in mechanics, cooling, vibration and noise of automotive engines', *Proc. I. Mech. E.*, Vol. 198D, No. 7, pp. 95–106

Radermacher K. (1982). 'The BMW eta engine concept', *Proc. I. Mech. E.*, Vol. 196

Rask R. B. (1981). 'Comparison of window, smoothed ensemble, and cycle-by-cycle data reduction techniques for laser Doppler anemometer measurements of in-cylinder velocity', in Morel T., Lohnmann R. P. and Rackley J. M. (eds), *Fluid Mechanics of Combustion Systems*, pp. 11–20, ASME

Rassweiler G. M. and Withrow L. (1938), 'Motion pictures of engine flame correlated with pressure cards', *SAE Paper 800131* (originally presented in January 1938)

Reid R. C., Prausnitz J. M. and Sherwood T. K. (1977). *The Properties of Gases and Liquids*, McGraw-Hill, New York

Rhodes D. B. and Keck J. C. (1985). 'Laminar burning speed measurements of indolene–air–diluent mixtures at high pressures and temperature', *SAE Paper 850047*

Ricardo H. R. and Hempson J. G. G. (1968). *The High Speed Internal Combustion Engine*, 5th edn, Blackie, London

Roark R. J. and Young W. C. (1976). '*Formulas for Stress and Strain*', 5th edn, McGraw-Hill, New York

Rogers G. F. C. and Mayhew Y. R. (1980a). *Engineering Thermodynamics, Work and Heat Transfer*, 3rd edn, Longman, London

Rogers G. F. C. and Mayhew Y. R. (1980b). *Thermodynamic and Transport Properties of Fluids, SI Units*, 3rd edn, Blackwell, Oxford

Rowe L. C. (1980). 'Automotive engine coolants: a review of their requirement and methods of evaluation', in Ailor W. H. (ed.), *Engine Coolant Testing: State of its Art*, ASTM STP 705, pp. 3–23

SAE (1987). 'SAE recommended practice, engine terminology and nomenclature', *SAE Handbook*, p. 24.01, SAE, Warrendale, Pennsylvania, USA

Scott A. W. (1991). *Development of the TWR Jaguar V12 Engine*, Autosports National Congress, NEC, Birmingham

Seizinger D. E., Marshall W. F. and Brooks A. L. (1985). 'Fuel influences on diesel particulates', *SAE Paper 850546*

Sher E. (1990). 'Scavenging the two-stroke engine', *Progress in Energy and Combustion Science*, Vol. 16, pp. 95–124

Sihling K. and Woschni G. (1979). 'Experimental investigation of the instantaneous heat transfer in the cylinder of a high speed diesel engine', *SAE Paper 790833*

Smith P. H. (1967). *Valve Mechanisms for High Speed Engines*, Foulis, Yeovil

Smith P. H. (1968). *Scientific Design of Exhaust and Intake Systems*, Foulis, Yeovil

de Soete G. C. (1983). 'Propagation behaviour of spark ignited flames in the early stages', *Int. Conf. on Combustion in Engineering*, Vol. 1, Mech. E. Conf. Proc., MEP, London

Soltau J. P. (1960). 'Cylinder pressure variations in petrol engines, *I. Mech. E. Conf. Proc.*, July 1960, pp. 96–116

Sorrell A, J. (1989). Spark ignition engine performance during warm-up. PhD Thesis, Brunel University, London

Spindt R. S. (1965). 'Air–fuel ratios from exhaust gas analysis', *SAE Paper* 650507

Stevens J. L. and Tawil, N. (1989). 'A new range of industrial diesel engines with emphasis on cost reduction techniques', Paper C372/028, *The Small Internal Combustion Engine*, pp. 137–144, I. Mech. E. Conf. Proc., MEP, London

Stillerud K. G. H. (1978). 'Aspects of two-stroke design viewpoints from outboard developments', *Design and Development of Small Internal Combustion Engines*, I. Mech. E. Conf. Publication 1978–5, MEP, London

Stone C. R. and Green-Armytage D. I. (1987). 'Comparison of methods for the calculation of mass fraction burnt from engine pressure–time diagrams', *Proc. I. Mech. E.*, Vol. 201, No. D1

Stone C. R. and Steele A. B. (1989). 'Measurement and modelling of ignition system energy and its effect on engine performance', *Proc. I. Mech. E.*, Vol. 203, Pt D, No. 4, pp. 277–86

Stone R. (1989a). *Motor Vehicle Fuel Economy*, Macmillan, London

Stone R. (1989b). *Review of Induction System Design*, Notes for Cranfield MSc Course in Automotive Product Engineering

Stone R. (1989c). 'Air flow measurement in internal combustion engines', *SAE Paper* 890242

Stone R. and Kwan E. (1989). 'Variable valve actuation mechanisms and the potential for their application', *SAE Paper* 890673, *SAE Trans.*, Vol. 98

Stone R. and Ladommatos N. (1992). 'Measurement and analysis of swirl in steady flow', to be published

Tabaczynski R. J. (1976). 'Turbulent combustion in spark-ignition engines', *Prog. Energy Combustion Sciences*, Vol. 2, pp. 143–65

Tabaczynski R. J. (1983). 'Turbulence measurements and modelling in reciprocating engines — an overview', *Int. Conf. on Combustion in Engineering*, Vol. I, I. Mech. E. Conf. Publication 1983–3, MEP, London

Tabaczynski R. J., Ferguson C. R. and Radhakrishnan K. (1977). 'A turbulent entrainment model for spark-ignition engine combustion', *SAE Paper* 770647, *SAE Trans.*, Vol. 86

Tabaczynski R. J., Trinker F. H. and Shannon B. A. S. (1980). 'Further refinement and validation of a turbulent flame propagation model for spark ignition engines', *Combustion and Flame*, Vol. 39, pp. 111–21

Tawfig M., Jager D. and Charlton S. J. (1990). 'In-cylinder measurement of mixture strength in a divided-chamber natural gas engine', *Fast Response FID Workshop*, September 1990, Brunel University, London

Taylor C. F. (1985a). *The Internal Combustion Engine in Theory and Practice*, Vol. I, MIT Press, Cambridge, Massachusetts

Taylor C. F. (1985b). *The Internal Combustion Engine in Theory and Practice*, Vol. II, MIT Press, Cambridge, Massachusetts

Timoney D. J. (1985). 'A simple technique for predicting optimum fuel air mixing conditions in a direct injection diesel engine with swirl', *SAE Paper* 851543

Toda T., Nohira H. and Kobashi K. (1976). 'Evaluation of burned gas ratio (BGR) as a predominant factor to NO_x', *SAE Paper* 760765, *SAE Trans.*, Vol. 85

Tomita E. and Hamamoto Y. (1988). 'The effect of turbulence on combustion in cylinder of spark ignition engine — evaluation of entrainment model', *SAE Paper* 880128

Trier C. J., Rhead M. M. and Fussey D. E. (1990). 'Evidence for the pyrosynthesis of parent polycyclic aromatic compounds in the combustion chamber of a diesel engine', Paper C394/003, *Automotive Power Systems — Environment and Conservation*, I. Mech. E. Conf. Proc., MEP, London

Tsuchiya K. and Hiramo S. (1975). 'Characteristics of 2-stroke motorcycle exhaust HC emission and effects of air-fuel ratio and ignition timing', *SAE Paper* 750908

Uchiyama H., Chiku T. and Sayo S. (1977). 'Emission control of two-stroke automobile engine', *SAE Paper* 770766, *SAE Trans.*, Vol. 86

Wade W. R., Havstad P. H., Ounsted E. J., Trinkler F. H. and Garwin I. J. (1984). 'Fuel economy opportunities with an uncooled DI diesel engine', *SAE Paper* 841286 (*also in SAE* SP-610)

Wakeman A. C., Ironside J. M., Holmes M., Edwards S. I. and Nutton D. (1987). 'Adaptive engine controls for fuel consumption and emissions reduction', *SAE Paper* 870083

Wallace F. J. (1986a). 'Engine simulation models with filling-and-emptying methods' in Horlock J. H. and Winterbone D. E. (eds), *The Thermodynamics and Gas Dynamics of Internal Combustion Engines*, Vol. II, Oxford University Press (*see also* Benson (1982))

Wallace F. J. (1986b). 'Filling and emptying methods', in Horlock J. H. and Winterbone D. E. (eds), *The Thermodynamics and Gas Dynamics of Internal Combustion Engines*, Vol. II, Oxford University Press (*see also* Benson (1982))

Wallace F. J., Tarabad M. and Howard D. (1983). 'The differential compound engine — a new integrated engine transmission system for heavy vehicles', *Proc. I. Mech. E.*, Vol. 197A

Wallace W. B. (1968). 'High-output medium-speed diesel engine air and exhaust system flow losses', *Proc. I. Mech. E.*, Vol. 182, Pt 3D, pp. 134–44

Walzer P. (1990). 'Automotive power systems for future environmental problems', Paper C394/036, *Automotive Power Systems — Environment and Conservation*, I. Mech. E. Conf. Proc., MEP, London

Walzer P., Heirrich H. and Langer M. (1985). 'Ceramic components in passenger-car diesel engines', *SAE Paper* 850567

Watanabe Y., Ishikawa H. and Miyahara M. (1987). 'An application study of evaporative cooling to heavy duty diesel engines', *SAE Paper* 870023

Watson N. (1979). *Combustion Gas Properties*, Internal Report, Dept of Mechanical Engineering, Imperial College of Science and Technology, London

Watson N. and Janota M. S. (1982). *Turbocharging the Internal Combustion Engine*, Macmillan, London

Watson N., Wijeyakumar S. and Roberts G. L. (1981). 'A microprocessor controlled test facility for transient vehicle engine system development', *3rd Int. Conf. on Automotive Electronics*, MEP, London

Weaving J. H. and Pouille J-P. (1990). 'Atmospheric pollution', in Weaving J. H. (ed.), *Internal Combustion Engineering, Science and Technology*, Elsevier Applied Science, London and New York

Weertman W. L. and Dean J. W. (1981). 'Chrysler Corporation's new 2.2 liter 4 cylinder engine', *SAE Paper* 810007

Wentworth J. T. (1971). 'Effect of combustion chamber surface temperature on exhaust hydrocarbon concentration', *SAE Paper* 710587

Whitehouse J. A. and Metcalfe J. A. (1956). 'The influence of lubricating oil on the power output and fuel consumption of modern petrol and compression ignition engines', *MIRA Report* 1956/2

Whitehouse N. D. (1987). 'An estimate of local instantaneous condition in a diesel engine' *1st Int. Conf. on Heat and Mass Transfer in Gasoline and Diesel Engines*, Dubrovnik

Whitehouse N. D. and Way R. J. B. (1968). 'Rate of heat release in diesel engines and its correlation with fuel injection data', *Proc. I. Mech. E.*, Vol. 177, No. 2, pp. 43–63

Whiteman P. R. (1989). 'PC based engine combustion analysis', *Session 24: Computer Aided Engineering for Powertrains*, Autotech, I. Mech. E., London

Wiebe I. (1967). 'Halbempirische Formel für die Verbrennungsgeschwindigkeit', *Verlag der Akademie der Wissenschaften der UdSSR*, Moscow

Wiedenmann H. M., Raff L. and Noack R. (1984). 'Heated zirconia oxygen sensor for stoichiometric and lean air–fuel ratios', *SAE Paper 840141*

Willumeit H. P. and Steinberg P. (1986). 'The heat transfer within the combustion engine', *MTZ*, Vol. 47, No. 1, pp. 9–14

Willumeit H. P., Steinberg P., Otting H., Scheibner B. and Lee W. (1984). 'New temperature control criteria for more efficient gasoline engines', *SAE Paper 841292*

Winterbone D. E. (1986). 'Transient performance', in Horlock J. H. and Winterbone D. E. (eds), *The Thermodynamics and Gas Dynamics of Internal Combustion Engines*, Vol. II, Oxford University Press (*see also* Benson (1982))

Winterbone D. E. (1990a). 'The theory of wave action approaches applied to reciprocating engines' in Weaving J. H. (ed.), *Internal Combustion Engineering*, Elsevier Applied Science, London and New York

Winterbone D. E. (1990b). 'Application of wave action techniques', in Weaving J. H. (ed.), *Internal Combustion Engineering*, Elsevier Applied Science, London and New York

Witze P. O. (1980). 'A critical comparison of hot-wire anemometry and laser Doppler velocimetry for IC engine applications', *SAE Paper 800132*

Wolf G. (1982). 'The large bore diesel engine', *Sulzer Technical Review*, Vol. 3

Worthen R. P. and Rauen D. G. (1986). 'Valve parameters measured in firing engine', *Automotive Engineering*, Vol. 94, No. 6, pp. 40–7

Woschni G. (1967). 'A universally applicable equation for the instantaneous heat transfer coefficient in the internal combustion engine', Paper 670931, *SAE Trans.*, Vol. 76, p. 3065

Wright S. D. (1990). 'The use of viscous flowmeters to measure instantaneous air flowrates in pulsating flow produced by internal combustion engines', *Year 4 Project Report*, Brunel University, London

Yamaguchi J. (1986). 'Optical sensor feeds back diesel ignition timing', *Automotive Engineering*, Vol. 94, No. 4, pp. 84–5

Yates D. (1988). 'The compact high-speed gas sampling valve for an internal combustion engine', *Experimental Methods in Engine Research and Development*, I. Mech. E. Seminar, MEP, London

Index

569